FORMULAS FROM ALGEBRA

Exponents

$$a^m a^n = a^{m+n}$$

$$(a^m)^n = a^{mn}$$

$$\frac{a^m}{a^n} = a^{m-n}$$

$$(ab)^n = a^n b^n$$

$$\left(\frac{a}{b}\right)^n = \frac{a^n}{b^n}$$

Radicals

$$(\sqrt[n]{a})^n = a$$

$$\sqrt[n]{a^n}$$

$$\sqrt[n]{\frac{a}{b}}$$

Logarithms

$$\log_a MN = \log_a M + \log_a N$$

$$\log_a (M/N) = \log_a M - \log_a N$$

$$\log_a (N^p) = p \log_a N$$

Factoring Formulas

$$x^2 - y^2 = (x - y)(x + y)$$

$$x^3 - y^3 = (x - y)(x^2 + xy + y^2)$$

$$x^3 + y^3 = (x + y)(x^2 - xy + y^2)$$

$$x^2 + 2xy + y^2 = (x + y)^2$$

$$x^2 - 2xy + y^2 = (x - y)^2$$

$$x^3 + 3x^2 y + 3xy^2 + y^3 = (x + y)^3$$

Binomial Formula

$$(x + y)^n = {}_nC_0 x^n y^0 + {}_nC_1 x^{n-1} y^1 + \cdots + {}_nC_{n-1} x^1 y^{n-1} + {}_nC_n x^0 y^n$$

Quadratic Formula

The solutions to $ax^2 + bx + c = 0$ are $x = \dfrac{-b \pm \sqrt{b^2 - 4ac}}{2a}$

Complex Numbers

Multiplication: $(a + bi)(c + di) = (ac - bd) + (ad + bc)i$

Polar form: $a + bi = r(\cos \theta + i \sin \theta)$ where $r = \sqrt{a^2 + b^2}$

Powers: $[r(\cos \theta + i \sin \theta)]^n = r^n(\cos n\theta + i \sin n\theta)$

Roots: $\sqrt[n]{r}\left[\cos\left(\dfrac{\theta + k \cdot 360°}{n}\right) + i \sin\left(\dfrac{\theta + k \cdot 360°}{n}\right)\right]$ $k = 0, 1, 2, \cdots, n - 1$

Plane
Trigonometry

SECOND EDITION

Plane Trigonometry

A PROBLEM-SOLVING APPROACH

Walter Fleming
Hamline University

Dale Varberg
Hamline University

Prentice Hall
Englewood Cliffs, New Jersey 07632

Library of Congress Cataloging-in-Publication Data

Fleming, Walter
    Plane trigonometry: a problem-solving approach / Walter
Fleming, Dale Varberg.—2nd ed.
        p.   cm.
    Includes index.
    ISBN 0-13-679051-8: $24.75 (est.)
        1. Trigonometry, Plane.   I. Varberg, Dale.   II. Title.
QA533.F53   1989
516.2′4—dc19

87-31014
CIP

Editorial/production supervision: Zita de Schauensee
Interior design: Christine Gehring-Wolf
Cover design: Maureen Eide
Manufacturing buyer: Paula Massenaro
Computer graphics cover art created by Genigraphics Corporation

Printed in the United States of America
10 9 8 7 6 5 4 3 2 1

Credits for quotations in text: Page 10: Harry M.
Davis, "Mathematical Machines," in *Scientific Amer-
ican*" (April 1949). Page 96: George Polya, *Mathematical Discovery* (New York: John Wiley & Sons, Inc.,
1962), vol. I, pp. 6–7. Page 215: Richard Courant and
Herbert Robbins, *What Is Mathematics?* (New York:
Oxford University Press, 1978), p. 198.

ISBN   0-13-679051-8

Prentice-Hall International (UK) Limited, *London*
Prentice-Hall of Australia Pty. Limited, *Sydney*
Prentice-Hall Canada Inc., *Toronto*
Prentice-Hall Hispanoamericana, S.A., *Mexico*
Prentice-Hall of India Private Limited, *New Delhi*
Prentice-Hall of Japan, Inc., *Tokyo*
Prentice-Hall of Southeast Asia Pte. Ltd., *Singapore*
Editora Prentice-Hall do Brasil, Ltda., *Rio de Janeiro*

Contents

4 Applications of Trigonometry 121

5 Exponential and Logarithmic Functions 169

6 Analytic Geometry 215

Appendix 275

Answers to Odd-Numbered Problems 289

Index of Teaser Problems 313

Index of Names and Subjects 315

Preface

It is now eight years since we wrote the preface for the first edition of this book. We referred then to two controversies affecting the teaching of trigonometry. One had to do with how to begin the subject. That controversy is still alive. Traditionalists plead for introducing the subject as it was done historically, in terms of angles and right triangles. That simple setting, they say, gives trigonometry its clearest and most intuitive foundation and allows an immediate study of significant practical problems.

Who needs triangles or even angles? ask the modernists. It is the analytic properties of the trigonometric functions together with the oscillatory character of their graphs that are important. And the quickest and surest way to get to these features is via the unit circle.

We have chosen a middle course. After an introductory section on right-triangle trigonometry, the properties of the trigonometric functions for angles and numbers are developed simultaneously, based on the unit circle.

Already in that earlier edition, we had chosen sides in another controversy. What is the proper attitude toward those electronic marvels—scientific calculators? Our answer, then as now, is that they should be used freely. While we do still include brief tables of the trigonometric functions (and an explanation, calling it what to do if your battery goes dead), we expect students to have and use calculators. Of course, most of the problems of trigonometry require neither tables nor a calculator. If a problem does involve a calculation, either a calculator or tables will yield the answer. In some cases, the calculator is so much to be preferred that we have marked the problem with © for calculator.

WRITING STYLE

We continue those features that led to a successful first edition. Our *writing is informal* but not sloppy. Results are stated correctly, explained carefully, and illustrated profusely. *Expertly drawn diagrams* appear in profusion throughout the book. We offer *many examples* in the text itself and then give more within each problem set, always accompanied by a related group of problems. To spark student interest, we open every section with *a vignette*, which may be an intriguing problem, an anecdote, a historical note, or a famous quote. Our *sections are of approximately equal length* and cover the amount of material that we think appropriate for one day's lesson. To avoid those long expanses of uninterrupted prose that mar some books (and frighten students), we divide each section into three or four short subsections. Finally, key words are printed in bold face and *important results are highlighted* by using italics or ruled boxes.

PROBLEM SOLVING

Since we consider problem solving to be the principal task of mathematics, our book poses good problems and gives clear guidance in how to solve them. There are over 2000 problems for students to work on, including numerous word problems. We apply trigonometry to many disciplines, especially the natural sciences. Our problem sets are in two parts. First, there is a set of what might be called *basic problems*. These problems are straightforward and are designed to cover the material of the section in a carefully ordered systematic way. Second, we offer a set of *miscellaneous problems* designed to cover the same material in a random way. The miscellaneous set always begins with easy problems but steadily gets more challenging and concludes with a *teaser*, a problem that we think will intrigue and challenge the very best students and maybe even the teacher. We have worked pretty hard (and had a lot of fun) constructing the teaser problems. Some are old nuggets that have been passed around for decades but there are some nice new ones too. We suggest that a teacher offer a prize to the student who succeeds in solving the most teasers.

The answer key at the back of the book gives answers to odd-numbered problems and to all problems of the review sets that end each chapter. Both authors worked all problems; we are confident that this has resulted in an accurate set of answers.

NEW IN THIS EDITION

We have gone through the book page by page looking for places where the exposition could be improved. Naturally, we have tried to respond to the suggestions and criticisms of reviewers and previous users. For example, we have expanded to a full section the material on double angle and half angle formulas; and we have given more attention to graphing. The sections on vectors now follow immediately after the sections on the law of sines and the law of cosines, this because of the close relationship between these topics. The suggestion that the treatment of inverse trigonometric functions should precede the study of trigonometric equations has been implemented. Other notable changes are as follows.

1. Because some mathematics departments prefer to include a study of exponential and logarithmic functions in their trigonometry course, we have added a chapter on these important functions (Chapter 5). This means that all the elementary transcendental functions can be treated together.

2. We have completely reworked the miscellaneous problems that occur at the end of each problem set to make them more interesting and more comprehensive. The result is a much larger collection of problems from which to select assignments. We have already mentioned the special teaser problem that concludes each problem set.

3. To help students recognize and avoid common errors, we have placed caution boxes at strategic places throughout the text.

4. We have rewritten the last chapter (now titled "Analytic Geometry"). At the request of several reviewers, we now give a complete treatment of the conic sections, approached via the standard definitions.

FLEXIBILITY

The first chapter, "Algebraic Preliminaries," is a review of those topics from algebra that will be needed in the course. It can be covered rapidly or omitted, at least in classes that have just completed a substantial algebra course. Then come two chapters, "The Trigonometric Functions" and "Trigonometric Identities and Equations," which must form the core of any trigonometry course. These are followed by a chapter called "Applications of Trigonometry" from which an instructor may pick those topics considered to be interesting or important. Next comes a chapter on the "Exponential and Logarithmic Functions," not strictly part of trigonometry but often studied along with it. Finally, there is a chapter on analytic geometry which highlights the way trigonometry can be used to study the conic sections and other important curves. The following dependence chart will help instructors design a syllabus that fits local conditions.

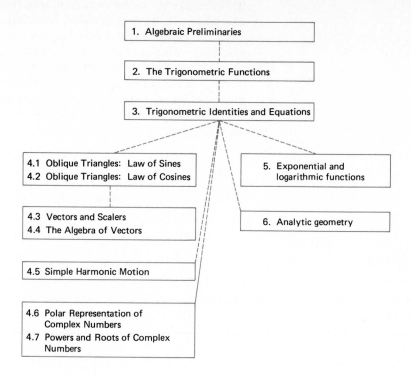

SUPPLEMENTARY MATERIALS

An extensive variety of instructional aids is available from Prentice Hall.

Instructor's Manual The instructor's manual was prepared by the authors of the textbook. It contains the following items.

 (a) Answers to all the even-numbered problems (answers to the odd problems appear at the end of the textbook).

 (b) Complete solutions to the last four problems in each problem set. This includes the teaser problem.

 (c) Six versions of a chapter test for each chapter together with an answer key for these tests.

 (d) A test bank of more than 1300 problems with answers designed to aid an instructor in constructing examinations.

 (e) A set of transparencies that illustrate key ideas.

Prentice Hall Test Generator The test bank of more than 1300 problems is available on floppy disk for the IBM PC. This allows the instructor to generate examinations by choosing individual problems, editing them, and if desired by creating completely new problems.

Student Solutions Manual This manual has worked-out solutions to every third problem (not including teaser problems).

Function Plotter Software A one-variable function plotter for the IBM PC is available with a qualified adoption. Contact your local Prentice Hall representative for details.

"How to Study Math" Designed to help your students overcome anxiety and to offer helpful hints regarding study habits, this useful booklet is available free with each copy sold. To request copies for your students in quantity, contact your local Prentice Hall representative.

ACKNOWLEDGMENTS

This and previous editions have profited from the warm praise and constructive criticism of many reviewers. We offer our thanks to the following people who gave helpful suggestions.

> Wayne Andrepont, *The University of Southwestern Louisiana*
> Steven Blasberg, *West Valley College*
> Paul Britt, *Louisiana State University*
> Donald Coram, *Oklahoma State University*
> Leonard Deaton, *California State University, Los Angeles*
> Margaret Gessaman, *The University of Nebraska at Omaha*
> Mark Hale, Jr., *University of Florida*
> D. W. Hall. *Michigan State University*
> James E. Hall, *University of Wisconsin, Madison*
> Allen Hesse, *Bowling Green State University*
> Robert McMillan, *Oklahoma State University*
> Eldon Miller, *University of Mississippi*
> John Pantano, *Santa Fe Community College*
> Wallace E. Parr, *University of Maryland, Baltimore County*
> Burla Sims, *University of Arkansas at Little Rock*
> Jean Smith, *Middlesex Community College*
> Carroll Wells, *Western Kentucky University*

The staff at Prentice-Hall is to be congratulated on another fine production job. The authors wish to express appreciation especially to Priscilla McGeehon (mathematics editor), Zita de Schauensee (project editor), and Christine Gehring-Wolf (designer) for their exceptional contributions.

Walter Fleming
Dale Varberg

"Thus under the Descartes-Fermat scheme points became pairs of numbers, and curves became collections of pairs of numbers subsumed in equations. The properties of curves could be deduced by algebraic processes applied to equations. With this development, the relation between number and geometry had come full circle. The classical Greeks had buried algebra in geometry, but now geometry was eclipsed by algebra. As the mathematicians put it, geometry was arithmetized."

Morris Kline

CHAPTER 1

Algebraic Preliminaries

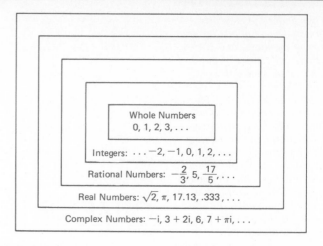

1-1 Numbers and Number Systems

Mathematics students are number crunchers. That popular image is both profoundly right and terribly wrong. Most mathematicians disdain adding up long columns of numbers, finding square roots, or doing long division. They have relegated such tasks to the electronic calculators that we shall discuss in the next section. But it is still true that numbers play a fundamental role in most of mathematics. Certainly this is true in trigonometry.

It would take too long to describe the long, tortuous road traveled by mankind in going from the whole numbers to the complex numbers. It is enough to call your attention to the five classes of numbers in our opening display, to suggest that they were developed roughly in the order listed. and to recall a few facts about the two largest of these classes.

THE REAL NUMBERS

Given a prescribed unit of length, we can (at least theoretically) measure the length of any line segment. The set of all numbers that can measure lengths, together with their negatives and zero, constitute the **real numbers**. The number $\sqrt{2}$ is a real number since it measures the hypotenuse of a right triangle with legs of unit length (Figure 1). So is π; it measures the circumference of a circle of unit diameter ($C = \pi d$). Every rational number (ratio of two integers) is a real number.

The best way to visualize the system of real numbers is as a set of labels for points on a line. Consider a horizontal line and select an arbitrary point to be labeled with 0. Then label a point one unit to the right with 1, a point one unit to the left with -1, and so on. The process is so familiar that we omit further details and draw a picture (Figure 2).

Of course, we cannot show all the labels, but we want you to imagine that

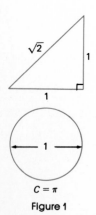

$C = \pi$

Figure 1

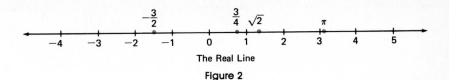

The Real Line

Figure 2

each point has a number label or **coordinate** that measures its distance to the right or left of 0. We refer to the resulting coordinate line as the **real line**.

If $b - a$ is positive, we say that a is less than b. We write $a < b$, which is called an **inequality**. On the real line, $a < b$ simply means that a is to the left of b (Figure 3). Similarly, if $b - a$ is positive or zero, we say a is less than or equal to b and write $a \leq b$. It is correct to say $5 < 6$, $5 \leq 6$, and $5 \leq 5$.

The symbol $|a|$, read the **absolute value** of a, is defined by

$$|a| = \begin{cases} a & \text{if } 0 \leq a \\ -a & \text{if } a < 0 \end{cases}$$

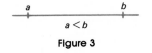

Figure 3

Geometrically, $|a|$ is the (undirected) distance from 0 to a. Some would say that $|a|$ is the magnitude of a without regard to its sign. Be careful with this: It is correct to say $|+5| = 5$ and $|-5| = 5$, but it is not necessarily true that $|x| = x$ (try $x = -2$). Finally, note that the distance between b and a on the real line is $|b - a|$; this is correct whether a is to the left or to the right of b, as we indicate in Figure 4.

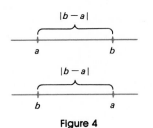

Figure 4

DECIMALS

There is another important way to describe the real numbers. We must first review a basic idea. Recall that

$$.4 = \frac{4}{10}$$

$$.42 = \frac{4}{10} + \frac{2}{100} = \frac{40}{100} + \frac{2}{100} = \frac{42}{100}$$

$$.731 = \frac{7}{10} + \frac{3}{100} + \frac{1}{1000} = \frac{700}{1000} + \frac{30}{1000} + \frac{1}{1000} = \frac{731}{1000}$$

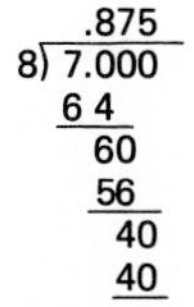

```
   .875
8) 7.000
   6 4
   ___
    60
    56
    ___
    40
    40
```

Figure 5

Clearly, each of these decimals represents a rational number.

Conversely, if we are given a rational number, we can find its decimal expansion by long division. For example, the division in Figure 5 shows that $\frac{7}{8} = .875$. When in Figure 6 we try the same procedure on $\frac{2}{11}$, something different happens. The decimal just keeps on going; it is a **nonterminating decimal.**

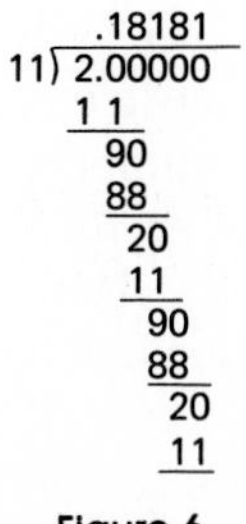

Actually, the **terminating decimal** .875 can be thought of as nonterminating if we adjoin zeros. Thus

```
    .18181
11) 2.00000
    1 1
    ___
     90
     88
     ___
     20
     11
     ___
     90
     88
     ___
     20
     11
```

Figure 6

$$\frac{7}{8} = .8750000 \ldots \qquad = .875\overline{0}$$

$$\frac{2}{11} = .181818 \ldots \qquad = .\overline{18}$$

$$\frac{2}{7} = .285714285714 \ldots = .\overline{285714}$$

Note that in each case, the decimal has a repeating pattern. This is indicated by putting a bar over the group of digits that repeat. Now we state a remarkable fact.

The rational numbers are precisely those numbers that can be represented as repeating nonterminating decimals (see Problems 13–24).

What about nonrepeating decimals like

$$.12112111211112 \ldots$$

They represent the **irrational numbers**, of which $\sqrt{2} = 1.41421356 \ldots$ and $\pi = 3.14159265 \ldots$ are the best-known examples. Together, the rational numbers and the irrational numbers make up the real numbers. Thus we may say that:

The real numbers are those numbers that can be represented as nonterminating decimals.

Rational Numbers (the repeating decimals)	Irrational Numbers (the nonrepeating decimals)

THE REAL NUMBERS

THE COMPLEX NUMBERS

The real numbers are the principal characters in this book. But if you should meet a complex number (as you will in Sections 4-6 and 4-7), we do not want you to faint. After all, complex numbers are simply pairs of real numbers hooked together in a special way.

Let i denote $\sqrt{-1}$, that is, a number whose square is -1. Of course, i is not a real number since the square of any (nonzero) real number is positive. We follow tradition and say that i is imaginary, though you should not conclude from this that i is less than a solid member of mathematical society. In fact, i plays the principal role in the enlargement of the real number system to the complex number system. **Complex numbers** are precisely those numbers that can be written in the form $a + bi$, where a and b are real numbers.

To have any genuine knowledge of the complex numbers requires that we understand how to add, subtract, multiply, and divide them. Here we have space for only the briefest description of these four operations.

$$(a + bi) + (c + di) = (a + c) + (b + d)i$$
$$(a + bi) - (c + di) = (a - c) + (b - d)i$$

$$(a + bi)(c + di) = (ac - bd) + (ad + bc)i$$

$$\frac{a + bi}{c + di} = \frac{(a + bi)(c - di)}{(c + di)(c - di)}$$

$$= \frac{ac + bd}{c^2 + d^2} + \frac{bc - ad}{c^2 + d^2}i$$

Example D (in the problem set) suggests a practical way to perform these operations.

Our first problem set gives you an opportunity to practice your skill at operating on various kinds of numbers and manipulating algebraic expressions. Do not disdain such practice. Success in trigonometry is highly dependent on strong algebraic skills.

Problem Set 1-1

Perform the indicated operations, removing all parentheses.

1. $3 + (7 - (2 - 3))$ 2. $5 + 2(3 - (6 - 2))$

3. $3(x - 2(x + 3))$ 4. $2x - 4(5 - 2(1 - x))$

EXAMPLE A (Operations on Fractions)   Perform the indicated operations.

(a) $\dfrac{3}{2}\left(\dfrac{2}{3} + \dfrac{3}{4}\right)$;   (b) $\dfrac{\frac{2}{5} - \frac{2}{3}}{\frac{7}{10}}$;   (c) $\dfrac{x^2 - x - 6}{x^2 - 16} \cdot \dfrac{x + 4}{x - 3}$

Solution.

(a) $\dfrac{3}{2}\left(\dfrac{2}{3} + \dfrac{3}{4}\right) = \dfrac{3}{2} \cdot \dfrac{2}{3} + \dfrac{3}{2} \cdot \dfrac{3}{4} = 1 + \dfrac{9}{8} = \dfrac{17}{8}$

(b) $\dfrac{\frac{2}{5} - \frac{2}{3}}{\frac{7}{10}} = \dfrac{\frac{6}{15} - \frac{10}{15}}{\frac{7}{10}} = \dfrac{\frac{-4}{15}}{\frac{7}{10}} = -\dfrac{4}{15} \cdot \dfrac{10}{7} = -\dfrac{8}{21}$

(c) Here we must remember how to factor.

$$\frac{x^2 - x - 6}{x^2 - 16} \cdot \frac{x + 4}{x - 3} = \frac{(x - 3)(x + 2)}{(x + 4)(x - 4)} \cdot \frac{x + 4}{x - 3} = \frac{x + 2}{x - 4}$$

In Problems 5–12, perform the indicated operations. Leave your answers as fractions in reduced form (that is, with no nontrivial common factors in numerator and denominator).

5. $\dfrac{2}{3}\left(\dfrac{5}{8} - \dfrac{2}{3}\right) + \dfrac{3}{4}$ 6. $\dfrac{3}{4}\left(\dfrac{3}{8} + \dfrac{2}{3}\right) - \dfrac{5}{16}$

7. $\dfrac{2 - \frac{3}{4}}{3 + \frac{1}{8}}$ 8. $\dfrac{\frac{1}{5} + \frac{3}{4}}{\frac{1}{8}}$

9. $\dfrac{x^2 + 3x - 10}{x^2 + 4x + 3} \cdot \dfrac{x^2 + 5x + 6}{x^2 - 4}$ 10. $\dfrac{x + 2}{2x - 3} \div \dfrac{x^2 - 4}{2x^2 - 3x}$

11. $\dfrac{x}{x^2 - 9} + \dfrac{1}{x - 3}$ 12. $\dfrac{1/a + 1/b}{a/b - b/a}$

Use long division to write each of the following rational numbers as a repeating decimal.

13. $\frac{5}{16}$ 14. $\frac{7}{8}$ 15. $\frac{3}{11}$

16. $\frac{13}{11}$ 17. $\frac{19}{17}$ 18. $\frac{5}{13}$

EXAMPLE B (Changing Repeating Decimals to Fractions)   Write
.171717 . . . = .$\overline{17}$ as a ratio of two integers.

Solution.   Let $x = .\overline{17}$. Then $100x = 17.\overline{17}$. When we subtract x from $100x$, we have the following result.

$$100x = 17.\overline{17}$$
$$\underline{x = .17}$$
$$99x = 17$$
$$x = \tfrac{17}{99}$$

Note that we multiplied x by 100 because the repeating group in the decimal for x contained two digits. For the decimal .$\overline{179}$, we would multiply by 1000.

Write each of the following decimals as a ratio of two integers in reduced form.

19. .$\overline{28}$ 20. .$\overline{34}$ 21. .$\overline{179}$

22. .$\overline{123}$ 23. .1$\overline{79}$ 24. .1$\overline{23}$

EXAMPLE C (Solving Equations)   Solve the following equations.

(a) $\frac{2}{3}(x - 1) = \frac{2}{5}x + 2$;   (b) $t^2 - 5t + 4 = 0$;
(c) $6x^2 + 2x - 3 = 0$

Solution.

(a) $15[\frac{2}{3}(x - 1)] = 15(\frac{2}{5}x + 2)$
$10(x - 1) = 6x + 30$
$10x - 10 = 6x + 30$
$4x = 40$
$x = 10$

(b) Here we factor $t^2 - 5t + 4$ and set each factor equal to zero.

$$t^2 - 5t + 4 = 0$$
$$(t - 1)(t - 4) = 0$$
$$t - 1 = 0 \qquad t - 4 = 0$$
$$t = 1 \qquad t = 4$$

(c) From the **quadratic formula**, we know the solutions to $ax^2 + bx + c = 0, a \neq 0$, are given by $(-b \pm \sqrt{b^2 - 4ac})/2a$. Letting

$a = 6$, $b = 2$, and $c = -3$, we have

$$\frac{-2 \pm \sqrt{4 + 72}}{12} = \frac{-2 \pm \sqrt{76}}{12}$$

$$= \frac{-2 \pm \sqrt{4}\sqrt{19}}{12} = \frac{2(-1 \pm \sqrt{19})}{12} = \frac{-1 \pm \sqrt{19}}{6}$$

Solve the equations in Problems 25–38.

25. $3(x - 2) = 9$

26. $-5(x + 4) = 30$

27. $\frac{2}{3}t + 4 = \frac{1}{2}t - 1$

28. $3(u - 2) = 2(u - 3) + u$

29. $\dfrac{2}{x - 3} + \dfrac{3}{x - 7} = \dfrac{7}{(x - 3)(x - 7)}$

30. $\dfrac{2}{x - 1} + \dfrac{3}{x + 1} = \dfrac{19}{x^2 - 1}$

31. $v^2 - 16 = 0$

32. $u^2 - 36 = 0$

33. $x^2 - 7x + 12 = 0$

34. $x^2 - 9x + 8 = 0$

35. $t^2 + 2t - 8 = 0$

36. $t^2 - 4t - 60 = 0$

37. $3u^2 + u - 1 = 0$

38. $2w^2 + 3w - 6 = 0$

39. A wire 130 centimeters long is bent into the shape of a rectangle that is 3 centimeters longer than it is wide. Find the width of the rectangle.

40. One leg of a right triangle is 2 inches longer than the other leg and the hypotenuse has length $\sqrt{174}$ inches. Find the lengths of the legs. *Hint:* Recall that in a right triangle, $c^2 = a^2 + b^2$, the Pythagorean theorem.

EXAMPLE D (Operations on Complex Numbers)   Perform the indicated operations and simplify.

(a) $(3 + 2i) + (-1 + 4i)$;   (b) $(3 + 2i)(-1 + 4i)$;

(c) $\dfrac{3 + 2i}{-1 + 4i}$

Solution.   Do not try to remember the formulas we gave in the text. Simply do what comes naturally, replacing i^2 by -1 wherever it appears. For division, there is a special technique.

(a) $(3 + 2i) + (-1 + 4i) = (3 - 1) + (2i + 4i) = 2 + 6i$

(b) $(3 + 2i)(-1 + 4i) = (-3 + 8i^2) + (12i - 2i)$

$$= (-3 + 8(-1)) + 10i = -11 + 10i$$

(c) To perform division, we multiply both numerator and denominator by the conjugate of the denominator. The **conjugate** of $a + bi$, symbolized by $\overline{a + bi}$, is $a - bi$.

$$\frac{3 + 2i}{-1 + 4i} = \frac{(3 + 2i)(-1 - 4i)}{(-1 + 4i)(-1 - 4i)}$$

$$= \frac{(-3 - 8i^2) + (-12i - 2i)}{1 + 16} = \frac{5 - 14i}{17} = \frac{5}{17} - \frac{14}{17}i$$

Perform the indicated operations in Problems 41–46. Write your answer in the form a + bi.

41. $(6 + 2i) - (4 + 3i)$ 42. $(3 - 2i) + (-4 + 6i)$

43. $(6 + 2i)(4 + 3i)$ 44. $(3 - 2i)(-4 + 6i)$

45. $\dfrac{6 + 2i}{4 + 3i}$ 46. $\dfrac{-4 + 6i}{3 - 2i}$

MISCELLANEOUS PROBLEMS

47. Write each of the following rational numbers both as a decimal and as a ratio of two integers.
    (a) $2.3 - \frac{9}{15}$ (b) $2.56 + (\frac{5}{6})$
    (c) $2(1.5)^2(\frac{8}{7})$ (d) $(3.25)^2 \div (\frac{3}{8})$

48. Write each of the following as a ratio of two integers in reduced form.
    (a) $.9\overline{36}$ (b) $.\overline{612}$
    (c) $(\frac{3}{22}) + .\overline{45}$ (d) $(3.3)(.\overline{23})$

49. Perform the indicated operations and simplify.
    (a) $\frac{2}{3} - \frac{3}{4}(\frac{3}{8} - \frac{5}{6})$ (b) $\dfrac{\frac{4}{5} - \frac{3}{4}}{\frac{4}{5} + \frac{3}{4}}$
    (c) $\dfrac{a^2(a - 1/a)^2}{(a + 1)^2}$ (d) $\left(\dfrac{2}{a + 1} + \dfrac{2}{1 - a}\right)(2 + 2a)$

50. Write each of the following in the form $a + bi$.
    (a) $\dfrac{(3 - 2i)(3i + 1)}{i^6}$ (b) $\dfrac{10i + 5}{4 - 3i}$
    (c) $(4 - i)^2 + 6i - 17$ (d) $\dfrac{(2 + i)^2}{3 + 4i} + \dfrac{3 - 4i}{2} + 2i - \dfrac{1}{2}$

51. Replace the box by $<$, $>$, or $=$ to make a true statement.
    (a) $.\overline{24}$ ☐ $\frac{1}{4}$ (b) $.\overline{666}$ ☐ $\frac{2}{3}$ (c) $\frac{41}{114}$ ☐ $.36$
    (d) $.\overline{24}$ ☐ $\frac{8}{33}$ (e) $\frac{44}{7}$ ☐ 2π (f) $6.\overline{9}$ ☐ 7

52. Find the number midway between each of the given two numbers. *Hint:* Find their average.
    (a) -2 and 15 (b) $\frac{3}{4}$ and $\frac{5}{3}$ (c) $.\overline{24}$ and $\frac{5}{11}$

53. In each case, find all values of x which satisfy the given equation.
    (a) $|x - 2| = 5$ (b) $|3x + 1| = 10$ (c) $|x^2 - 8| = 4$

54. The inequality $|x - 3| < 1.5$ is equivalent to $-1.5 < x - 3 < 1.5$, which in turn is equivalent to $1.5 < x < 4.5$. Write each of the following in the form $a < x < b$.
    (a) $|x - 2| < 5$ (b) $|2x + 3| < 2$ (c) $|x - \frac{5}{6}| < \frac{7}{3}$

55. The inequality $2 < x < 11$ is equivalent to $|x - 6.5| < 4.5$. (Note that 6.5 is midway between 2 and 11 and 4.5 is half of $11 - 2$.) Write each of the following as a single inequality involving an absolute value.
    (a) $11 < x < 23$ (b) $-5 < x < \frac{11}{4}$ (c) $3.12 \leq x \leq 7.76$
    (d) $(2 + xi) \cdot (1 + i) = -1 + 5i$
    (e) $(2 - 3i)(3 + i) = 1 + i + x(2 + 2i)$

56. Given that π is an irrational number, prove that $\pi + 1.5$ and $(1.5)\pi$ are also irrational. *Hint:* Try a proof by contradiction.

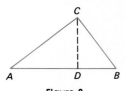

Figure 7

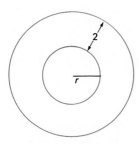

Figure 8

57. Solve each of the following equations.
    (a) $\frac{3}{4}x - \frac{2}{3} = \frac{5}{12}$ (b) $2x^2 + 9x - 5 = 0$
    (c) $x^3 - 4x^2 - 12x = 0$ (d) $(x + 3)^2 + (x - 5)^2 = 50$
    (e) $\dfrac{4}{x + 1} + \dfrac{2}{x - 1} = 2$ (f) $2x = \dfrac{x - 3}{x - 2}$

58. The sum of the squares of three consecutive positive even integers is 440. Find the smallest of these integers.

59. The sum of a number and its reciprocal is $13/6$. Find the number.

60. Two triangles are **similar** if corresponding angles are equal. From geometry, we know that the ratios of corresponding sides of similar triangles are equal. Thus, for the similar triangles of Figure 7,

$$\frac{a}{d} = \frac{b}{e} = \frac{c}{f}$$

Suppose one triangle has sides of length 4, 5, and 6 and that the longest side of a similar triangle measures 7.5. Find the lengths of the other two sides.

61. The triangle in Figure 8 is a right triangle with $\overline{AB} = 10$, $\overline{BC} = 6$, and $\overline{AC} = 8$. The dotted segment CD is perpendicular to AB. Show that triangles ADC and CDB are both similar to triangle ACB and find $\overline{CD}$, $\overline{AD}$, and $\overline{BD}$.

62. A square of side length $10\sqrt{2}$ is inscribed in a circle (the circle goes through all four vertices).
    (a) Find the radius of the circle.
    (b) Find the area of the region outside the square but inside the circle. *Recall*: The area of a circle of radius r is πr^2.

63. Three circles of radius r are mutually tangent. Find the area of the small region (a curved triangle) trapped between the three circles.

64. Suppose that the ring (annulus) of Figure 9 is 2 centimeters wide and that this ring has the same area as the smaller circle. Find r, the radius of the smaller circle.

65. A rectangular garden is twice as long as wide. It is surrounded by a walk 2 feet wide. If the walk has an area of 256 square feet, find the dimensions of the garden.

66. **TEASER** Figure 10 represents the end view of a humidifier, a rotating disk that is partially submerged in water. We wish to determine the area A of the wet part of the disk (an annulus) but cannot measure either its inner or its outer radius because of the device that holds the wheel. However, it is easy to measure a and b. Express A in terms of a and b. Note that this shows that all annuli (no matter what their radii) have the same area if they intersect a given line with the same measurements a and b.

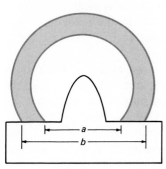

Figure 9

Figure 10

A new revolution is taking place in technology today. It both parallels and completes the Industrial Revolution that started a century ago. The first phase of the Industrial Revolution meant the mechanization, then the electrification of brawn. The new revolution means the mechanization and electrification of brains.

Harry M. Davis

1-2 Pocket Calculators

Most people can do arithmetic if they have to. But few do it with either enthusiasm or accuracy. Frankly, it is rather a dull subject. The spectacular sales of pocket electronic calculators demonstrate both our need to do arithmetic and our distaste for it.

Pocket calculators vary greatly in what they can do. The simplest perform only the four arithmetic operations and sell for under $10. The most sophisticated are programmable and can automatically perform dozens of operations in sequence. For this course, a standard scientific calculator is ideal. In addition to the four arithmetic operations, it will calculate values for the exponential and logarithmic functions and the trigonometric and inverse trigonometric functions.

Two kinds of logic are used in pocket calculators, **reverse Polish** logic and **algebraic** logic. The former avoids the use of parentheses but is slightly tricky to learn. Algebraic logic uses parentheses and mimics the procedures of ordinary algebra. The Texas Instruments Model 30 shown earlier uses algebraic logic and will serve to illustrate our discussion of calculators. You will need to study the instruction book for any other calculator to understand the minor modifications required to use that calculator. And if you already are proficient with a calculator, skim the rest of this section. But try a few of the problems for practice.

ENTERING DATA

The first thing to understand is what kind of numbers a calculator will accept as data. Those nonterminating decimals of Section 1-1 obviously will not fit; they must be rounded off to some finite sequence of digits (8 digits on many calculators). A decimal such as 238.75142 fits fine; to enter it, simply press the keys

$$238 \boxed{\cdot} 75142$$

in succession. To enter the negative of this number, press the same keys and then the key to change the sign, $\boxed{+/-}$.

What we have said takes care of numbers with eight digits or less. But what about very small or very large numbers like .00000000000123 or 593,800,000,000? These numbers must first be written in scientific notation. Recall that a positive number is in **scientific notation** when it is in the form $c \times 10^n$, where $1 \le c < 10$ and n is an integer. Thus

$$.00000000000123 = 1.23 \times 10^{-12}$$

$$593,800,000,000 = 5.938 \times 10^{11}$$

To enter 5.938×10^{11}, press the keys

$$5 \boxed{\cdot} 938 \boxed{EE} 11$$

The key $\boxed{EE}$, which stands for *enter exponent*, controls the two places at the extreme right of the display, which are reserved for the exponent on 10. After pressing the indicated keys, the display will show

$$\boxed{5.938 \quad 11}$$

If you press

$$1 \boxed{\cdot} 23 \boxed{EE} 12 \boxed{+/-}$$

it will show

$$\boxed{1.23 \quad -12}$$

This entry represents 1.23×10^{-12}. Finally, to represent -1.23×10^{-12}. press

$$1 \boxed{\cdot} 23 \boxed{+/-} \boxed{EE} 12 \boxed{+/-}$$

In making a calculation, you may enter some numbers in standard notation and others in scientific notation. The calculator understands either form and makes the proper translations. If any of the entered data is in scientific notation, it will display the answer in this format. Also, if the answer is too large or too small for standard format, the calculator will automatically convert the result of a calculation to scientific notation.

DOING ARITHMETIC

The five keys $\boxed{+}$, $\boxed{-}$, $\boxed{\times}$, $\boxed{\div}$, and $\boxed{=}$ are the work horses for arithmetic in any calculator using algebraic logic. To perform the calculation

$$175 + 34 - 18$$

simply press the following keys.

$$175 \boxed{+} 34 \boxed{-} 18 \boxed{=}$$

The answer 191 will appear in the display.

Or consider (175)(14)/18. Press

$$175 \boxed{\times} 14 \boxed{\div} 18 \boxed{=}$$

and the calculator will display 136.11111.

An expression involving additions (or subtractions) and multiplications (or divisions) may be ambiguous. For example, $2 \times 3 + 4 \times 5$ could have several meanings depending on which operations are performed first.

1. $2 \times (3 + 4) \times 5 = 70$
2. $(2 \times 3) + (4 \times 5) = 26$
3. $((2 \times 3) + 4) \times 5 = 50$
4. $2 \times (3 + (4 \times 5)) = 46$

Parentheses are used in mathematics and in calculators to indicate the order in which operations are to be performed. To do Calculation 1, press

$$2 \boxed{\times} \boxed{(} 3 \boxed{+} 4 \boxed{)} \boxed{\times} 5 \boxed{=}$$

Similarly to do Calculation 3, press

$$\boxed{(} \boxed{(} 2 \boxed{\times} 3 \boxed{)} \boxed{+} 4 \boxed{)} \boxed{\times} 5 \boxed{=}$$

Recall that in arithmetic, we have an agreement that multiplications and divisions are done before additions and subtractions when no parentheses are used. Thus,

$$2 \times 3 + 4 \times 5$$

is interpreted as $(2 \times 3) + (4 \times 5)$. The same convention is used in most calculators. Pressing

$$2 \boxed{\times} 3 \boxed{+} 4 \boxed{\times} 5 \boxed{=}$$

will yield the answer 26. Similarly for Calculation 3, pressing

$$\boxed{(} 2 \boxed{\times} 3 \boxed{+} 4 \boxed{)} \boxed{\times} 5 \boxed{=}$$

will yield 50 since, within the parentheses, the calculator will do the multiplication first. However, when in doubt, it is always safest to use parentheses; without them, it is easy to make errors.

SPECIAL FUNCTIONS

Most scientific calculators have keys for finding powers and roots of a number. On such a calculator, the $\boxed{y^x}$ key is used to raise a number y to the xth power. For example, to calculate $2.75^{-.34}$, press

$$2 \;\boxed{\cdot}\; 75 \;\boxed{y^x}\; \boxed{\cdot}\; 34 \;\boxed{+/-}\; \boxed{=}$$

and the correct result, .70896841, will appear in the display.

Finding a root is the inverse of raising a number to a power. For example, taking a cube root is the inverse of cubing a number. Thus, to calculate $\sqrt[3]{17}$, press 17 $\boxed{\text{INV}}$ $\boxed{y^x}$ 3 $\boxed{=}$ and you will get 2.5712816. To use the $\boxed{y^x}$ key, y must be positive. However, x may be either positive or negative. Some calculators have a special key for this process, $\boxed{\sqrt[x]{y}}$. To use this key, press 17 $\boxed{\sqrt[x]{y}}$ 3 $\boxed{=}$.

Square roots occur so often that on many calculators there is a special key for them. Thus to calculate $\sqrt{17}$, simply press 17 $\boxed{\sqrt{}}$ and you will immediately get 4.1231056. On other calculators, you will need to press 17 $\boxed{\text{INV}}$ $\boxed{x^2}$.

Similarly, there is a special key for both common logarithms, $\boxed{\log}$, and natural logarithms, $\boxed{\ln x}$. To obtain log 56.34, simply press

$$56 \;\boxed{\cdot}\; 34 \;\boxed{\log}$$

and the number 1.7508168 will appear in the display. To calculate

$$15 + 3 \log 56.34$$

press

$$15 \;\boxed{+}\; 3 \;\boxed{\times}\; 56 \;\boxed{\cdot}\; 34 \;\boxed{\log}\; \boxed{=}$$

Note that it was not necessary to use any parentheses. Just as one would do in algebra, the calculator first finds the logarithm, then multiplies by 3, and finally adds 15. You should check the instruction book for your calculator to make sure you understand the order in which it does operations when no parentheses are used.

The real power of calculators for trigonometry will be evident when we study the trigonometric functions sine, cosine, and tangent. Then, rather than looking up these functions in tables, we may simply press the keys $\boxed{\sin}$, $\boxed{\cos}$, and $\boxed{\tan}$. Not only is this much faster, but it also gives us eight-digit accuracy rather than the four- or five-digit accuracy of a table. But we are getting ahead of ourselves; that comes in Chapter 2.

SIGNIFICANT DIGITS AND ACCURACY OF ANSWERS

When working with measurements (which are always approximate), it is important to know what rules we should follow in making calculations and reporting answers. Suppose, for example, that we wish to calculate the area of

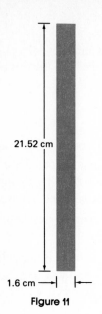

21.52 cm

1.6 cm →|

Figure 11

a rectangular metal strip whose length and width are recorded as 21.52 centimeters and 1.6 centimeters, respectively (Figure 11). Multiplying these numbers, we get 34.432. Can we honestly claim 34.432 square centimeters as the area of the metal strip? Hardly, since to say that 21.52 is the length really means that the length is between 21.515 and 21.525. Similarly, the width is between 1.55 and 1.65. Thus the area is somewhere between $(21.515)(1.55) = 33.34815$ and $(21.525)(1.65) = 35.51625$. If we are to report one number as the area, what should it be? To answer, we need the concept of significant digits.

We illustrate what we mean by **significant digits** with four examples.

NUMBER	SIGNIFICANT DIGITS
.0024	2, 4
1.205	1, 2, 0, 5
1.4	1, 4
1.40	1, 4, 0

In each case, the string of significant digits starts with the first nonzero digit and ends with the last digit definitely specified.

Now we can state a rule to use in calculations involving approximate numbers.

RULE 1

In any calculation involving multiplication, division, raising to powers, or extracting roots of numbers obtained as measurements, give the answer with the same number of significant digits as the measurement with fewest significant digits.

In using this rule, you should carry at least one more significant digit than is present in the least accurate measurement through all intermediate calculations and then round at the final step. Incidentally, we round up if the first neglected digit is 5 or more and down otherwise.

Here is how we use the rule in the example involving the area of a metal strip.

	LENGTH	WIDTH
Measurement	21.52	1.6
Number of significant digits	4	2
Measurement rounded off	21.5	1.6

Area: $(21.5)(1.6) = 34.40$
Area to two significant digits: 34

Suppose that the width had been measured as 1.61. Then we would have calculated

$$(21.52)(1.61) = 34.6472$$

and reported 34.6 as the area.

Though it will be needed less often in this book, there is also a rule for additions and subtractions.

RULE 2

In a calculation involving additions or subtractions of numbers obtained as measurements, the reported answer should not show more decimal places than the measurement with fewest decimal places.

For example, suppose we are to find the perimeter of a triangle whose sides were measured as 4.123, 5.2, and 3.49. While these numbers have a sum of 12.813, we should round this off to 12.8 in reporting the answer, since the least precise of the data, 5.2, has only one decimal place.

Whether the final zeros of a number are significant digits or whether they simply serve to place the decimal point must sometimes be determined from the context. If the population of a city is given as 490,000 in a list that clearly gives populations to the nearest thousand, then the first zero is significant but the others are not. We can avoid any ambiguity by always writing approximate numbers in scientific notation. The population of that city would be written as 4.90×10^5. If it were correct to the nearest hundred, we would write 4.900×10^5.

Problem Set 1-2

Use your calculator to perform the following calculations. Assume all numbers are exact and give your answers with the maximum accuracy that your calculator allows. To catch errors caused by pressing the wrong keys or failing to use parentheses properly, we suggest that you make a quick mental estimate of the answer. For example, the answer to Problem 3 might be estimated as $(3 - 6)(14 \times 50) = -2100$.

1. $34.1 - 49.95 + 64.2$

2. $7.465 + 3.12 - .0156$

3. $(3.42 - 6.71)(14.3 \times 51.9)$

4. $(21.34 + 2.37)(74.13 - 26.3)$

5. $\dfrac{514 + 31.9}{52.6 - 50.8}$

6. $\dfrac{547.3 - 832.7}{.0567 + .0416}$

7. $\dfrac{(6.34 \times 10^7)(537.8)}{1.23 \times 10^{-5}}$

8. $\dfrac{(5.23 \times 10^{16})(.0012)}{1.34 \times 10^{11}}$

9. $\dfrac{6.34 \times 10^7}{.00152 + .00341}$

10. $\dfrac{3.134 \times 10^{-8}}{5.123 + 6.1457}$

11. $\dfrac{532 + 1.346}{34.91}(1.75 - 2.61)$

12. $\dfrac{39.95 - 42.34}{15.76 - 16.71}(5.31 \times 10^4)$

13. $(1.214)^3$

14. $(3.617)^{-2}$

15. $\sqrt[3]{1.215}$

16. $\sqrt{1.5789}$

17. $\dfrac{(1.34)(2.345)^3}{\sqrt{364}}$

18. $\dfrac{(14.72)^{12}(59.3)^{11}}{\sqrt{17.1}}$

19. $\dfrac{\sqrt{130} - \sqrt{5}}{15^6 - 4^8}$

20. $\dfrac{\sqrt{143.2} + \sqrt{36.1}}{(234.1)^4 - (11.2)^2}$

21. $(\log 4.156)^2$

22. $(\log 9.132)^3$

23. $\dfrac{\sqrt[3]{4.31} + 2\ln 5.79}{1.413 \times 10^{-4}}$

24. $\dfrac{(\log 459.3)^{.23}}{74.12 - \sqrt{69.3}}$

Figure 12

Each of the following problems involves measurements; therefore your answers should be rounded according to the rules given in the text.

25. Find the length of the hypotenuse of a right triangle whose legs were measured to be 4.134 centimeters and 59.62 centimeters, respectively. *Note:* In a right triangle, $a^2 + b^2 = c^2$.

26. Find the area of a circle whose diameter was measured to be 5.123 inches. *Note:* $A = \pi r^2$.

27. Find the length of a running track if it has the shape of a square capped on opposite ends by semicircles (Figure 12). The square measures 42.55 meters on a side. *Note:* For a circle, $C = 2\pi r$.

28. Find the area of the region enclosed by the track in Problem 27.

29. Light travels at 2.998×10^5 kilometers per second and 1 mile is 1.609 kilometers. How long will it take a light ray to travel from the sun to the earth $(9.30 \times 10^7$ miles)?

30. It is 2.51×10^{13} miles to our nearest star. How long does it take light from there to reach us (see Problem 29)?

MISCELLANEOUS PROBLEMS

31. Calculate $\dfrac{4}{(2.68)^2} + \sqrt{96.28} - \dfrac{\pi}{\sqrt{.092}}$.

32. Calculate $\dfrac{3.2 \times 10^4}{(4.9 + \sqrt{6.32})^3}$.

33. Thomas A. Doolittle lived a long life of 92 years, 131 days, 5 hours, and 6 minutes. During this time his heart beat an average of 72 beats per minute. How long did he live in heartbeats? Assume that a year has 365.24 days.

34. Last year the ABC Company had a total payroll of $1,843,338. It had 141 employees who worked an average of 38.22 hours per week. On the average, how much per hour did the company pay its workers? Assume that a year has 52 weeks and give your answer to the nearest penny.

To do the remaining problems, you may need some of the formulas from geometry listed inside the front cover. All data are to be considered to be approximate and therefore answers should be rounded according to the rules given in this section.

35. Find the volume V and surface area S of a rectangular box whose dimensions are 2.62, 4.91, and 3.19, all in feet.

36. Find the volume V and surface area S of a sphere of radius 9.62 inches.

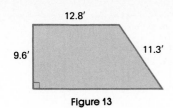

12.8'

9.6'   11.3'

Figure 13

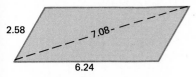

2.58

7.08

6.24

Figure 14

37. The radius of a spherical balloon decreased from 42.6 meters to 37.8 meters. Calculate the decrease in volume.

38. Find the slant height s and the volume V of a right circular cone of base radius 4.92 centimeters and height 12.61 centimeters.

39. Find the perimeter P and the area A of the trapezoid shown in Figure 13.

40. According to Isaac Asimov, the radius of the observable universe is 1.23×10^{28} centimeters.
    (a) Find the volume of the observable universe.
    (b) About how many protons (each of volume 1.7×10^{-40} cubic centimeters) could the observable universe hold, assuming protons can be packed tightly together?

41. The area A of a triangle with sides of length a, b, and c and with semiperimeter $s = (a + b + c)/2$ is given by the formula (Heron's formula)

$$A = \sqrt{s(s - a)(s - b)(s - c)}$$

Find the area of a triangle with sides of lengths 12.4, 18.7, and 25.9.

42. Find the area of the parallelogram shown in Figure 14.

43. The mean $\bar{x}$ and standard deviation s of a set of values $x_1, x_2, \ldots, x_n$ are given by the formulas

$$\bar{x} = \frac{1}{n}(x_1 + x_2 + \cdots + x_n) \qquad s = \left[\frac{1}{n}\left(x_1^2 + x_2^2 + \cdots + x_n^2\right) - \bar{x}^2\right]^{1/2}$$

Five men have heights 68.7, 71.8, 74.6, 70.8, and 65.9, all in inches. Find $\bar{x}$ and s for this set of values.

44. **TEASER** Driving across the plains of North Dakota, I noticed that I would soon be passing an immensely long freight train, which was proceeding on tracks parallel to the road I was traveling. Guessing that the train would be traveling at a constant rate, I determined to measure the length of this gargantuan train. To do it, I made use of both my car's speedometer and its odometer. Traveling at the speed of 55 miles per hour, I found it took me 3.55 miles to pass the train, while at the speed of 59 miles per hour, it took me 3.10 miles to pass the train. From this I deduced the length of the train. See if you can do so.

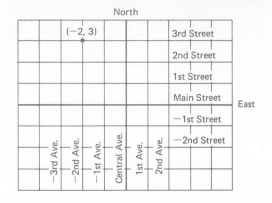

[Coordinate geometry], far more than any of his metaphysical speculations, immortalized the name of Descartes, and constitutes the greatest single step ever made in the progress of the exact sciences.

John Stuart Mill

1-3   Cartesian Coordinates

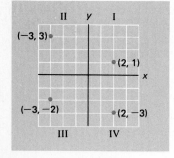

Figure 15

Two Frenchmen, René Descartes and Pierre de Fermat, are responsible for the idea in this section. Although Descartes is better known and usually gets credit for the idea, Fermat introduced the concept earlier.

Imagine the plane laid out like a giant city map. Main Street, the chief east-west artery, appears as a horizontal line. Central Avenue, the principal north-south road, is a vertical line. Every other road gets its name from its position relative to these two chief roads. First Avenue is one block east of Central Avenue; Minus Second Avenue is two blocks west. Similarly, Third Street is three blocks north of Main Street, while Minus Fourth Street is four blocks south. Any intersection can be specified by giving two numbers—first the number of the avenue and then the number of the street. For example, $(-2, 3)$ specifies the intersection of Minus Second Avenue and Third Street.

It's just a small step from this to the grand idea of Descartes and Fermat. Take two real lines and place them on the plane, one horizontal and the other vertical. Call the first the **x-axis** and the second the **y-axis**; call their intersection point the **origin**. Now any point in the plane is uniquely specified by giving two coordinates (x, y). The first, called the **x-coordinate** or **abscissa**, measures the directed distance from the vertical axis. The second, called the **y-coordinate** or **ordinate**, measures the directed distance from the horizontal axis. A few examples are shown in Figure 15. Note that the two axes divide the plane into four quadrants; they are traditionally numbered I, II, III, and IV, as shown.

Why is the introduction of coordinates such an important idea? You will see.

THE DISTANCE FORMULA

The first important consequence is a simple formula for the distance between any two points. It is based on the **Pythagorean theorem**, which says that if a and b measure the two legs of a right triangle and c measures its hypotenuse

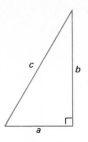

Figure 16

(Figure 16), then

$$a^2 + b^2 = c^2$$

The relationship between the three sides of a triangle holds only for a right triangle.

Now consider two points P and Q, with coordinates (x_1, y_1) and (x_2, y_2), respectively. Together with $R(x_2, y_1)$, they are vertices of a right triangle (Figure 17). The lengths of PR and RQ are $|x_2 - x_1|$ and $|y_2 - y_1|$, respectively. When we apply the Pythagorean Theorem and take the square root of both sides, we obtain the following expression for $d(P, Q)$, the distance between P and Q.

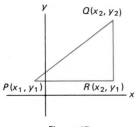

Figure 17

$$d(P, Q) = \sqrt{(x_2 - x_1)^2 + (y_2 - y_1)^2}$$

This is called the **distance formula**.

As an example, the distance between $(-2, 3)$ and $(4, -1)$ is

$$\sqrt{(4 - (-2))^2 + (-1 - 3)^2} = \sqrt{36 + 16} = \sqrt{52} \approx 7.21$$

The formula is correct even if the two points lie on the same horizontal or vertical line. For example, the distance between $(-2, 2)$ and $(6, 2)$ is

$$\sqrt{(-2 - 6)^2 + (2 - 2)^2} = \sqrt{64} = 8$$

THE EQUATION OF A CIRCLE

It is a small step from the distance formula to the equation of a circle. **A circle** is the set of points which lie at a fixed distance r (called the radius) from a fixed point (called the center). Consider, for example, a circle of radius 3 with center at $(0, 0)$. As in Figure 18, let (x, y) denote any point on this circle (the key step). By the distance formula,

$$\sqrt{(x - 0)^2 + (y - 0)^2} = 3$$

When we square both sides, we obtain

$$x^2 + y^2 = 9$$

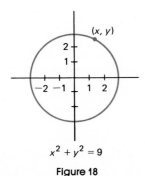

$x^2 + y^2 = 9$

Figure 18

which we call the *equation* of this circle.

What is significant here is that we have transformed a geometric object (a circle) into an algebraic object (an equation). That and the reverse process (graphing an equation) constitute the giant idea to which John Stuart Mill referred in our opening quotation.

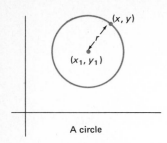

A circle

Figure 19

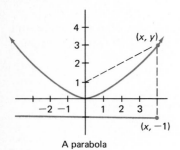

A parabola

Figure 20

We can handle the general circle (Figure 19) of radius r centered at (x_1, y_1) just as easily. Its equation, called the **standard equation of the circle**, is

$$(x - x_1)^2 + (y - y_1)^2 = r^2$$

A PARABOLA

As a second example of writing an equation for a curve, consider the set of all points (x, y) that are equidistant from the point $(0, 1)$ and the horizontal line one unit below the x-axis (Figure 20). From the distance formula we have

$$\sqrt{(x - 0)^2 + (y - 1)^2} = \sqrt{(x - x)^2 + (y + 1)^2}$$

Squaring both sides yields

$$x^2 + y^2 - 2y + 1 = y^2 + 2y + 1$$

which simplifies to $x^2 = 4y$ or $y = \frac{1}{4}x^2$. This type of equation will receive more attention in the next section; it is the equation of a parabola.

Problem Set 1-3

In Problems 1–6, plot the points P and Q in the Cartesian plane and find the distance between them.

1. $P(1, 5)$, $Q(-2, 1)$ 2. $P(-1, 12)$, $Q(4, 0)$
3. $P(2, -2)$, $Q(4, 3)$ 4. $P(-1, 5)$, $Q(2, 6)$
C 5. $P(\pi, -1.517)$, $Q(-\sqrt{2}, \sqrt{3})$ C 6. $P(2.134, 2.612)$, $Q(\sqrt{5}, \sqrt{7})$

7. Plot the points $A(1, 1)$, $B(6, 3)$, $C(4, 8)$, and $D(-1, 6)$ and show that they are vertices of a rectangle. *Hint:* It is enough to show that the diagonals are of equal length.

8. Show that $A(1, 3)$, $B(2, 6)$, $C(4, 7)$, and $D(3, 4)$ are vertices of a parallelogram.

9. $A(2, -1)$ and $C(8, 7)$ are two of the vertices of a rectangle with sides parallel to the coordinate axes. Find the coordinates of the other two vertices.

10. Show that $A(2, -4)$, $B(8, -2)$, and $C(4, 0)$ are vertices of a right triangle.

11. Write the equation of the circle with center at $(1, -2)$ and radius 3.

12. Write the equation of the circle with center $(-\pi, 1)$ and radius $\sqrt{2}$.

EXAMPLE (More on Circles) When $(x - x_1)^2 + (y - y_1)^2 = r^2$ is expanded, it takes the form $x^2 + y^2 + Ax + By = C$. We therefore expect an equation of the latter form to represent a circle. Find the center and radius of the circle with equation

$$x^2 + y^2 - 2x + 6y = 6$$

Solution. We rearrange the terms and then complete the squares by adding the same number to both sides of the equation.

$$(x^2 - 2x \quad) + (y^2 + 6y \quad) = 6$$

$$(x^2 - 2x + 1) + (y^2 + 6y + 9) = 6 + 1 + 9$$

$$(x - 1)^2 + (y + 3)^2 = 16$$

The last equation is in standard form; it is the equation of a circle centered at $(1, -3)$ with radius 4.

Use the method just described to find the centers and radii of circles with the following equations.

13. $x^2 + y^2 - 8x + 5y = 1$

14. $x^2 + y^2 - 3x + 9y = 0$

15. $2x^2 + 2y^2 - 8x + 16y = 4$

16. $3x^2 + 3y^2 - 6x + 9y = -1$

17. $x^2 + y^2 + \pi x - 2y = 0$

18. $x^2 + y^2 - 6x + 4y = -13$

MISCELLANEOUS PROBLEMS

19. Find the equation of the circle that has the line segment connecting $(-1, 3)$ and $(7, 9)$ as a diameter. *Hint:* The coordinates of the midpoint of a line segment are found by averaging the corresponding coordinates of the endpoints.

20. Find the equation of the circle that has center $(-3, 4)$ and is tangent to the x-axis.

21. Find two points on the x-axis that are 6 units from the point $(5, 2)$.

22. Find the equation of the circle with center on the x-axis that goes through $(4, 3)$ and $(8, 1)$.

23. The points $(0, 0)$, $(2, 2)$, and (a, b) with $a < 0$ are the coordinates of an equilateral triangle. Find a and b.

24. Find the two points where the line $y = x$ intersects the circle with equation $(x + 3)^2 + (y - 2)^2 = 17$.

25. How far is it between the centers of the circles with equations $x^2 + y^2 + 4x - 8y = 16$ and $x^2 + y^2 - 6x + 10y = 15$?

26. How far is it from the point $(8, 3)$ to the circle with equation $x^2 + y^2 - 4x + 2y = 11$?

27. A belt fits tightly around the two circular wheels with equations $(x - 2)^2 + (y - 3)^2 = 9$ and $(x - 8)^2 + (y + 1)^2 = 9$. How long is this belt?

28. Find the equations of the two circles of radius 5, tangent to the vertical line $x = 3$, with centers on the line $y = x$.

29. The common chord of two intersecting circles has length 10. The circles have radii 13 and $\sqrt{41}$, respectively. How far apart are their centers?

30. There are two circles with center at $(1, -4)$ that are tangent to the circle $(x + 1)^2 + (y - 4)^2 = 16$. (One of these circles has the given circle in its interior.) Find the equations of the two circles.

31. The x-axis is along the ground. A tricycle has the hub of its rear wheel at $(0, 4)$ and the hub of its front wheel at $(20, 10)$. Jamie rode the tricycle to the right along the x-axis so that the front wheel made 5 complete revolutions.
    (a) How many revolutions did the rear wheel make?
    (b) What are the new coordinates of the hub of the front wheel?

32. Point B is 12 miles downstream and on the opposite side of a river from point A. The river is straight and $\frac{1}{2}$ mile wide. Starting at A, Tricia ran 7 miles along the shore to point C and then swam on a straight line to B.
   (a) Find the total length of Tricia's path.
   (b) How long did it take if Tricia ran at 7 miles per hour and swam at 4 miles per hour?

33. A point moves in the xy-plane in such a way that its distance from the point $(3, 2)$ is always equal to its distance from the vertical line $x = -5$. Derive the equation of its path. *Hint:* See the example just before this problem set.

34. A point moves so that its distance from the point $(0, 0)$ is always three times its distance from the point $(0, 4)$. Derive the equation of its path and identify this curve.

35. Let (a, b) be any point above the x-axis on the circle $x^2 + y^2 = r^2$. Show that the points (a, b), $(-r, 0)$, and $(r, 0)$ are the vertices of a right triangle.

36. Let $a > b > 0$. Consider the four points (a, b), $(5a, -b)$, $(3a, 5b)$, and $(-a, 7b)$. Show that they are vertices of a parallelogram and find the length of its shortest diagonal.

37. Starting at $(1, 0)$, I walked continuously around the circle $x^2 + y^2 = 1$ in a clockwise manner until I had traveled $19\pi/4$ units. How far was I then from my starting point?

38. **TEASER** Let the xy-plane be oriented so the positive y-axis points north and the positive x-axis points east. Starting at $(0, 0)$, Arnold walked 16 miles east, 8 miles south, 4 miles west, 2 miles north, 1 mile east, $\frac{1}{2}$ mile south, and so on, ad infinitum. How far is his destination from his starting point?

Start with a curve

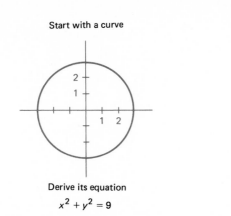

Derive its equation

$$x^2 + y^2 = 9$$

$$y = x^2 - 3$$

Draw its graph

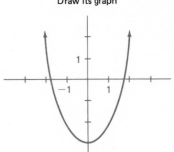

1-4 Graphing Equations

The Cartesian program runs in both directions. Not only can we go from a curve to its equation, as we did in the last section for circles and parabolas, we can also go from an equation to its graph. The former turns geometry into algebra; the latter transforms algebra into geometry.

What do we mean by the **graph** of an equation in x and y? We mean simply the set of all points P whose coordinates (x, y) satisfy the equation. How shall we draw the graph of an equation? This is best explained by means of an example.

THE GRAPH OF $y = x^2 - 3$

To obtain a graph, we follow a definite procedure.

1. Obtain the coordinates of a few points.
2. Plot those points in the plane.
3. Connect the points with a smooth curve in the order of increasing x-values.

The best way to do Step 1 is to make a **table of values**. Assign values to one of the variables, say x, determine the corresponding values of the other variable, and list the pairs of values in tabular form. The whole three-step procedure is illustrated for $y = x^2 - 3$ in Figure 21 on the next page.

Of course, you need to use common sense and even a little faith. When you connect the points that you have plotted with a smooth curve, you are assuming that the curve behaves nicely between consecutive points, which is faith. That is why you should plot enough points so the outline of the curve

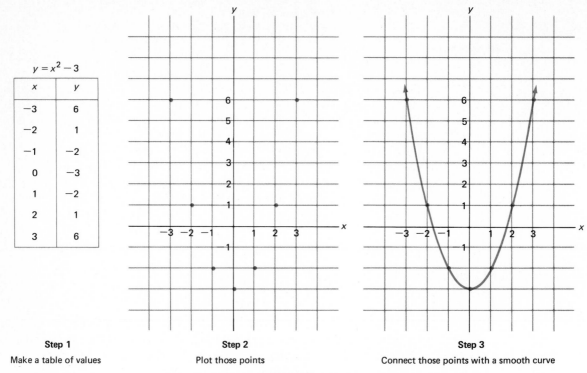

$y = x^2 - 3$	
x	y
-3	6
-2	1
-1	-2
0	-3
1	-2
2	1
3	6

Step 1

Make a table of values

Step 2

Plot those points

Step 3

Connect those points with a smooth curve

Figure 21

seems very clear; the more points you plot, the less faith you will need. Also, you should recognize that you can seldom display the whole curve. In our example, the curve has infinitely long arms opening wider and wider. But our graph does show the essential features. This is our goal in graphing: show enough of the graph so the essential features are visible.

SYMMETRY OF A GRAPH

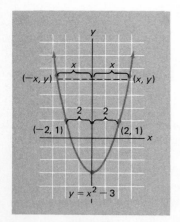

Figure 22

The graph of $y = x^2 - 3$, drawn above and again in Figure 22, has a nice property of symmetry. If the coordinate plane were folded along the y-axis, the two branches would coincide. For example, $(3, 6)$ would coincide with $(-3, 6)$, $(2, 1)$ would coincide with $(-2, 1)$, and, more generally, (x, y) would coincide with $(-x, y)$. Algebraically, this corresponds to the fact that we may replace x by $-x$ in the equation $y = x^2 - 3$ without changing the equation.

Whenever an equation is unchanged by replacing (x, y) with $(-x, y)$, the graph of the equation is said to be **symmetric with respect to the y-axis**. Likewise, if the equation is unchanged when (x, y) is replaced by $(x, -y)$, its graph is **symmetric with respect to the x-axis**. The equation $x = 1 + y^2$ is of the latter type; its graph is shown in Figure 23.

A third type of symmetry is **symmetry with respect to the origin**. It

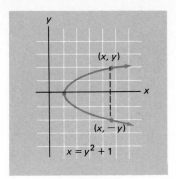

Figure 23

occurs whenever replacing (x, y) by $(-x, -y)$ produces no change in the equation. The equation $y = x^3$ is a good example, since $-y = (-x)^3$ is equivalent to $y = x^3$. Note that the dotted line segment from $(-x, -y)$ to (x, y) in the graph of $y = x^3$ is bisected by the origin (Figure 24).

In graphing $y = x^3$, we used a smaller scale on the y-axis than on the x-axis. This made it possible to show a larger portion of the graph (it also distorted the graph by flattening it out). We suggest that before putting scales on the two axes, you should examine your table of values. Choose scales so that all or most of your points can be plotted and still keep your graph of reasonable size.

Graphing an equation is an extremely important operation. It gives us a picture to look at. Most of us can absorb qualitative information from a picture much more easily than from symbols. But if we want precise quantitative information, then equations are better; they are easier to manipulate. This is why we need both geometry and algebra, that is, both pictures and equations.

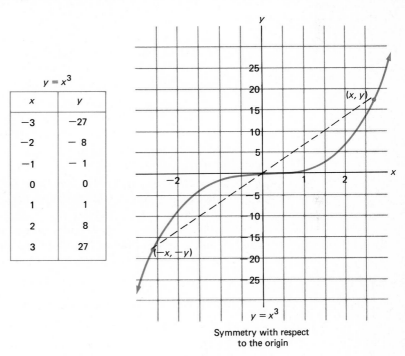

$y = x^3$

x	y
-3	-27
-2	-8
-1	-1
0	0
1	1
2	8
3	27

Symmetry with respect
to the origin

Figure 24

Problem Set 1-4

In Problems 1-12, sketch the graphs of the given equations. Be sure to check for possible symmetries as a first step.

1. $y = 3x - 2$

2. $y = -2x + 1$

3. $y = -x^2 + 4$

4. $y = 2x^2 - x$

5. $x = y^2 - 2$

6. $x = -y^2 + 4$

7. $y = x^3 - 3x$

8. $y = x^3 + 1$

9. $y = 1/(x^2 + 1)$

10. $y = x/(x^2 + 1)$

11. $2x^2 + y^2 = 8$ *Note:* This is equivalent to $y = \pm \sqrt{8 - 2x^2}$.

12. $x^2 - y^2 = 1$

EXAMPLE A (The Graph of $y = ax + b$) Problems 1 and 2 above illustrate curves with equations of the form $y = ax + b$. Show that the graph of such an equation is always a straight line.

Solution. First note that the point $(0, b)$ is on the graph; we call b the **y-intercept**. Next take any other point (x, y) on the graph (Figure 25) and consider the ratio of rise to run in moving from $(0, b)$ to (x, y).

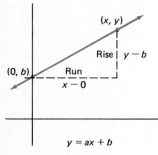

$y = ax + b$

Figure 25

$$\frac{\text{rise}}{\text{run}} = \frac{y - b}{x - 0} = \frac{ax + b - b}{x} = a$$

Since this ratio is constant, that is, it does not depend on (x, y), the graph must be a line. The number a is called the **slope** of the line.

13. Determine the slope and y-intercept for $y = 2x - 5$ and sketch the graph.

14. Determine the slope and y-intercept for $y = -3x + 2$ and sketch the graph. Note that this graph falls from left to right, corresponding to its negative slope.

15. Determine the slope and y-intercept for $3y - 4x = 7$. *Hint:* Solve for y.

16. Sketch the graph of the line through $(-2, 3)$ that is parallel to the line with equation $2x + 4y = 6$.

17. Sketch the graphs of $x = -2$ and $x = 3$. Convince yourself that they are vertical lines. (Slope is not defined for such lines, since rise over run involves division by zero.)

18. The equation $Ax + By = C$ always represents a line if A and B are not both zero. Show this by changing the equation to the form:
    (a) $y = ax + b$ if $B \neq 0$;
    (b) $x = c$ if $B = 0$.

EXAMPLE B (The Graph of $y = ax^2 + bx + c$) Determine the characteristics of the graph of $y = ax^2 + bx + c$.

Solution. We sketched the graph of $y = x^2 - 3$ in the text. You sketched $y = -x^2 + 4$ and $y = 2x^2 - x$ in Problems 3 and 4. All of these curves are cup-shaped curves called **parabolas**. The critical point on such a curve is the low point if the cup opens upward and the high point if the cup opens downward. We call this point the **vertex**. To find the x-coordinate of the vertex of $y = ax^2 + bx + c$, we use the process of completing the square.

$$y = ax^2 + bx + c$$

$$y = a\left(x^2 + \frac{b}{a}x \qquad\qquad\right) + c$$

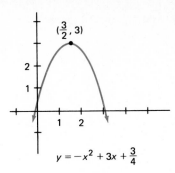

$(\frac{3}{2}, 3)$

$y = -x^2 + 3x + \frac{3}{4}$

Figure 26

$$y = a\left[x^2 + \frac{b}{a}x + \left(\frac{b}{2a}\right)^2\right] + c - a\left(\frac{b}{2a}\right)^2$$

$$y = a\left(x + \frac{b}{2a}\right)^2 + \frac{4ac - b^2}{4a}$$

If $a > 0$, this will be a parabola opening upward with vertex occurring where the squared term is smallest, namely at $x = -b/2a$. If $a < 0$, the parabola opens downward; again, its vertex occurs where $x = -b/2a$. The graph of $y = -x^2 + 3x + \frac{3}{4}$ is shown in Figure 26.

In each of the following, determine the coordinates of the vertex and then sketch the graph.

19. $y = 2x^2 - 4x + 3$
20. $y = x^2 - 4x + 5$
21. $y = -x^2 - 6x$
22. $y = -2x^2 + 8x - 2$

MISCELLANEOUS PROBLEMS

23. Sketch the graphs of $y = x^2 + 2x$ and $y = 4x + 3$ using the same coordinate axes.

24. Determine the points of intersection of the two graphs in Problem 23. *Hint:* Solve the two equations simultaneously by solving $x^2 + 2x = 4x + 3$.

C 25. Sketch the graphs of $x^2 + y^2 = 4$ and $y = 4x + 3$ using the same coordinate axes and find the points of intersection (accurate to four decimal places). *Hint:* You will need the quadratic formula $x = (-b \pm \sqrt{b^2 - 4ac})/2a$.

26. Use the same axes to sketch the graphs of $y = x^2 - 5x + 5$ and $y = -x^2 + 2x + 2$ and find their points of intersection.

27. Find the value(s) of m such that the line $y = mx - 1$ is tangent to the parabola $y = 2x^2 - 2x + 1$. *Hint:* A tangent line and the parabola will have only one point in common.

28. Show that the equation of the line of slope m which goes through the point (x_1, y_1) can be written in the form $y - y_1 = m(x - x_1)$. This is called the **point-slope form** for the equation of a line. Use this to find the equation of each of the following lines.
    (a) The line through $(-2, 3)$ with slope 4.
    (b) The line through the two points $(2, 4)$ and $(-1, 8)$.
    (c) The line through $(5, 2)$ that is parallel to the line $y = 2x + 1$.
    (d) The line through $(-1, 2)$ that is parallel to the line $2x + 3y = 7$.

29. It can be shown that lines with slopes m_1 and m_2 are perpendicular if and only if $m_1 m_2 = -1$. Use this fact and Problem 28 to find the equation of each of the following lines.
    (a) The line through $(2, 3)$ that is perpendicular to the line $y = -2x + 3$.
    (b) The line through $(-4, 1)$ that is perpendicular to the line $3x - 2y = 1$.
    (c) The line through the intersection of the lines $4x - 3y = 2$ and $2x + 5y = -12$ that is perpendicular to the first line.

30. Write the equation of the circle that has the origin and the point on the graph of $y = x^3 + 5$, where $x = 1$, as the endpoints of a diameter.

31. Suppose that a ball thrown straight up has height y feet (above the ground) after t seconds, where $y = -16t^2 + 80t + 96$.
    (a) When does the ball reach maximum height?
    (b) What is the maximum height?
    (c) When does the ball hit the ground?

32. Sketch the graphs of $y = 2x + 6$ and $x^2 + (y - 1)^2 = 20$ using the same axes and find their points of intersection.

33. Use the same axes to sketch the graphs of $x^2 + y^2 = 25$ and $(x - 2)^2 + (y + 1)^2 = 26$ and find their points of intersection.

34. Recall that $|x| = x$ if $x \geq 0$ and $|x| = -x$ if $x < 0$. Use the same axes to sketch the graphs of $y = |x|$, $y = |x - 2|$, and $y = |x + 2|$.

35. Sketch the graph of $y = |x + 2| + |x| + |x - 2| - 5$. Begin by establishing that the graph is symmetric with respect to the y-axis.

36. Determine the symmetries and sketch the graphs of $y = 2^x + 2^{-x}$ and $y = 2^x - 2^{-x}$.

37. Sketch the graphs of $|x| + |y| = 1$, $x^2 + y^2 = 1$, and $x^4 + y^4 = 1$ using the same axes.

38. **TEASER** Sketch the graphs of each of the following.
    (a) $y + |y| + x + |x| = 0$
    (b) $|x + y| + |x - y| = 2$

The word *function*, together with its synonyms *mapping* and *transformation*, makes its appearance in all brances of mathematics. The definition we give is due to the French mathematician P. G. Lejeune Dirichlet (1805–1859).

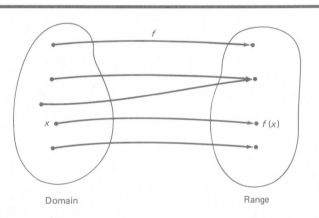

Domain   Range

1-5 Functions and Their Inverses

Think of a function as a gun. It takes ammunition from one set, called the *domain*, and fires it at a target set, called the *range*. Each shell hits a *single* target point, but it could happen that several shells land on the same point. We can state the definition more formally and introduce some notation at the same time.

A **function** f is a rule that associates with each object x in one set, called the **domain**, a unique value $f(x)$ from a second set. The set of values so obtained is called the **range** of the function.

The definition puts no restrictions on the domain and range sets. The domain might consist of the people in a psychology class and the range the set of chairs they occupy the first day of class. The function then associates with each person the chair occupied by that person.

More relevant in trigonometry will be examples where both the domain and range consist of real numbers. In such a case, we can often give a formula for the function. To illustrate, let the domain be the set of all real numbers and the range be the set of real numbers greater than or equal to 1; let f be specified by the formula $f(x) = x^2 + 1$. Then $f(1) = 1^2 + 1 = 2, f(-2) = (-2)^2 + 1 = 5, f(\frac{1}{2}) = (\frac{1}{2})^2 + 1 = \frac{5}{4}$, and so on.

GRAPHS OF FUNCTIONS

When both the domain and range of a function consist of numbers, we can picture the function by drawing its graph on a coordinate plane. The **graph** of the function f is simply the graph of the equation $y = f(x)$, As examples, consider the three functions described below.

1. $f(x) = x^2 + 1$ domain: all real numbers
2. $g(x) = x^3$ domain: all real numbers
3. $h(x) = 2/(x - 1)$ domain: all real numbers except 1

In each case, we have taken the domain to be the largest set of real numbers for which the formula makes sense and gives real number values. We call this the **natural domain** of a function.

Following the procedure described in Section 1-4 (make a table of values, plot the corresponding points, connect these points with a smooth curve), we obtain the graphs in Figure 27.

Pay special attention to the graph of h; it points out an oversimplification that we have made and now need to correct. When connecting the plotted

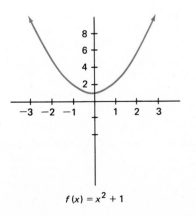

$f(x) = x^2 + 1$

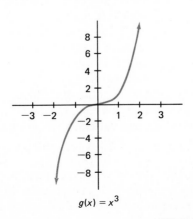

$g(x) = x^3$

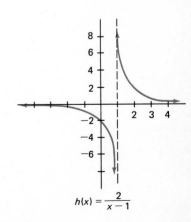

$h(x) = \dfrac{2}{x - 1}$

Figure 27

points by a smooth curve, do not do so in a mechanical way that ignores special features that may be apparent from the formula of the function. In the case of $h(x) = 2/(x - 1)$, it is clear that something dramatic must happen near $x = 1$. In fact, the values of $|h(x)|$ increase without bound. (For example, $h(.99) = -200$ and $h(1.001) = 2000$.) We have indicated this by drawing a dotted vertical line, called an **asymptote**, at $x = 1$. As x approaches 1, the shape of the graph looks more and more like this line, though this line itself is not part of the graph. Actually the graph of h also has a horizontal asymptote, namely the x-axis.

EVEN AND ODD FUNCTIONS

We can predict the symmetry of the graph of a function by inspecting its formula. If $f(-x) = f(x)$, then the graph is symmetric with respect to the y-axis. Such a function is called an **even function**, probably because a function whose formula involves only even powers of x is even. The function $f(x) = x^2 + 1 = x^2 + x^0$ is even, so is $f(x) = 3x^{14} - 2x^6 + 11$.

If $f(-x) = -f(x)$, the graph of f is symmetric with respect to the origin. We call such a function an **odd function**. A function that gives $f(x)$ as a sum of odd powers of x is odd. Thus $f(x) = x^3$ and $f(x) = x^{99} - 11x^5 + x$ are formulas for odd functions.

Notice that the function with formula $h(x) = 2/(x - 1)$ is neither even nor odd. To see this, observe that $h(-x) = 2/(-x - 1)$, which is not equal to either $h(x)$ or $-h(x)$. The graph of h was shown earlier. Note that it is neither symmetric with respect to the y-axis nor the origin. (The graph of h does have symmetry with respect to the point $(1, 0)$, but this is not immediately obvious from the formula for $h(x)$.)

INVERSE FUNCTIONS

The three functions graphed earlier in this section allow us to illustrate another important idea. For some functions, two distinct values of x are always associated with two distinct values of y (that is, $x_1 \neq x_2$ implies $f(x_1) \neq f(x_2)$). Such functions are called **one-to-one** functions. Examples of one-to-one functions are $g(x) = x^3$ and $h(x) = 2/(x - 1)$. The function $f(x) = x^2 + 1$ is not one-to-one, since two different values of x (for example, -1 and 1) yield the same value of y. A simple graphical criterion for deciding whether a function is one-to-one is this: If every horizontal line intersecting the graph of a function intersects it in exactly one point, then the corresponding function is one-to-one.

One-to-one functions are important because they can be reversed. If f is a one-to-one function taking x to $f(x)$, then we have another function f^{-1} (called the **inverse function**) which takes $f(x)$ right back to x, that is,

$$f^{-1}(f(x)) = x$$

To put it another way, f is a gun which picks up a shell at x and fires it to the target point y. Because f is one-to-one, there is only one shell that lands at y. Therefore, f^{-1} can, without ambiguity, pick up that shell and fire it back to x (Figure 28).

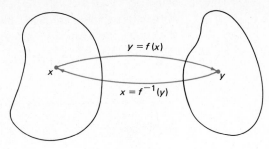

$y = f(x)$

$x = f^{-1}(y)$

Figure 28

A problem that may be somewhat sticky is finding a formula for f^{-1}. The key idea is to rewrite the equation that expresses y in terms of x in such a way that it expresses x in terms of y. The result is $x = f^{-1}(y)$. For example, consider the function g determined by the formula $y = g(x) = x^3$. Solving $y = x^3$ for x yields $x = \sqrt[3]{y}$. Thus $g^{-1}(y) = \sqrt[3]{y}$. Similarly, suppose $y = 2/(x - 1)$ is the formula for h. If we multiply both sides by $x - 1$, we have

$$(x - 1)y = 2$$

If we next remove parentheses, giving

$$xy - y = 2$$

and then solve for x, we get

$$x = \frac{2 + y}{y}$$

Thus $h^{-1}(y) = (2 + y)/y$.

There is another consideration. We often want the domain variable for a function to be called x. For g^{-1} and h^{-1} above, it is called y. But that is easily fixed; whether we say $g^{-1}(y) = \sqrt[3]{y}$, $g^{-1}(u) = \sqrt[3]{u}$, or $g^{-1}(x) = \sqrt[3]{x}$ is simply a matter of taste. As formulas for functions, they say exactly the same thing. We therefore offer the following procedure for finding the formula for the inverse of a one-to-one function.

1. Solve $y = f(x)$ for x in terms of y.
2. Use the symbol $f^{-1}(y)$ to name the resulting expression in y.
3. Replace y by x to get $f^{-1}(x)$.

Example A (in the problem set) explains how the graphs of f and f^{-1} are related.

We make one last remark. Suppose we try the three-step procedure on a function that is not one-to-one, such as $f(x) = x^2 + 1$. Let's see where it breaks down. If we solve $y = x^2 + 1$ for x, we obtain

$$x = \pm \sqrt{y - 1}$$

As expected, there are *two* values of x for each value of y. This, of course, does not determine a function. Example B explains a procedure that is often used in a case of this type.

Problem Set 1-5

In Problems 1–6, let $f(x) = \sqrt{x-1}$ and $g(x) = 1/(x+3)$. Calculate each of the following.

1. $f(5) + g(5)$ 2. $f(1) - g(1)$ 3. $g(-2.9)$

4. $g(-2.99)$ 5. $g(-3.001)$ 6. $g(f(1))$

7. What are the natural domains for the functions f and g above?

8. If $f(u) = \sqrt{u^2 - 1}/u$, what is the natural domain for f?

9. The function f squares a number t, adds 2, and then cubes the result. Write a formula for $f(t)$.

10. A window has the shape of a square topped by a semicircle. Write formulas for its area $A(x)$ and perimeter $P(x)$ if the width of the window is x.

In Problems 11–16, sketch the graph of the given function. Indicate whether the function is even, odd, or neither, and whether or not it is one-to-one.

11. $f(x) = 3x$ 12. $f(x) = -x^2 + 2x$

13. $f(t) = 2/(t+2)$ 14. $f(x) = x/(x+3)$

15. $f(x) = 2x^2/(x^2 + 1)$ 16. $f(u) = \sqrt[3]{u}$

17. Calculate $f^{-1}(2)$ and $f^{-1}(4)$ for the function of Problem 13.

18. Calculate $f^{-1}(2)$ and $f^{-1}(-3)$ for the function of Problem 16.

19. Following the three-step method suggested in the text, find a formula for $f^{-1}(x)$ if $f(x) = 2x/(x+2)$.

20. Find a formula for $g^{-1}(x)$ if $g(x) = x/(x-3)$.

EXAMPLE A (The Graph of f^{-1})   Assuming that f is one-to-one so that it has an inverse, how is the graph of $y = f^{-1}(x)$ related to the graph of $y = f(x)$? Illustrate with $f(x) = x^3$.

Solution.   The graphs of $y = f(x)$ and $x = f^{-1}(y)$ are identical. To get the graph of $y = f^{-1}(x)$ from that of $x = f^{-1}(y)$, we interchange the roles of x and y, which amounts to a reflection in the line $y = x$. This is illustrated in Figure 29 for $f(x) = x^3$.

In Problems 21–26, sketch the graph of $y = f(x)$. Then use the reflection principle of Example A to draw the graph of $y = f^{-1}(x)$. Finally obtain a formula for $f^{-1}(x)$.

21. $f(x) = \sqrt[5]{x}, \ x \geq 0$ 22. $f(x) = 2x + 1$

23. $f(x) = x/(x+1)$ 24. $f(x) = 2/x$

25. $f(x) = 2^x$ 26. $f(x) = \log x$

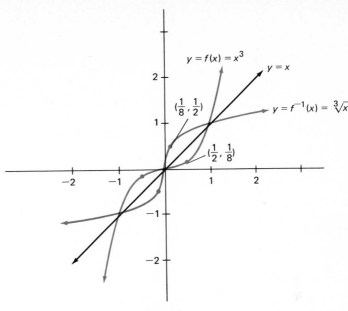

Figure 29

EXAMPLE B (Restricting the Domain)   While it is true that $f(x) = x^2$ does not have an inverse if we use its natural domain, it can be made to have an inverse if its domain is properly restricted. Show one way that this can be done.

Solution.   We restrict the domain to the set of real numbers x such that $x \geq 0$, which is denoted by $\{x: x \geq 0\}$. This has the effect of cutting off the left branch of the graph (see Figure 30). Then f is one-to-one and has an inverse. Its formula is $f^{-1}(x) = \sqrt{x}$. If we had chosen instead to use the left branch (corresponding to $x \leq 0$), then the formula would have been $f^{-1}(x) = -\sqrt{x}$.

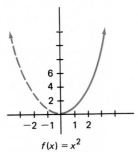

$f(x) = x^2$
Domain: $\{x : x \geq 0\}$

Figure 30

In Problems 27–30, restrict the domain so f has an inverse and find a formula for $f^{-1}(x)$. Since this can be done in more than one way, we suggest that you make the domain as large as possible using the right-hand part of the x-axis.

27. $f(x) = x^4$

28. $f(x) = 4 + x^2$

29. $f(x) = 2x^2 - 3$

30. $f(x) = x^2 - 2x$

EXAMPLE C (Translations)   Show how the graph of $y = f(x - c)$ is related to the graph of $y = f(x)$.

Solution.   The graph of $y = f(x - c)$ has the same shape as the graph of $y = f(x)$, but it is translated (moved rigidly) c units to the right if c is positive and $|c|$ units to the left if c is negative. We illustrate in Figure 31 for $f(x) = x^2$.

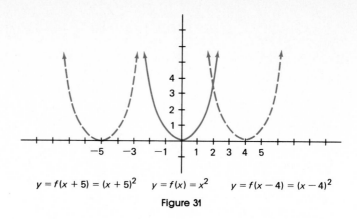

$$y = f(x + 5) = (x + 5)^2 \qquad y = f(x) = x^2 \qquad y = f(x - 4) = (x - 4)^2$$

Figure 31

In Problems 31–34, sketch both graphs. Draw the second graph by translating the first.

31. $f(x) = -x^2 + 1,\qquad g(x) = f(x - 3) = -(x - 3)^2 + 1$
32. $f(x) = x^3,\qquad g(x) = f(x + 4) = (x + 4)^3$
33. $f(x) = 2^x,\qquad g(x) = f(x + 3) = 2^{x+3}$
34. $f(x) = 1/x,\qquad g(x) = f(x - 4) = 1/(x - 4)$

MISCELLANEOUS PROBLEMS

35. Let $f(x) = 1/(x - 3)$ and $g(x) = 2x - 1$. Determine each of the following.
    (a) $f(2.5)$ (b) $f(2.9)$ (c) $f(-1)g(-1)$
    (d) $[f(-2)]^2/g(.6)$ (e) $f(g(\frac{1}{2}))$ (f) $g(f(\frac{1}{2}))$
    (g) $[g(x^2)]^2$ (h) $g^{-1}(5)$ (i) $f^{-1}(2)$

36. Let $f(x) = 2/x$. Show that $f(f(x)) = x$. What can you deduce from this about $f^{-1}(x)$?

37. Which of the following functions are even, which are odd, and which are neither?
    (a) $f(x) = 5$ (b) $f(x) = \pi x$ (c) $f(x) = x|x|$
    (d) $f(x) = 2x^2 + x$ (e) $f(x) = x^7(2x^3 - x)$ (f) $f(x) = x^7/(x^2 + 1)$

38. Suppose that $f(x)$ is an even function and that $g(x)$ is an odd function. Which of the following functions are even, which are odd, and which are neither?
    (a) $f(x) \cdot g(x)$ (b) $f(x) + g(x)$ (c) $[g(x)]^2$
    (d) $xf(x)$ (e) $g(x)/(x^2 + 1)$ (f) $f(g(x))$

39. Show that $f(x) = x/(x - 1)$ is self-inverse, that is, show that $f^{-1}(x) = f(x)$.

40. Let $f(x) = (x - 3)/(x + 1)$. Show that $f^{-1}(x) = f(f(x))$.

41. Let $f(x) = x^2 + 6x$. Restrict the domain to $x \geq -3$ and then determine the formula for $f^{-1}(x)$.

42. In each column of the accompanying table, work down the column, performing the given operation on the result of the line above.
    (a) Write the formula for $f(x)$.
    (b) Write the formula for $g(x)$.
    (c) Show that f and g are inverse to each other, that is, show that $f(g(x)) = x$ and $g(f(x)) = x$.

Choose a number x	Choose a number x
Subtract 1	Subtract 5
Cube	Divide by 2
Multiply by 2	Take the cube root
Add 5	Add 1
Call the result $f(x)$	Call the result $g(x)$

43. Let $f(x)$ denote the sum of the squares of three consecutive even integers, x being the middle integer. Write the formula for $f(x)$.

44. Let x denote the length of a side of an equilateral triangle. Write the formula for the area $A(x)$ of the triangle.

45. Let x denote the length of the diagonal of a square. Write the formulas for its area $A(x)$ and its perimeter $P(x)$.

46. A piece of cardboard 10 centimeters by 12 centimeters is made into an open box by cutting squares of width x centimeters from each of the four corners and turning up the sides. Write a formula for $V(x)$, the volume of the resulting box.

47. A wire 120 centimeters long is cut into two pieces. The first piece, of length x centimeters, is formed into a square and the second is formed into a circle. Write a formula for $A(x)$, the total area of the two regions enclosed.

48. Using the same axes, sketch the graphs of $f(x) = \sqrt{x}$, $g(x) = \sqrt{x-2}$, and $h(x) = \sqrt{x+4}$.

49. Let (x) denote the distance to the integer nearest x. For example, $(2.3) = .3$ and $(2.8) = .2$. Sketch the graph of $y = (x)$.

50. We call a function f **periodic** if there is a positive constant p such that $f(x + p) = f(x)$ for all x in the domain of f. The smallest such positive constant is the **period** of f. (The premier examples of periodic functions are the sine and cosine functions, both with period 2π; see Section 2-6). Convince yourself that $f(x) = (x)$ is periodic and determine its period.

51. Sketch the graph of $f(x) = (2x)$ and determine its period.

52. Sketch the graph of $f(x) = (\frac{1}{3}x)$ and determine its period.

53. Let $f(n)$ be the nth digit in the decimal expansion of $3/13$. Convince yourself that f is periodic and determine its period.

54. **TEASER** Let

$$f(x) = \begin{cases} 1 & \text{if } x = 0 \\ 1/q & \text{if } x = p/q \text{ is a rational number in reduced form and } q > 0 \\ 0 & \text{if } x \text{ is irrational} \end{cases}$$

(a) Sketch the graph of f as best you can.

(b) Show that f is periodic and determine its period.

Chapter Summary

The fundamental numbers in this book are the **real numbers**. These numbers consist of the **rational numbers** (such as $\frac{3}{4}$, $-\frac{9}{10}$, $.\overline{95}$) and the **irrational numbers** (such as π, $\sqrt{2}$, and $\sqrt{3}$). Every real number can be represented as a (nonterminating) **decimal**. And each such number is the **coordinate** of a point on the **real line**. In a few instances, we shall need a larger number system, the **complex numbers**. These are numbers of the form $a + bi$, where a and b are real numbers and $i^2 = -1$.

Calculations with numbers are greatly facilitated by use of scientific calculators. These electronic devices handle decimals with a specified number of digits (usually 8 or 10) and hence can give only approximate answers to many problems. Very large or very small numbers must be entered in **scientific notation**. When calculations involve numbers resulting from measurements of physical quantities, answers should not be given with more accuracy than the data warrant. This involves the notion of **significant digits**.

Points in the plane can be uniquely specified by giving two numbers, the **x-** and **y-coordinates** of the point. This allows us to give a simple expression, called the **distance formula**, for the distance between two points. More importantly, it leads to the concepts of the **equation** of a curve and the **graph** of an equation. Graphing an equation is usually accomplished by making a **table of values**, plotting the corresponding points, and connecting these points with a smooth curve. However, qualitative features, such as properties of **symmetry**, can simply the process.

The key idea of **function** is introduced as a rule which associates with each object from one set (the **domain**) an object from a second set (the **range**). A function f is best visualized by drawing its **graph**, the graph of the equation $y = f(x)$. The concepts of **even** and **odd** functions are mirrored by the symmetries of the graphs. Some functions are **one-to-one**. Associated with such a function is another function, called its **inverse function**, which undoes the work the original function accomplished.

Chapter Review Problem Set

1. Simplify and write as a fraction in reduced form.
   (a) $\dfrac{4}{11}\left(\dfrac{7}{8} - \dfrac{5}{12}\right)$ (b) $1 - \dfrac{\frac{4}{3} + \frac{2}{3}}{13}$ (c) $.1\overline{27}$

2. Write as a decimal.
   (a) $(1.01)^2/5$ (b) $\frac{14}{9}$

3. Solve the equation.
   (a) $2t^3 - 10t = 0$ (b) $t^2 - 3t = 1$

4. Write in the form $a + bi$.
   (a) $(3 + 2i)(1 - 3i)$ (b) $(3 + 2i)/(1 - 3i)$

5. Write the inequality $|x - 2| \le 2.5$ in the form $a \le x \le b$.

🖸 6. Use a scientific calculator to evaluate

$$\frac{(\sqrt{24.31} + \log 57.31)^3}{(4.31 \times 10^{-3})}$$

7. Find the distance from $(2, -8)$ to the center of the circle with equation $x^2 + 8y + y^2 = 20$.

8. Write the equation of a circle of radius 5, if it is in the second quadrant and is tangent to both axes.

9. Sketch the graph of $y = x^2 - x - 6$. What are the coordinates of the lowest point on this graph?

10. Find the distance between the two points where the graph of $y = x^2 - x - 6$ meets the graph of $y = 1$.

11. Find the x-coordinate of the point in the first quadrant that lies on the line $y = x - 2$ and is 4 units from the origin.

12. If $f(x) = 2/(x - 3)$, evaluate each expression.
    (a) $f(0)$ (b) $f(2)$ (c) $f(\frac{31}{10})$

13. What is the natural domain of $f(x) = \sqrt{x - 3}$?

14. Sketch the graph of $f(x) = \sqrt{x - 3}$.

15. If $f(x) = \sqrt{x - 3}$, determine a formula for $f^{-1}(x)$.

16. State whether each function is even, odd, or neither.
    (a) $f(x) = 1/(x^2 + 1)$ (b) $f(x) = x/(x^2 + 1)$
    (c) $f(x) = (x^2 + x)^2$ (d) $f(x) = (x^3 + x)^2$

17. A cylindrical pipe is 50 centimeters long, has outer diameter of 6 centimeters, and is made of material x centimeters thick. Write a formula for $V(x)$, the volume of the interior of the pipe.

The great book of Nature lies open before our eyes and true philosophy is written in it. . . . But we cannot read it unless we have first learned the language and characters in which it is written. . . . It is written in mathematical language and the characters are triangles, circles, and other geometrical figures.

Galileo

CHAPTER 2

The Trigonometric Functions

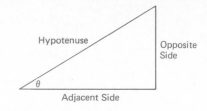

2-1   Right-Triangle Trigonometry

A triangle is called a *right triangle* if one of its angles is a right angle, that is, a 90° angle. The other two angles are necessarily acute angles (less than 90°) since the sum of all three angles in a triangle is 180°. Let θ (the Greek letter theta) denote one of these acute angles. We may label the three sides relative to θ: adjacent side, opposite side, and hypotenuse, as shown in the diagram above. In terms of these sides, we introduce the three fundamental ratios of trigonometry, sine θ, cosine θ, and tangent θ. Using obvious abbreviations, we give the following definitions.

$$\sin \theta = \frac{\text{opp}}{\text{hyp}}$$

$$\cos \theta = \frac{\text{adj}}{\text{hyp}}$$

$$\tan \theta = \frac{\text{opp}}{\text{adj}}$$

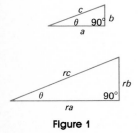

Figure 1

Thus with every acute angle θ, we associate three numbers, sin θ, cos θ, and tan θ. A careful reader might wonder whether these numbers depend only on the size of θ, or if they also depend on the lengths of the sides of the right triangle with which we started. Consider two different right triangles, each with the same angle θ (as in Figure 1). You may think of the lower triangle as a magnification of the upper one. Each of its sides has length r times that of the corresponding side in the upper triangle. If we calculate sin θ from the lower triangle, we get

$$\sin \theta = \frac{\text{opp}}{\text{hyp}} = \frac{rb}{rc} = \frac{b}{c}$$

which is the same result we get using the upper triangle. We conclude that for a given θ, sin θ has the same value no matter which right triangle is used to compute it. So do cos θ and tan θ.

SPECIAL ANGLES

We can use the Pythagorean theorem ($a^2 + b^2 = c^2$) to find the values of sine, cosine, and tangent for the special angles 30°, 45°, and 60°. Consider the two right triangles of Figure 2, which involve these angles.

Figure 2

To see that the indicated values of a are correct, note that in the first triangle, $a^2 + a^2 = 2^2$, which gives $a = \sqrt{2}$. In the second, which is half of an equilateral triangle, $a^2 + 1^2 = 2^2$, or $a = \sqrt{3}$.

From these triangles, we obtain the following important facts.

$$\sin 45° = \frac{\sqrt{2}}{2} \qquad \cos 45° = \frac{\sqrt{2}}{2} \qquad \tan 45° = 1$$

$$\sin 30° = \frac{1}{2} \qquad \cos 30° = \frac{\sqrt{3}}{2} \qquad \tan 30° = \frac{1}{\sqrt{3}} = \frac{\sqrt{3}}{3}$$

$$\sin 60° = \frac{\sqrt{3}}{2} \qquad \cos 60° = \frac{1}{2} \qquad \tan 60° = \sqrt{3}$$

OTHER ANGLES

When you need the sine, cosine, or tangent of an angle other than the special ones just considered, you may do one of two things. If you have a scientific calculator, you may simply push two or three keys and have your answer correct to eight or more significant digits. Otherwise, you will need to use Table A in the Appendix.

Several facts about Table A should be noted. First, it gives answers usually to four decimal places. Second, angles are measured in degrees and tenths of degrees. By interpolation (see first section of Appendix), it is possible to consider angles measured to the nearest hundredth of a degree. Finally, notice that the left column of the table lists angles from 0° to 45°. For angles from 45° to 90°, use the right column; you must then also use the bottom captions. To make sure that you are reading the table (or your calculator) correctly, check that you get each of the following answers.

$$\tan 33.1° = .6519 \qquad \sin 26.9° = .4524$$

$$\cos 54.3° = .5835 \qquad \tan 82° = 7.115$$

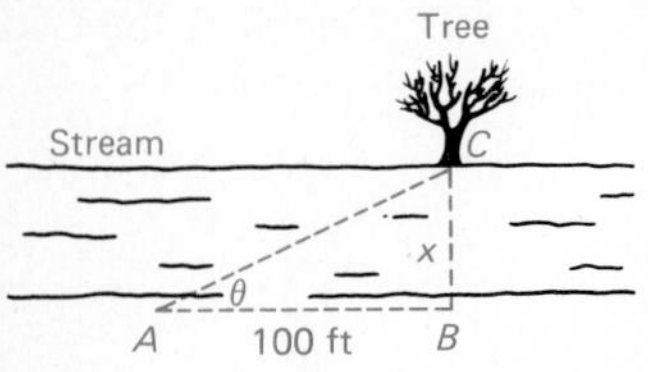

APPLICATIONS

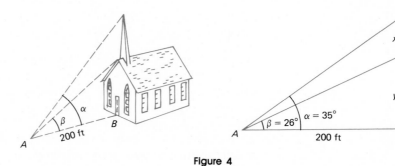

Figure 3

Suppose that you wish to measure the distance across a stream but do not want to get your feet wet. Here is how you might proceed.

Pick out a tree at C on the opposite shore and set a stone at B directly across from it on your shore (Figure 3). Set another stone at A, 100 feet up the shore from B. With an angle measuring device (for example, a protractor or a transit), measure angle θ between AB and AC. Then x, the length of BC, satisfies the following equation.

$$\tan \theta = \frac{\text{opp}}{\text{adj}} = \frac{x}{100}$$

or

$$x = 100 \tan \theta$$

For example, if θ measures 29°, you find from your scientific calculator or Table A that tan 29° = .5543. Then x = 100(.5543) = 55.43 feet. Since you used stones and trees for points, this suggests that you should not give your answer with such accuracy. It would be better to say that the distance x is approximately 55 feet.

As a more difficult example, consider a church with a steeple, as shown in Figure 4. The problem is to calculate the height of the steeple while standing on the ground. To find the height, mark a point B on the ground directly below the steeple and another point A 200 feet away on the ground. At A, measure the *angles of elevation* α and β to the top and bottom of the steeple. This is all the information you will need, provided you know your trigonometry.

Figure 4

Let x be the height of the steeple and y be the distance from the ground to the bottom of the steeple. Suppose that α = 35° and β = 26°. Then

$$\tan 35° = \frac{x + y}{200}$$

$$\tan 26° = \frac{y}{200}$$

If you solve for y in the second equation and substitute the value in the first, you will get the following sequence of equations.

$$\tan 35° = \frac{x + 200 \tan 26°}{200}$$

$$200 \tan 35° = x + 200 \tan 26°$$

$$x = 200 \tan 35° - 200 \tan 26°$$

$$= 200(.7002 - .4877)$$

$$x = 42.5 \text{ feet}$$

Problem Set 2-1

In Problems 1–6, use Table A to evaluate each expression. If you have a scientific calculator, use it as a check.

1. sin 41.3°   2. tan 54.4°   3. cos 49.2°
4. sin 89.3°   5. tan 72.3°   6. cos 38.7°

In Problems 7–12, use Table A to find θ. We suggest you also do these problems on a calculator. For example, to do Problem 7 on many calculators, press .2164 INV sin . This will give the inverse sine of .2164, that is, the angle whose sine is .2164.

7. sin θ = .2164   8. tan θ = .3096   9. tan θ = 2.311
10. cos θ = .9354   11. cos θ = .3535   12. sin θ = .7302

Each of the remaining problems in this problem set involves a considerable amount of arithmetic that you can do by hand (using tables) or by using a calculator. (If you use a calculator to find values for the trigonometric functions, be sure that it is in the degree mode.) In Problems 13–18, find x.

13.

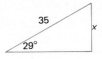

14.

15.

16.

17.

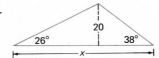

18.

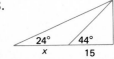

EXAMPLE A (Solving a Right Triangle Given an Angle and a Side) To solve a triangle means to determine all its unknown parts. Solve the right triangle which has hypotenuse of length 14.6 and an angle measuring 33.2°.

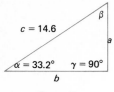

c = 14.6

β

a

α = 33.2°   γ = 90°

b

Figure 5

Solution. First, we draw the triangle labeling the known parts and assigning letters to the unknown parts. Our convention is to use the first three Greek letters, α, β, and γ (alpha, beta, and gamma) for the angles and a, b, and c for the lengths of the respective sides opposite these angles (see Figure 5). We need to find β, a, and b.

(i) $\beta = 90° - 33.2° = 56.8°$

(ii) $\sin 33.2° = a/14.6$, so

$$a = 14.6 \sin 33.2° = (14.6)(.5476) \approx 7.99$$

(iii) $\cos 33.2° = b/14.6$, so

$$b = 14.6 \cos 33.2° = (14.6)(.8368) \approx 12.2$$

Notice that we gave the answers to three significant digits since the given data have three significant digits.

Solve each of the following triangles. First draw the triangle, labeling it as in the example with $\gamma = 90°$

19. $\alpha = 42°$, $c = 35$ 20. $\beta = 29°$, $c = 50$

21. $\beta = 56.2°$, $c = 91.3$ 22. $\alpha = 69.9°$, $c = 10.6$

23. $\alpha = 39.4°$, $a = 120$ 24. $\alpha = 40.6°$, $b = 163$

EXAMPLE B (Solving a Right Triangle Given Two Sides) Solve the right triangle which has legs $a = 42.8$ and $b = 94.1$.

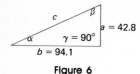

c

β

a = 42.8

α   γ = 90°

b = 94.1

Figure 6

Solution. First, we draw the triangle and label its parts (Figure 6). We must find α, β, and c.

(i) $\tan \alpha = \dfrac{42.8}{94.1} \approx .4548$

Now we can find α by using Table A backwards, that is, by looking under tangent in the body of the table for .4548 and determining the corresponding angle. Or better, we can use the $\boxed{\text{INV}}\ \boxed{\text{tan}}$ keys on a scientific calculator. On many calculators, press

$$\boxed{(}\ 42.8\ \boxed{\div}\ 94.1\ \boxed{)}\ \boxed{\text{INV}}\ \boxed{\text{tan}}$$

In either case, the result is $\alpha \approx 24.5°$.

(ii) $\beta = 90° - \alpha \approx 90° - 24.5° = 65.5°$

(iii) We could find c by using $c^2 = a^2 + b^2$. Instead, we use $\sin \alpha$.

$$\sin \alpha = \sin 24.5° = \frac{42.8}{c}$$

$$c = \frac{42.8}{\sin 24.5°} = \frac{42.8}{.4147} \approx 103$$

Solve the right triangles satisfying the given information in Problems 25–32, assuming that c is the hypotenuse. You can do them either with tables or a calculator.

25. $a = 9, b = 12$
26. $a = 24, b = 10$
27. $a = 40, c = 50$
28. $c = 41, a = 40$
29. $a = 14.6, c = 32.5$
30. $a = 243, c = 419$
31. $a = 9.52, b = 14.7$
32. $a = .123, b = .456$

33. A straight path leading up a hill rises 26 feet per 100 horizontal feet. What angle does it make with the horizontal?

34. A 20-foot ladder leans against a wall, making an angle of 76° with the level ground. How high up the wall is the upper end of the ladder?

35. Find the angle of elevation of the sun if a woman 5 feet 9 inches tall casts a shadow 46.8 feet long. (The *angle of elevation* is the upward angle made with the horizontal.)

36. A guy wire to a pole makes an angle of 69° with the level ground and is 14 feet from the pole at the ground. How high above the ground is the wire attached to the pole?

37. Suppose that the woman in Problem 35 is walking with her daughter Sue, who is 3 feet 10 inches tall. How long is Sue's shadow?

38. Find the length of the supporting wire in Problem 36.

MISCELLANEOUS PROBLEMS

The following problems can be solved using either a calculator or tables. We recommend using a calculator.

39. Calculate each value.
    (a) tan 14.5°
    (b) 24.6 cos 74.3°
    (c) $15.6 (\sin 14°)^2 / \cos 87°$

40. Find θ in each case.
    (a) $\sin \theta = .6691$
    (b) $\cos \theta = .5519$
    (c) $\tan \theta = 5.396$

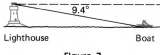

Lighthouse   Boat

Figure 7

41. From the top of a lighthouse 120 feet above sea level, the *angle of depression* (the downward angle from the horizontal) to a boat adrift on the sea is 9.4° (Figure 7). How far from the foot of the lighthouse is the boat?

42. Solve the right triangle in which $b = 67.3$ and $c = 82.9$.

43. Find x in Figure 8.

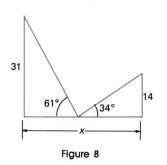

Figure 8

44. When the *angle of elevation* (the upward angle from the horizontal) of the sun is 28.4°, the Eiffel Tower in Paris casts a horizontal shadow 1822 feet long. How high is the tower?

45. With her hands 5 feet above the ground, Sally is pulling on a kite. If the kite is 200 feet above the ground and the kite string makes an angle of 32.4° with the horizontal, how many feet of string are out?

46. A plane is flying directly away from a ground observer at a constant rate, maintaining an elevation of 15,000 feet. At a certain instant, the observer measures the angle of elevation as 44° and 15 seconds later as 31°. How fast is the plane flying in miles per hour?

47. From a window in an office building, I am looking at a television tower that is 600 meters away (horizontally). The angle of elevation of the top of the tower

is 19.6° and the angle of depression of the base of the tower is 21.3°. How tall is the tower?

48. The vertical distance from first to second floor of a certain department store is 28 feet. The escalator, which has a horizontal reach of 96 feet, takes 25 seconds to carry a person between floors. How fast does the escalator travel?

49. The Great Pyramid is about 480 feet high and its square base measures 760 feet on a side. Find the angle of elevation of one of its edges, that is, find β in Figure 9.

50. Find the angle between a principal diagonal and a face diagonal of a cube.

51. A regular hexagon (6 equal sides) is inscribed in a circle of radius 4. Find the perimeter P and area A of this hexagon.

52. A regular decagon (10 equal sides) is inscribed in a circle of radius 12. What percent of the area of the circle is the area of the decagon?

53. Find the area of the regular 6-pointed Star of David that is inscribed in a circle of radius 1 (Figure 10).

54. **TEASER** Find the area of the regular 5-pointed star (the pentagram) that is inscribed in a circle of radius 1 (Figure 11).

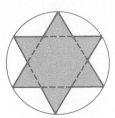
Figure 10

Figure 11

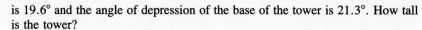
Figure 9

The Dynamic View of Angles

In geometry, we take a static view of angles. An angle is simply the union of two rays with a common endpoint (the vertex). In trigonometry, angles are thought of in a dynamic way. An angle is determined by rotating a ray about its endpoint from an initial position to a terminal position.

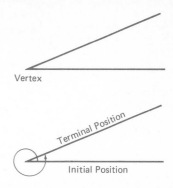

2-2 Angles and Arcs

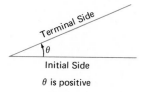

θ is positive

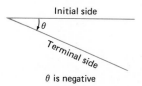

θ is negative

Figure 12

For the solution of right triangles (which involve acute angles), we required only the familiar and simple notion of angle from high-school geometry. But for the broader development of trigonometry, we need the new perspective on angles suggested by our opening display. Not only do we allow arbitrarily large angles, but we also distinguish between positive and negative angles. If an angle is generated by a counterclockwise rotation, it is positive; if generated by a clockwise rotation, it is negative (Figure 12). To know an angle, in trigonometry, is to know how the angle came into being. It is to know the initial side, the terminal side, and the kind of rotation that produced the angle.

DEGREE MEASUREMENT

Take a circle and divide its circumference into 360 equal parts. The angle with vertex at the center determined by one of these parts has measure one **degree** (written 1°). This way of measuring angles is due to the ancient Babylonians and is so familiar that we used it in Section 2-1 without comment. There is a refinement, however, that we avoid. The Babylonians divided each degree into 60 minutes and each minute into 60 seconds; some people still follow this cumbersome practice. If we need to measure angles to finer accuracy than a degree, we will use decimal parts. Thus we write 40.5° rather than 40°30′.

It is important that we be familiar with measuring both positive and negative angles, as well as angles resulting from large rotations. Three angles are shown in Figure 13. Note that all three have the same initial and terminal sides.

RADIAN MEASUREMENT

The best way to measure angles is in radians. Take a circle of radius r. The familiar formula $C = 2\pi r$ tells us that the circumference has 2π (about 6.28) arcs of length r around it. The angle with vertex at the center of a circle

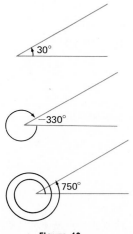

Figure 13

θ measures
one radian
(about 57.3°)

Figure 14

determined by an arc of length equal to its radius measures one **radian** (Figure 14). Thus an angle of size 360° measures 2π radians and an angle of size 180° measures π radians. We abbreviate the latter by writing

$$180° = \pi \text{ radians}$$

To convert from degrees to radians, all one needs to remember is the result in the box. By dividing by 2, 3, 4, and 6, respectively, we get the conversions for several special angles.

$$90° = \frac{\pi}{2} \text{ radians}$$

$$60° = \frac{\pi}{3} \text{ radians}$$

$$45° = \frac{\pi}{4} \text{ radians}$$

$$30° = \frac{\pi}{6} \text{ radians}$$

If we divide the boxed formula by 180, we get

$$1° = \frac{\pi}{180} \text{ radians}$$

and if we divide by π, we get

$$\frac{180°}{\pi} = 1 \text{ radian}$$

The following rules thus hold.

To convert from degrees to radians, multiply by $\pi/180$.
To convert from radians to degrees, multiply by $180/\pi$.

For example,

$$22° = 22\left(\frac{\pi}{180}\right) \text{ radians} \approx .38397 \text{ radians}$$

and

$$2.3 \text{ radians} = 2.3\left(\frac{180}{\pi}\right)° \approx 131.78°$$

Some scientific calculators have a key that makes these conversions automatically.

ARC LENGTH AND AREA

Radian measure is almost invariably used in calculus because it is an intrinsic measure. The division of a circle into 360 parts was quite arbitrary; its division into parts of radius length (2π parts) is more natural. Because of this, formulas using radian measure tend to be simple, while those using degree measure are often complicated. As an example, consider arc length. Let t be the radian measure of an angle θ with vertex at the center of a circle of radius r. This angle cuts off an arc of length s which satisfies the simple formula

$$s = rt$$

This follows directly from the fact that an angle of one radian ($t = 1$) cuts off an arc of length r (see Figure 15).

A second nice formula is that for the area of the sector cut off from a circle by a central angle of t radians (Figure 16). Note that the area A of this sector is to the area of the whole circle as t is to 2π, that is, $A/\pi r^2 = t/2\pi$. Thus

$$A = \frac{1}{2}r^2 t$$

THE UNIT CIRCLE

The formula for arc length takes a particularly simple form when $r = 1$, namely, $s = t$. We emphasize its meaning. *On the unit circle, the length of an arc is the same as the radian measure of the angle it determines.*

Someone is sure to point out a difficulty in what we have just said. What happens when t is greater than 2π or when t is negative? To understand our meaning, imagine an infinitely long string on which the real number scale has been marked. Think of wrapping this string around the unit circle as shown in Figure 17.

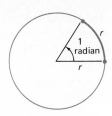

$s = 2r$

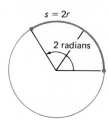

2 radians

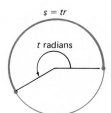

$s = tr$

t radians

Figure 15

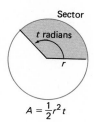

Sector

t radians

$A = \frac{1}{2}r^2 t$

Figure 16

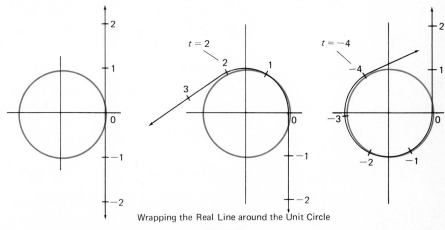

Wrapping the Real Line around the Unit Circle

Figure 17

Now if we think of the directed length (that is, the signed length) of a piece of the string, the formula $s = t$ holds no matter what t is. For example, the length of string corresponding to an angle of 8π radians is 8π. That piece of string wraps counterclockwise around the unit circle exactly 4 times. A piece of string corresponding to an angle of -3π radians would wrap clockwise around the unit circle one and a half times, its directed length being -3π.

Problem Set 2-2

Convert each of the following to radians. You may leave π in your answer.

1. 120°
2. 225°
3. 240°
4. 150°
5. 210°
6. 330°
7. 315°
8. 300°
9. 540°
10. 450°
11. −420°
12. −660°
13. 160°
14. 200°
15. $(20/\pi)°$
16. $(150/\pi)°$

Convert each of the following to degrees. Give your answer correct to the nearest tenth of a degree.

17. $\dfrac{4}{3}\pi$ radians

18. $\dfrac{5}{6}\pi$ radians

19. $-\dfrac{2\pi}{3}$ radians

20. $-\dfrac{7\pi}{4}$ radians

21. 3π radians

22. 3 radians

ⓒ 23. 4.52 radians

ⓒ 24. $\dfrac{11}{4}$ radians

ⓒ 25. $\dfrac{1}{\pi}$ radians

ⓒ 26. $\dfrac{4}{3\pi}$ radians

27. Find the radian measure of the angle at the center of a circle of radius 6 inches which cuts off an arc of length

   (a) 12 inches;   (b) 18.84 inches.

28. Find the length of the arc cut off on a circle of radius 3 feet by an angle at the center of

   (a) 2 radians;   (b) 5.5 radians;

   (c) $\dfrac{\pi}{4}$ radians;   (d) $\dfrac{5\pi}{6}$ radians.

29. Find the radius r for each of the following.

(a)

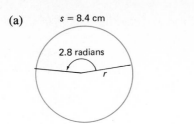

s = 8.4 cm

2.8 radians

r

(b)

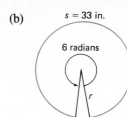

s = 33 in.

6 radians

r

30. Through how many radians does the minute hand of a clock turn in 1 hour? The hour hand in 1 hour? The minute hand in 5 hours?

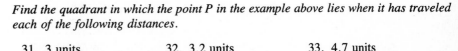

EXAMPLE A (Locating a Point on the Unit Circle)   Figure 18 shows a unit circle with center at the origin. Suppose that a point P moves in a counterclockwise direction around the circle starting at $(1, 0)$. In which quadrant is P when it has traveled a distance of 4 units? Of 40 units?

Solution.   Keep in mind that the distance P travels equals the radian measure of the angle through which OP turns. A distance of 4 units puts P in quadrant III since $\pi < 4 < 3\pi/2$. Once around the circle is $2\pi \approx 6.28$ units. If you divide 40 by 6.28, you get

$$40 = 6(6.28) + 2.32$$

Since 2.32 is between $\pi/2$ and π, traveling 40 units around the unit circle will put P in quadrant II.

Figure 18

Find the quadrant in which the point P in the example above lies when it has traveled each of the following distances.

31. 3 units

32. 3.2 units

33. 4.7 units

34. 4.8 units

35. $\left(\dfrac{5\pi}{2} + 1\right)$ units

36. $\left(\dfrac{9\pi}{2} - 1\right)$ units

37. 100 units

38. 200 units

EXAMPLE B (Angular Velocity)   A formula closely related to the arc length formula $s = rt$ is the formula

$$v = r\omega$$

which connects the speed (velocity) of a point on the rim of a wheel of radius r with the angular velocity ω at which the wheel is turning. Here ω is measured in radians per unit of time. Use this formula to determine the angular velocity in radians per second of a bicycle wheel of radius 16 inches if the bicycle is being ridden down the road at 30 miles per hour.

Solution.   We must use consistent units. You can check that the speed of a point on the rim of the wheel (30 miles per hour) translates to 44 feet per second and that the radius of the wheel is $\frac{4}{3}$ feet. Thus

$$44 = \frac{4}{3}\omega$$

or

$$\omega = \frac{3}{4}(44) = 33 \text{ radians per second}$$

39. Sally is pedaling her tricycle so the front wheel (radius 8 inches) turns at 4 revolutions per second. How fast is she moving down the sidewalk in feet per second? *Hint:* Four revolutions per second is 8π radians per second.

40. Suppose that the tire on a car has an outer diameter 2.5 feet. How many revolutions per minute does the tire make when the car is traveling 60 miles per hour?

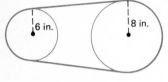

41. A dead fly is stuck to a belt that passes over two pulleys 6 inches and 8 inches in radius, as shown in Figure 19. Assuming no slippage, how fast is the fly moving when the larger pulley turns at 20 revolutions per minute?

42. How fast (in revolutions per minute) is the smaller wheel in Problem 41 turning?

Figure 19

MISCELLANEOUS PROBLEMS

43. Convert to radians.
    (a) $-1440°$ (b) $2\frac{1}{2}$ revolutions   (c) $(60/\pi)°$

44. Convert to degrees.
    (a) $23\pi/36$ radians   (b) -4.63 radians   (c) $3/(2\pi)$ radians

45. Find the length of arc cut off on a circle of radius 4.25 centimeters by each central angle.
    (a) 6 radians   (b) $(18/13\pi)°$ (c) $17\pi/6$ radians

46. The front wheel of Tony's tricycle has a diameter of 20 inches. How far did he travel in pedaling through 60 revolutions?

47. The pedal sprocket of Maria's bicycle has radius 12 centimeters, the rear wheel sprocket has radius 3 centimeters, and the wheels have radius 40 centimeters. How far did Maria travel if she pedaled continuously for 30 revolutions of the pedal sprocket?

48. A belt traveling at the rate of 60 feet per second drives a pulley (a wheel) at the rate of 900 revolutions per minute. Find the radius of the pulley.

49. Assume that the earth is a sphere of radius 3960 miles. How fast (in miles per hour) is a point on the equator moving as a result of the earth's rotation about its axis?

© 50. The orbit of the earth about the sun is an ellipse that is nearly circular with radius 93 million miles. Approximately, what is the earth's speed (in miles per hour) in its path around the sun? You will need the fact that a complete orbit takes 365.25 days.

51. The angle subtended by the sun at the earth (93 million miles away) is .0093 radians. Find the diameter of the sun.

© 52. A nautical mile is the length of 1 minute ($\frac{1}{60}$ of a degree) of arc on the equator of the earth. How many miles are there in a nautical mile?

53. One of the authors (Dale Varberg) lives at exactly 45° latitude north (see Figure 20). How long would it take him to fly to the North Pole at 600 miles per hour (assuming the earth is a sphere of radius 3960 miles)?

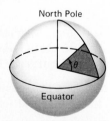

North Pole

Equator

θ measures latitude north

Figure 20

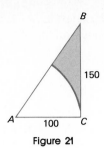

Figure 21

54. New York City is located at 40.5° latitude north. How far is it from there to the equator?

55. Oslo, Norway, and Leningrad, Russia, are both located at 60° latitude north. Oslo is at longitude 6° east (of the prime meridian) whereas Leningrad is at 30° east. How far apart are these two cities along the 60° parallel?

☐c 56. Find the area of the shaded region of the right triangle *ABC* shown in Figure 21.

57. The minute hand and hour hand of a clock are both 6 inches long and reach to the edge of the dial. Find the area of the pie-shaped region between the two hands at 5:40.

58. A cone has radius of base *R* and slant height *L*. Find the formula for its lateral surface area. *Hint:* Imagine the cone to be made of paper, slit it up the side, and lay it flat in the plane.

59. Find the area of the polar rectangle shown in Figure 22. The two curves are arcs of concentric circles.

60. **TEASER** Consider two circles both of radius *r* and with the center of each lying on the rim of the other. Find the area of the common part of the two circles.

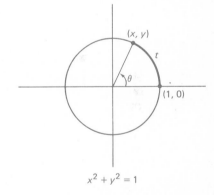

Figure 22

Definitions of Sine and Cosine

Place an angle θ, whose radian measure is t, in **standard position**, that is, put θ in the coordinate plane so that its vertex is at the origin and its initial side is along the positive x-axis. Let (x,y) be the coordinates of the point of intersection of the terminal side with the unit circle. We define both $\sin \theta$ (sine of θ) and $\sin t$ by

$$\sin \theta = \sin t = y.$$

Similarly,

$$\cos \theta = \cos t = x.$$

2-3 The Sine and Cosine Functions

In Section 2-1, we defined the sine and cosine for positive acute angles. The definitions in our opening display are more general and hence more widely applicable. They should be studied carefully. Notice that we have defined the sine and cosine for any angle θ and also for the corresponding number t. Both concepts are important. In geometric situations, angles play a central role; thus we are likely to need sines and cosines of angles. But in most of pure mathematics and in many scientific applications, it is the trigonometric functions of numbers that are important. In this connection, we emphasize that the number *t* may be positive or negative, large or small. And we may think of it as the radian measure of an angle, as the directed length of an arc on the unit circle, or simply as a number.

CONSISTENCY WITH EARLIER DEFINITIONS

Do the definitions given in Section 2-1 for the sine and cosine of an acute angle harmonize with those given here? Yes. Take a right triangle *ABC* with an acute angle θ. Place θ in standard position, thus determining a point $B'(x, y)$ on the unit circle and a point $C'(x, 0)$ directly below it on the *x*-axis (see Figure 23).

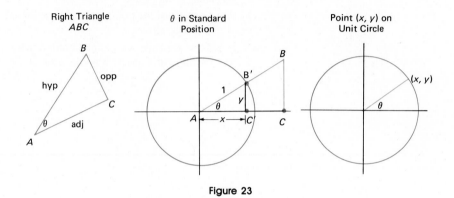

Figure 23

Notice that triangles *ABC* and *AB'C'* are similar. It follows that

$$\frac{\text{opp}}{\text{hyp}} = \frac{BC}{AB} = \frac{B'C'}{AB'} = \frac{y}{1} = y$$

$$\frac{\text{adj}}{\text{hyp}} = \frac{AC}{AB} = \frac{AC'}{AB'} = \frac{x}{1} = x$$

On the left are the old definitions of sin θ and cos θ; on the right are the new ones. They are consistent.

SPECIAL ANGLES

In Section 2-1, we learned that

$$\cos 45° = \frac{\sqrt{2}}{2} \qquad \sin 45° = \frac{\sqrt{2}}{2}$$

$$\cos 30° = \frac{\sqrt{3}}{2} \qquad \sin 30° = \frac{1}{2}$$

$$\cos 60° = \frac{1}{2} \qquad \sin 60° = \frac{\sqrt{3}}{2}$$

Making use of the consistency of the old and new definitions of sine and cosine, we conclude that the point on the unit circle corresponding to $\theta = 45° = \pi/4$ radians must have coordinates $(\sqrt{2}/2, \sqrt{2}/2)$. Similarly, the point corresponding to $\theta = 30° = \pi/6$ radians has coordinates $(\sqrt{3}/2, 1/2)$ and the point corresponding to $\theta = 60° = \pi/3$ radians has coordinates $(1/2, \sqrt{3}/2)$.

Now we can make use of obvious symmetries to find the coordinates of many other points on the unit circle. In the two diagrams of Figure 24, we show a number of these points, noting first the radian measure of the angle and then the coordinates of the corresponding point on the unit circle.

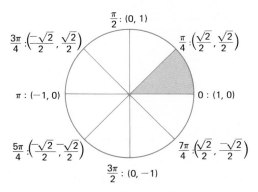

Some multiples of $\pi/4$

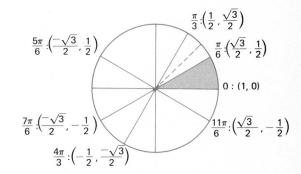

Some multiples of $\pi/6$

Figure 24

Notice, for example, how the coordinates of the points corresponding to $t = 5\pi/6$, $7\pi/6$, and $11\pi/6$ are related to the point corresponding to $t = \pi/6$. You should have no trouble seeing other relationships.

Once we know the coordinates of a point on the unit circle, we can state the sine and cosine of the corresponding angle. In particular, we get the values in the table in Figure 25. They are used so often that you should memorize them.

t	0	$\dfrac{\pi}{6}$	$\dfrac{\pi}{4}$	$\dfrac{\pi}{3}$	$\dfrac{\pi}{2}$	π	$\dfrac{3\pi}{2}$
$\cos t$	1	$\dfrac{\sqrt{3}}{2}$	$\dfrac{\sqrt{2}}{2}$	$\dfrac{1}{2}$	0	-1	0
$\sin t$	0	$\dfrac{1}{2}$	$\dfrac{\sqrt{2}}{2}$	$\dfrac{\sqrt{3}}{2}$	1	0	-1

Figure 25

PROPERTIES OF SINES AND COSINES

Think of what happens to x and y as t increases from 0 to 2π in Figure 26, that is, as P travels all the way around on the unit circle. For example, x steadily decreases until it reaches its smallest value of -1 at $t = \pi$; then it starts to

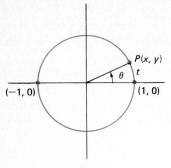

Figure 26

increase until it is back to 1 at $t = 2\pi$. We have just described the behavior of $\cos t$ (or $\cos\theta$) as t increases from 0 to 2π. You should trace the behavior of $\sin t$ in the same way. Notice that both x and y are always between -1 and 1 (inclusive). It follows that

$$-1 \le \sin t \le 1$$
$$-1 \le \cos t \le 1$$

Since P is on the unit circle, $x^2 + y^2 = 1$, and $x = \cos t$ and $y = \sin t$, it follows that

$$(\sin t)^2 + (\cos t)^2 = 1$$

It is conventional to write $\sin^2 t$ instead of $(\sin t)^2$ and $\cos^2 t$ instead of $(\cos t)^2$. Thus we have

$$\sin^2 t + \cos^2 t = 1$$

This is an identity; it is true for all t. Of course we can just as well write

$$\sin^2\theta + \cos^2\theta = 1$$

We have established one basic relationship between the sine and the cosine; here are two others, valid for all t.

$$\sin\left(\frac{\pi}{2} - t\right) = \cos t$$

$$\cos\left(\frac{\pi}{2} - t\right) = \sin t$$

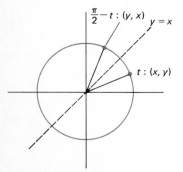

Figure 27

These relationships are easy to see when $0 < t < \pi/2$. Notice that t and $\pi/2 - t$ are measures of complementary angles (two angles with measures totaling $90°$ or $\pi/2$). That means that t and $\pi/2 - t$ determine points on the unit circle which are reflections of each other about the line $y = x$ (see Figure 27). Thus if one point has coordinates (x, y), the other has coordinates (y, x). The result given above follows from this fact.

Finally, we point out that $t, t \pm 2\pi, t \pm 4\pi, \ldots$ all determine the same point on the unit circle and thus have the same sine and cosine. This repetitive behavior puts the sine and cosine into a special class of functions, for which we give the following definition. A function f is **periodic** if there is a positive number p such that

$$f(t + p) = f(t)$$

for every t in the domain of f. The smallest such p is called the **period** of f. Thus we say that sine and cosine are periodic functions with period 2π and write

$$\boxed{\begin{array}{c} \sin(t + 2\pi) = \sin t \\ \cos(t + 2\pi) = \cos t \end{array}}$$

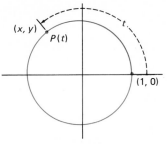

Figure 28

THE TRIGONOMETRIC POINT $P(t)$

We have introduced $\cos t$ and $\sin t$ as the x- and y-coordinates of the point on the unit circle whose directed distance from $(1, 0)$ along the unit circle is t. This point is called a **trigonometric point** and will be denoted by $P(t)$ (Figure 28). We may regard $P(t)$ to be a function of t, since for each t there is a unique point $P(t)$. This function, moreover, is periodic with period 2π—that is,

$$P(t + 2\pi) = P(t)$$

It follows that

$$P(t + k2\pi) = P(t)$$

for any integer k, a fact that allows us to find the coordinates of $P(t)$ for any t, no matter how large t is. Suppose, for example, that we wish to find the coordinates of $P(16\pi/3)$, shown in Figure 29. Since

$$\frac{16\pi}{3} = \frac{4\pi}{3} + 4\pi$$

it follows that

$$P\left(\frac{16\pi}{3}\right) = P\left(\frac{4\pi}{3}\right)$$

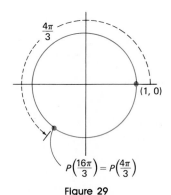

Figure 29

From the special angle diagrams on page 55, $P(4\pi/3)$ has coordinates $(-1/2, -\sqrt{3}/2)$. We conclude that $P(16\pi/3)$ also has these coordinates. This means that

$$\cos\left(\frac{16\pi}{3}\right) = -\frac{1}{2} \qquad \sin\left(\frac{16\pi}{3}\right) = -\frac{\sqrt{3}}{2}$$

Problem Set 2-3

In Problems 1–8, find the coordinates of the trigonometric point P(t) for the indicated value of t. Hint: Begin by drawing a unit circle and locating P(t) on it; then relate P(t) to the diagrams on page 55.

1. $t = \dfrac{13\pi}{6}$ 2. $t = \dfrac{19\pi}{6}$ 3. $t = \dfrac{19\pi}{4}$

4. $t = \dfrac{15\pi}{4}$ 5. $t = 24\pi + \dfrac{5\pi}{4}$ 6. $t = 16\pi + \dfrac{5\pi}{6}$

7. $t = -\dfrac{7\pi}{6}$ 8. $t = \dfrac{13\pi}{4}$

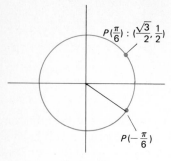

$P(\frac{\pi}{6}) : (\frac{\sqrt{3}}{2}, \frac{1}{2})$

$P(-\frac{\pi}{6})$

Figure 30

EXAMPLE A (Using $P(t)$ to Find Sine and Cosine Values)   Find
(a) $\sin(-\pi/6)$; (b) $\cos(29\pi/4)$.

Solution.

(a) We locate $P(-\pi/6)$ on the unit circle (Figure 30) and note that its
y-coordinate is $-\frac{1}{2}$ because of its position relative to $P(\pi/6)$. Thus
$\sin(-\pi/6) = -\frac{1}{2}$.

(b) We simplify the problem by removing a large multiple of 2π—that
is, by noting that

$$\frac{29\pi}{4} = 6\pi + \frac{5\pi}{4}$$

from which we conclude $P(29\pi/4) = P(5\pi/4)$. Then we refer to
the diagrams on page 293, or better yet, we simply observe that
$P(5\pi/4)$, being diametrically opposite from $P(\pi/4)$ on the unit
circle, has coordinates $(-\sqrt{2}/2, -\sqrt{2}/2)$ (see Figure 31). We conclude that

$$\cos\left(\frac{29\pi}{4}\right) = -\frac{\sqrt{2}}{2}$$

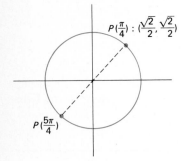

$P(\frac{\pi}{4}) : (\frac{\sqrt{2}}{2}, \frac{\sqrt{2}}{2})$

$P(\frac{5\pi}{4})$

Figure 31

Using the method of Example A, find the value of each of the following.

9. $\sin(-\pi/4)$	10. $\sin(-5\pi/4)$	11. $\sin(9\pi/4)$
12. $\sin(15\pi/4)$	13. $\cos(13\pi/4)$	14. $\cos(-7\pi/4)$
15. $\cos(10\pi/3)$	16. $\cos(25\pi/6)$	17. $\sin(5\pi/2)$
18. $\cos 7\pi$	19. $\sin(-4\pi)$	20. $\cos(7\pi/2)$
21. $\cos(19\pi/6)$	22. $\sin(14\pi/3)$	23. $\cos(-\pi/3)$
24. $\sin(-5\pi/6)$	25. $\cos(125\pi/4)$	26. $\cos(-13\pi/6)$
27. $\sin 510°$	28. $\sin(-390°)$	29. $\cos 840°$
30. $\cos(-720°)$	31. $\cos(-210°)$	32. $\sin 900°$

EXAMPLE B (Sine and Cosine of $-t$)   Show that for all t

$$\sin(-t) = -\sin t$$
$$\cos(-t) = \cos t$$

that is, sine is an odd function and cosine is an even function.

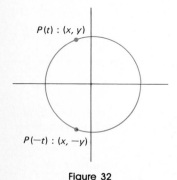

$P(t) : (x, y)$

$P(-t) : (x, -y)$

Figure 32

Solution.   The points $P(-t)$ and $P(t)$ are symmetric with respect to the x-axis
(Figure 32). Thus if $P(t)$ has coordinates (x, y), $P(-t)$ has coordinates
$(x, -y)$ and so

$$\sin(-t) = -y = -\sin t$$

$$\cos(-t) = x = \cos t$$

33. If $\sin 1.87 = .95557$ and $\cos 1.87 = -0.29476$, find $\sin(-1.87)$ and
$\cos(-1.87)$.

34. If $\sin 15.2° = 0.2622$ and $\cos 15.2° = 0.9650$, find $\sin(-15.2°)$ and $\cos(-15.2°)$.

35. Given $P(t)$ with coordinates $(1/\sqrt{5}, -2/\sqrt{5})$.
    (a) What are the coordinates of $P(-t)$?
    (b) What are the values of $\sin(-t)$ and $\cos(-t)$?

36. If t is the radian measure of an angle in quadrant III and $\sin t = -\frac{3}{5}$, evaluate each expression.
    (a) $\sin(-t)$
    (b) $\cos t$ Hint: Use the fact that $\sin^2 t + \cos^2 t = 1$.
    (c) $\cos(-t)$

37. Note that $P(t)$ and $P(\pi + t)$ are symmetric with respect to the origin (Figure 33). Use this to show that
    (a) $\sin(\pi + t) = -\sin t$;   (b) $\cos(\pi + t) = -\cos t$.

38. Note that $P(t)$ and $P(\pi - t)$ are symmetric with respect to the y-axis. Use this fact to find identities analogous to those in Problem 37 for $\sin(\pi - t)$ and $\cos(\pi - t)$.

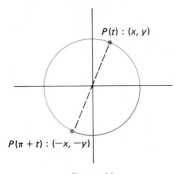

$P(t) : (x, y)$

$P(\pi + t) : (-x, -y)$

Figure 33

MISCELLANEOUS PROBLEMS

39. Use the unit circle to determine the sign (plus or minus) of each of the following.
    (a) $\cos 2$
    (b) $\sin(-3)$
    (c) $\cos 428°$
    (d) $\sin 21.4$
    (e) $\sin(23\pi/32)$
    (f) $\sin(-820°)$

40. Find the coordinates of $P(t)$ for the indicated values of t.
    (a) $t = \pi$
    (b) $t = -3\pi/2$
    (c) $t = -3\pi/4$
    (d) $t = 5\pi/6$
    (e) $t = 44\pi/3$
    (f) $t = -93.5\pi$

41. Let $P(t)$ have coordinates $(x, -1/2)$.
    (a) Find the two possible values of x.
    (b) Find the corresponding values of t.

42. With initial point $(0, -1)$, a string of length $4\pi/3$ is wound clockwise around the unit circle. What are the coordinates of the terminal point?

43. For what values of t satisfying $0 \le t < 2\pi$ are the following true?
    (a) $\sin t = \cos t$
    (b) $\frac{1}{2} < \sin t < \frac{\sqrt{3}}{2}$
    (c) $\cos^2 t \ge .25$
    (d) $\cos^2 t > \sin^2 t$

44. Find the four smallest positive solutions to the following equations.
    (a) $\sin t = 1$
    (b) $|\cos t| = \frac{1}{2}$
    (c) $\cos t = -\frac{\sqrt{3}}{2}$
    (d) $\sin t = -\frac{\sqrt{2}}{2}$

45. In each case, assume that θ is an angle in standard position with terminal side in the fourth quadrant. Use $\sin^2 \theta + \cos^2 \theta = 1$ to determine the indicated value.
    (a) $\cos \theta$ if $\sin \theta = -\frac{4}{5}$
    (b) $\sin \theta$ if $\cos \theta = \frac{24}{25}$

46. Use the unit circle to find identities for $\sin(2\pi - t)$ and $\cos(2\pi - t)$.

47. If $P(t)$ has coordinates $(\frac{4}{5}, -\frac{3}{5})$, evaluate each of the following. *Hint:* For (e) and (f), see Problems 37 and 38.
    (a) $\sin(-t)$
    (b) $\sin(\frac{\pi}{2} - t)$
    (c) $\cos(2\pi + t)$
    (d) $\cos(2\pi - t)$
    (e) $\sin(\pi + t)$
    (f) $\cos(\pi - t)$

sin t	cos t	sin($t + \pi$)	cos($t + \pi$)	sin($\pi - t$)	sin($2\pi - t$)	Least positive value of t
$\sqrt{3}/2$	$-\frac{1}{2}$					
	$\sqrt{2}/2$	$\sqrt{2}/2$				
$-\frac{1}{2}$			$-\sqrt{3}/2$			
-1						
			$\sqrt{3}/2$		$\frac{1}{2}$	
	0				-1	

48. Fill in all the blanks in the chart above.

49. Recall that [ ] and ( ) denote "the greatest integer in" and "the distance to the nearest integer," respectively. Determine which of the following functions are periodic and, if so, specify the period.
    (a) $f(x) = (x)$ (b) $f(x) = (3x)$
    (c) $f(x) = [x]$ (d) $f(x) = x - [x]$

50. Suppose that $f(x)$ is periodic with period 2 and that $f(x) = 4 - x^2$ for $0 \le x < 2$. Evaluate each of the following.
    (a) $f(2)$ (b) $f(4.5)$
    (c) $f(-.5)$ (d) $f(8.8)$

51. Evaluate

$$\sin 1° + \sin 2° + \sin 3° + \cdots + \sin 357° + \sin 358° + \sin 359°$$

52.   Evaluate

$$\sin^2 1° + \sin^2 2° + \sin^2 3° + \cdots + \sin^2 357° + \sin^2 358° + \sin^2 359°$$

New Functions from Old Ones	
tangent:	$\tan t = \dfrac{\sin t}{\cos t}$
cotangent:	$\cot t = \dfrac{\cos t}{\sin t}$
secant:	$\sec t = \dfrac{1}{\cos t}$
cosecant:	$\csc t = \dfrac{1}{\sin t}$

"Strange as it may sound, the power of mathematics rests on its evasion of all unnecessary thought and on its wonderful saving of mental operations."

Ernst Mach

2-4 Four More Trigonometric Functions

Without question, the sine and cosine are the most important of the six trigonometric functions. Not only do they occur most frequently in applications, but the other four functions can be defined in terms of them, as our opening box shows. This means that if you learn all you can about sines and cosines, you will automatically know a great deal about tangents, cotangents, secants, and cosecants. Ernst Mach would say that it is a way to evade unnecessary thought.

Look at the definitions in the opening box again. Naturally, we must rule out any values of t for which a denominator is zero. For example, $\tan t$ is not defined for $t = \pm\pi/2, \pm3\pi/2, \pm5\pi/2$, and so on. Similarly, $\csc t$ is not defined for such values as $t = 0, \pm\pi$, and $\pm2\pi$.

PROPERTIES OF THE NEW FUNCTIONS

The wisdom of the opening paragraph will now be demonstrated. Recall the identity $\sin^2 t + \cos^2 t = 1$. Out of it come two new identities.

$$1 + \tan^2 t = \sec^2 t$$
$$1 + \cot^2 t = \csc^2 t$$

To show that the first identity is correct, we take its left side, express it in terms of sines and cosines, and do a little algebra.

$$1 + \tan^2 t = 1 + \left(\frac{\sin t}{\cos t}\right)^2$$

$$= 1 + \frac{\sin^2 t}{\cos^2 t}$$

$$= \frac{\cos^2 t + \sin^2 t}{\cos^2 t}$$

$$= \frac{1}{\cos^2 t}$$

$$= \left(\frac{1}{\cos t}\right)^2$$

$$= \sec^2 t$$

The second identity is verified in a similar fashion.

Suppose we wanted to know whether cotangent is an even or an odd function (or neither). We simply recall that $\sin(-t) = -\sin t$ and $\cos(-t) = \cos t$ and write

$$\cot(-t) = \frac{\cos(-t)}{\sin(-t)} = \frac{\cos t}{-\sin t} = -\frac{\cos t}{\sin t} = -\cot t$$

Thus cotangent is an odd function.

In a similar vein, recall the identities

(i) $$\sin\left(\frac{\pi}{2} - t\right) = \cos t$$

(ii) $$\cos\left(\frac{\pi}{2} - t\right) = \sin t$$

From them, we obtain

$$\text{(iii)} \qquad \tan\left(\frac{\pi}{2} - t\right) = \frac{\sin(\pi/2 - t)}{\cos(\pi/2 - t)} = \frac{\cos t}{\sin t} = \cot t$$

These three identities are examples of what are called **cofunction identities.** Sine and cosine are confunctions; so are tangent and cotangent; as are secant and cosecant. Notice that identities (i), (ii), and (iii) all have the form

$$\text{function}\left(\frac{\pi}{2} - t\right) = \text{cofunction}(t)$$

With cosecant as the function, we have

$$\csc\left(\frac{\pi}{2} - t\right) = \sec t$$

ALTERNATIVE DEFINITIONS OF THE TRIGONOMETRIC FUNCTIONS

There is another approach to trigonometry favored by some authors. Let θ be an angle in standard position and suppose that (a, b) is any point on its terminal side at a distance r from the origin (Figure 34). Then

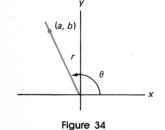

Figure 34

$$\sin \theta = \frac{b}{r} \qquad \cos \theta = \frac{a}{r}$$

$$\tan \theta = \frac{b}{a} \qquad \cot \theta = \frac{a}{b}$$

$$\sec \theta = \frac{r}{a} \qquad \csc \theta = \frac{r}{b}$$

To see that these definitions are equivalent to those we gave earlier, consider first an angle θ with terminal side in quadrant I (see Figure 35).

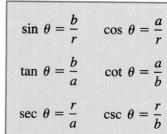

Figure 35

By similar triangles,

$$\frac{b}{r} = \frac{y}{1} \quad \text{and} \quad \frac{a}{r} = \frac{x}{1}$$

Actually these ratios are equal no matter in which quadrant the terminal side of θ is, since b and y always have the same sign, as do a and x. The first two formulas in the box now follow from our original definitions, which say that

$$\sin \theta = y \quad \text{and} \quad \cos \theta = x$$

The others are a consequence of the fact that the remaining four functions can be expressed in terms of sines and cosines.

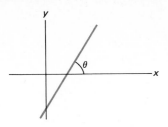

Figure 36

THE TANGENT FUNCTION AND SLOPE

Recall that the slope m of a line is the ratio of rise to run. In particular, if the line goes through the point (a, b) and also the origin, its slope is b/a. But this number b/a is also the tangent of the nonnegative angle θ that the line makes with the positive x-axis (see Figure 34).

In general, the smallest nonnegative angle θ that a line makes with the positive x-axis is called the **angle of inclination** of the line (Figure 36). It follows that for any nonvertical line, the slope m of the line satisfies

$$m = \tan \theta$$

As an example, suppose that a line has angle of inclination $120°$ and goes through the point $(1, 2)$. Then its slope is $m = \tan 120° = -\sqrt{3}$ and the line has equation

$$y - 2 = -\sqrt{3}(x - 1)$$

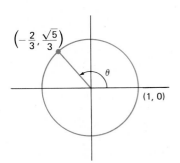

Figure 37

Problem Set 2-4

1. If $\sin t = \frac{4}{5}$ and $\cos t = -\frac{3}{5}$, evaluate each function.
   (a) $\tan t$ (b) $\cot t$ (c) $\sec t$ (d) $\csc t$
2. If $\sin t = -1/\sqrt{5}$ and $\cos t = 2/\sqrt{5}$, evaluate each function.
   (a) $\tan t$ (b) $\cot t$ (c) $\sec t$ (d) $\csc t$
3. Find the values of $\tan \theta$ and $\csc \theta$ for the angle θ of Figure 37.
4. Find $\cot \alpha$ and $\sec \alpha$ for α as shown in Figure 38.

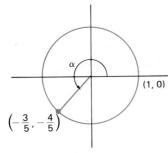

Figure 38

Keeping in mind what you know about the sines and cosines of special angles, find each of the values in Problems 5–22.

5. $\tan(\pi/6)$	6. $\cot(\pi/6)$	7. $\sec(\pi/6)$
8. $\csc(\pi/6)$	9. $\cot(\pi/4)$	10. $\sec(\pi/4)$
11. $\csc(\pi/3)$	12. $\sec(\pi/3)$	13. $\sin(4\pi/3)$
14. $\cos(4\pi/3)$	15. $\tan(4\pi/3)$	16. $\sec(4\pi/3)$
17. $\tan \pi$	18. $\sec \pi$	19. $\tan 330°$
20. $\cot 120°$	21. $\sec 600°$	22. $\csc(-150°)$

23. For what values of t on $0 \le t \le 4\pi$ is each of the following undefined?
    (a) $\sec t$ (b) $\tan t$ (c) $\csc t$ (d) $\cot t$
24. For which values of t on $0 \le t \le 4\pi$ is each of the following equal to 1?
    (a) $\sec t$ (b) $\tan t$ (c) $\csc t$ (d) $\cot t$

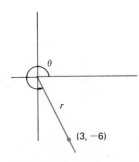

Figure 39

EXAMPLE (Using the a, b, r Definitions)   Suppose that the point $(3, -6)$ is on the terminal side of an angle in standard position (Figure 39). Find $\sin \theta$, $\tan \theta$, and $\sec \theta$.

Solution. First we find r.

$$r = \sqrt{3^2 + (-6)^2} = \sqrt{45} = 3\sqrt{5}$$

Then

$$\sin \theta = \frac{b}{r} = \frac{-6}{3\sqrt{5}} = -\frac{2}{\sqrt{5}}$$

$$\tan \theta = \frac{b}{a} = \frac{-6}{3} = -2$$

$$\sec \theta = \frac{r}{a} = \frac{3\sqrt{5}}{3} = \sqrt{5}$$

In Problems 25–28, find sin θ, tan θ, and sec θ, assuming that the given point is on the terminal side of θ.

25. $(5, -12)$ 26. $(7, 24)$ 27. $(-1, -2)$

28. $(-3, 2)$

29. If $\tan \theta = \frac{3}{4}$ and θ is an angle in the first quadrant, find $\sin \theta$ and $\sec \theta$. *Hint:* The point $(4, 3)$ is on the terminal side of θ.

30. If $\tan \theta = \frac{3}{4}$ and θ is an angle in the third quadrant, find $\cos \theta$ and $\csc \theta$. *Hint:* The point $(-4, -3)$ is on the terminal side of θ.

31. If $\sin \theta = \frac{5}{13}$ and θ is an angle in the second quadrant, find $\cos \theta$ and $\cot \theta$. *Hint:* A point with y-coordinate 5 and $r = 13$ is on the terminal side of θ. Thus the x-coordinate must be -12.

32. If $\cos \theta = \frac{4}{5}$ and $\sin \theta < 0$, find $\tan \theta$.

33. Where does the line from the origin to $(5, -12)$ intersect the unit circle?

34. Where does the line from the origin to $(-6, 8)$ intersect the unit circle?

35. Find the angle of inclination of the line $5x + 2y = 6$.

36. Find the equation of the line with angle of inclination $75°$ that passes through $(-2, 4)$.

MISCELLANEOUS PROBLEMS

37. Evaluate without use of a calculator.
    (a) $\sec(7\pi/6)$ (b) $\tan(-2\pi/3)$ (c) $\csc(3\pi/4)$
    (d) $\cot(11\pi/4)$ (e) $\csc(570°)$ (f) $\tan(180.045°)$

[c] 38. Calculate.
    (a) $\tan(\sin 2.4)$ (b) $\cot(\tan 1.49)$ (c) $\csc(\sin 11.8°)$
    (d) $\sec^2(\tan 91.2°)$ (e) $\csc(\tan \pi)$ (f) $\tan[\tan(\tan 1.5)]$

39. If $\csc t = 25/24$ and $\cos t < 0$, find each of the following.
    (a) $\sin t$ (b) $\cos t$ (c) $\tan t$
    (d) $\sec(\frac{\pi}{2} - t)$ (e) $\cot(\frac{\pi}{2} - t)$ (f) $\csc(\frac{\pi}{2} - t)$

40. Show that each of the following are identities.
    (a) $\tan(-t) = -\tan t$ (b) $\sec(-t) = \sec t$ (c) $\csc(-t) = -\csc t$

41. Find the two smallest positive values of t that satisfy each of the following.
    (a) $\tan t = -1$ (b) $\sec t = \sqrt{2}$ (c) $|\csc t| = 1$

42. Find the angle of inclination of the line that is perpendicular to the line $4x + 3y = 9$.

43. Write each of the following in terms of sines and cosines and simplify.

(a) $\dfrac{\sec \theta \csc \theta}{\tan \theta + \cot \theta}$

(b) $(\tan \theta)(\cos \theta - \csc \theta)$

(c) $\dfrac{(1 + \tan \theta)^2}{\sec^2 \theta}$

(d) $\dfrac{\sec \theta \cot \theta}{\sec^2 \theta - \tan^2 \theta}$

(e) $\dfrac{\cot \theta - \tan \theta}{\csc \theta - \sec \theta}$

(f) $\tan^4 \theta - \sec^4 \theta$

44. Let θ be a first quadrant angle. Express each of the other five trigonometric functions in terms of $\sin \theta$ alone.

45. Use the identities of Problem 37 in Section 2-3, namely,

$$\sin(t + \pi) = -\sin t \quad \text{and} \quad \cos(t + \pi) = -\cos t$$

to establish each of the following identities.

(a) $\tan(t + \pi) = \tan t$

(b) $\cot(t + \pi) = \cot t$

(c) $\sec(t + \pi) = -\sec t$

(d) $\csc(t + \pi) = -\csc t$

Note: From (a) and (b), we conclude that tangent and cotangent are periodic with period π.

46. Show that $|\sec t| \geq 1$ and $|\csc t| \geq 1$ for all t for which these functions are defined.

47. If $\tan \theta = \frac{5}{12}$ and $\sin \theta < 0$, evaluate $\cos^2 \theta - \sin^2 \theta$.

48. A wheel of radius 5, centered at the origin, is rotating counterclockwise at a rate of 1 radian per second. At $t = 0$, a speck of dirt on the rim is at $(5, 0)$. What are the coordinates of the speck at time t?

49. At $t = 2\pi/3$, the speck in Problem 44 came loose and flew off along the tangent line. Where did it hit the x-axis?

50. Find the coordinates of P in Figure 40.

51. The face of a clock is in the xy-plane with center at the origin and 12 on the positive y-axis. Both hands of the clock are 5 units long.
(a) Find the slope of the minute hand at 2:24.
(b) Find the slope of the line through the tips of both hands at 12:50.

52. From an airplane h miles above the surface of the earth (a sphere of radius 3960 miles), I can just see a bright light on the horizon d miles away. If I measure the angle of depression of the light as 2.1°, help me determine d and h.

53. A wheel of radius 20 centimeters is used to drive a wheel of radius 50 centimeters by means of a belt that fits around the wheels. How long is the belt if the centers of the two wheels are 100 centimeters apart?

54. **TEASER** Express the length L of the crossed belt that intersects in angle 2α and fits around wheels of radius r and R (Figure 41) in terms of r, R, and α.

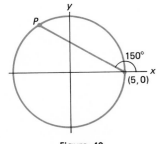

Figure 40

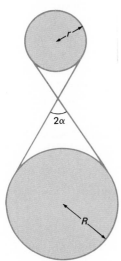

Figure 41

t (rad.)	Sin t	Tan t	Cot t	Cos t
.40	.38942	.42279	2.3652	.92106
.41	.39861	.43463	2.3008	.91712
.42	.40776	.44657	2.2393	.91309
.43	.41687	.45862	2.1804	.90897
.44	.42594	.47078	2.1241	.90475
.45	.43497	.48306	2.0702	.90045
.46	.44395	.49545	2.0184	.89605
.47	.45289	.50797	1.9686	.89157
.48	.46178	.52061	1.9208	.88699
.49	.47063	.53339	1.8748	.88233
.50	.47943	.54630	1.8305	.87758

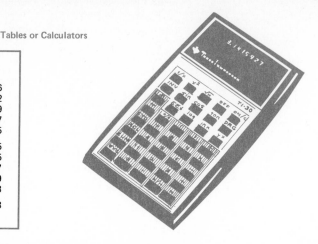

2-5 Finding Values of the Trigonometric Functions

In order to make significant use of the trigonometric functions, we will have to be able to calculate their values for angles other than the special angles we have considered. The simplest procedure is to press the right key on a calculator and read the answer. About the only thing to remember is to make sure the calculator is in the right mode, degree or radian, depending on what we want.

Even though calculators are becoming standard equipment for most mathematics and science students, we think you should also know how to use tables. That is the subject we take up now. We might call it "what to do when your battery goes dead."

The opening display gives a small portion of a five-place table of values for sin t, tan t, cot t, and cos t. (The complete table appears as Table B at the back of the book.) From it we read the following:

$$\sin .44 = .42594 \qquad \tan .44 = .47078$$
$$\cot .44 = 2.1241 \qquad \cos .44 = .90475$$

These results are not exact; they have been rounded off to five significant digits. Keep in mind that you can think of sin .44 in two ways, as the sine of the number .44 or, if you like, as the sine of an angle of radian measure .44.

Table B appears to have two defects. First, t is given only to 2 decimal places. If we need sin .44736, we have to round or perhaps to interpolate (see first section of Appendix).

$$\sin .44736 \approx \sin .45 = .43497$$

A more serious defect appears to be the fact that values of t go only to 2.00.

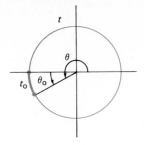

Figure 42

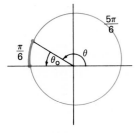

Figure 43

This limitation evaporates once we learn about reference angles and reference numbers, our next topic.

REFERENCE ANGLES AND REFERENCE NUMBERS

Let θ be any angle in standard position and let t be its radian measure. Associated with θ is an acute angle θ_0, called the **reference angle** and defined to be the smallest positive angle between the terminal side of θ and the x-axis (Figure 42). The radian measure t_0 of θ_0 is called the **reference number** corresponding to t. For example, the reference number for $t = 5\pi/6$ is $t_0 = \pi/6$ (Figure 43). Once we know t_0, we can find $\sin t$, $\cos t$, and so on, no matter what t is. Here is how we do it.

Examine the four diagrams in Figure 44.

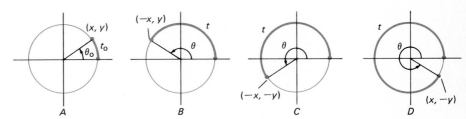

Figure 44

Each angle θ in B, C, and D has θ_0 as its reference angle and, of course, each t has t_0 as its reference number. Now we make a crucial observation. In each case, the point on the unit circle corresponding to t has the same coordinates, except for sign, as the point corresponding to t_0. It follows from this that

$$\sin t = \pm\sin t_0 \qquad \cos t = \pm\cos t_0$$

with the $+$ or $-$ sign being determined by the quadrant in which the terminal side of the angle falls. For example,

$$\sin \frac{5\pi}{6} = \sin \frac{\pi}{6} \qquad \cos \frac{5\pi}{6} = -\cos \frac{\pi}{6}$$

or, in degree notation,

$$\sin 150° = \sin 30° \qquad \cos 150° = -\cos 30°$$

We chose the plus sign for the sine and the minus sign for the cosine because in the second quadrant the sine function is positive, whereas the cosine function is negative.

What we have just said applies to all six trigonometric functions. If T stands for any one of them, then

$$\boxed{T(t) = \pm T(t_0) \quad \text{and} \quad T(\theta) = \pm T(\theta_0)}$$

with the plus or minus sign being determined by the quadrant in which the terminal side of θ lies. Of course $T(t_0)$ itself is always nonnegative since $0 \leq t_0 \leq \pi/2$.

EXAMPLES

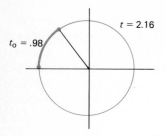

Figure 45

If we wish to calculate cos 2.16 using tables, we must first find the reference number for 2.16. Approximating π by 3.14, we find that (see Figure 45)

$$t_0 = 3.14 - 2.16 = .98$$

and thus, using Table B,

$$\cos 2.16 = -\cos .98 = -.55702$$

Notice we chose the minus sign because the cosine is negative in quadrant II.

To calculate tan 24.95 is slightly more work. First we remove as large a multiple of 2π as possible from 24.95. Using 6.28 for 2π, we get

$$24.95 = 3(6.28) + 6.11$$

The reference number for 6.11 is (see Figure 46)

$$t_0 = 6.28 - 6.11 = .17$$

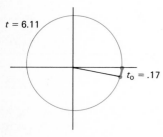

Figure 46

Thus

$$\tan 24.95 = \tan 6.11 = -\tan .17 = -.17166$$

We choose the minus sign because the tangent is negative in quadrant IV.

Now use your pocket calculator to find tan 24.95 the easy way. Be sure you put it in radian mode. You will get $-.18480$ instead of $-.17166$, a rather large discrepancy. Whom should you believe? We suggest that you trust your calculator. The reason we were so far off is that 6.28 is a rather poor approximation for 2π, and multiplying it by 3 made matters worse. Had we used 6.2832 for 2π, we would have obtained $t_0 = .1828$ and tan 24.95 $= -.18486$.

Problem Set 2-5

Find the value of each of the following using Table A or Table B.

1. sin 1.38   2. cos .67   3. cos 42.8°   4. tan 18.0°
5. cot .82   6. tan 1.11   7. sin 68.3°   8. cot 49.6°

EXAMPLE A (Finding Reference Numbers)   Find the reference number t_0 for each of the following values of t.

(a) $t = 20.59$ (b) $t = \dfrac{5\pi}{2} - .92$

$$\begin{array}{r} 3 \\ 6.28\overline{)20.59} \\ \underline{18.84} \\ 1.75 \end{array}$$

Figure 47

Solution.

(a) To get rid of the irrelevant multiples of 2π, we divide 20.59 by 6.28 ($2\pi \approx 6.28$), obtaining 1.75 as remainder (Figure 47). Since 1.75 is between $\pi/2$ and π, we subtract it from π. Thus

$$t_0 \approx \pi - 1.75 \approx 3.14 - 1.75 = 1.39$$

(b) Since $5\pi/2 - .92 = 2\pi + \pi/2 - .92$, it follows that

$$t_0 = \frac{\pi}{2} - .92 \approx 1.57 - .92 = .65$$

Find the reference number t_0 if t has the given value. Use 3.14 for π.

9. 1.84	10. 2.14	11. 3.54	12. 3.74
13. 5.18	14. 6.08	15. 10.48	16. 8.38
17. −1.12	18. −1.86	19. −2.64	20. −4.24

Find the reference number for each of the following. You may leave your answer in terms of π.

21. $13\pi/8$	22. $37\pi/36$	23. $40\pi/3$	24. $-11\pi/5$
25. $3\pi + .24$	26. $3\pi/2 + .17$	27. $3\pi - .24$	28. $3\pi/2 - .17$
29. $11\pi/2$	30. 26π		

Find the value of each of the following using Table B and $\pi = 3.14$. Calculators will give slightly different results because of this crude approximation to π.

31. cos 1.42	32. sin .97	33. tan 1.39	34. cot .08
35. sin 2.14	36. cos 3.08	37. cot 5.62	38. tan 4.11
39. cos(−2.54)	40. sin(−4.18)		

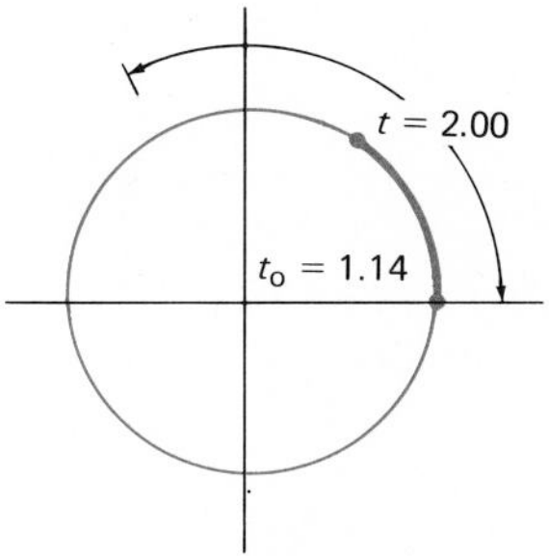

EXAMPLE B (Finding t When sin t or cos t Is Given) Find 2 values of t between 0 and 2π for which (a) sin $t = .90863$; (b) cos $t = -.95824$.

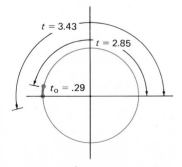

Figure 48

Figure 49

Solution.

(a) We get $t = 1.14$ directly from Table B (or using a calculator). Since the sine is also positive in quadrant II, we seek a value of t between $\pi/2$ and π for which 1.14 is the reference number (Figure 48). Only one number fits the bill:

$$\pi - 1.14 \approx 3.14 - 1.14 = 2.00$$

(b) We know that cos $t_0 = .95824$ and so $t_0 = .29$. Now the cosine is negative in quadrants II and III. Thus we are looking for two numbers between $\pi/2$ and $3\pi/2$ with .29 as reference number (Figure 49). One is $\pi - .29 \approx 3.14 - .29 = 2.85$, and the other $\pi + .29 \approx 3.14 + .29 = 3.43$.

Find two values of t between 0 and 2π for which the given equality holds.

41. sin $t = .94898$	42. cos $t = .72484$	43. cos $t = -.08071$

44. $\sin t = -.48818$ 45. $\tan t = 4.9131$ 46. $\cot t = 1.4007$

47. $\tan t = -3.6021$ 48. $\cot t = -.47175$

Find the reference angle (in degrees) for each of the following angles. For example, the reference angle for $\theta = 124.1°$ is $\theta_0 = 180° - 124.1° = 55.9°$.

49. $139.6°$ 50. $218.1°$ 51. $348.7°$

52. $375.4°$ 53. $-99.8°$ 54. $-224.4°$

EXAMPLE C (Finding sin θ, cos θ, and so on, When θ Is Any Angle Given in Degrees)   Find the value of each of the following.

(a) $\cos 214.6°$ (b) $\cot 658°$

Solution.   So far, we have used Table A to find the sine, cosine, and so on, of positive angles measuring less than $90°$. Here we do this for angles of arbitrary (degree) measure.

(a) The reference angle is

$$214.6° - 180° = 34.6°$$

$$\cos 214.6° = -\cos 34.6° = -.8231$$

We used the minus sign since cosine is negative in quadrant III.

(b) First we reduce our angle by $360°$

$$658° = 360° + 298°$$

The reference angle for $298°$ is $360° - 298°$, or $62°$. In the column with cot at the bottom and $62°$ at the right, we find $.5317$. Therefore $\cot 658° = -.5317$.

CAUTION

$\cos 99° = \cos 81°$ ~~= .1564~~   *wrong*
$\cos 99° = -\cos 81°$ $= -.1564$ Be sure to assign the correct sign.

Find the value of each of the following.

55. $\sin 156.1°$ 56. $\cos 138.7°$ 57. $\tan 348.9°$ 58. $\cot 224.9°$

59. $\cos(-66.1°)$ 60. $\sin 487°$ 61. $\cos 441.3°$ 62. $\sin 180.2°$

63. $\cot(-134°)$ 64. $\tan 311.6°$

Find two different degree values of θ between $0°$ and $360°$ for which the given equality holds.

65. $\sin \theta = .3633$ 66. $\cos \theta = .9907$ 67. $\tan \theta = .4942$

68. $\cot \theta = 1.2799$ 69. $\cos \theta = -.9085$ 70. $\sin \theta = -.2045$

MISCELLANEOUS PROBLEMS

71. Use Table A or B to find each of the following. You may approximate π by 3.14.

    (a) $\cos 5.63$ (b) $\sin 10.34$ (c) $\tan 8.42$

    (d) $\sin 311.3°$ (e) $\tan(-411°)$ (f) $\cos 1989°$

72. Use Tables A *and* B to calculate.

    (a) $\sin(\cos 134°)$ (b) $\sin[(\tan 1.5)°]$ (c) $\tan(-5.4°) + \tan(-5.4)$

© 73. Calculate.

    (a) $\cos(\sin 2.42°)$ (b) $\cos^3(\sin^2 2.42)$ (c) $\sqrt{\tan 4.21 + \ln(\sin 7.12)}$

74. Use Table B and $\pi = 3.14$ to find two values of t between 0 and 2π for which each of the following is true.
(a) $\sin t = .62879$ (b) $\cos t = -.90045$ (c) $\tan t = -4.4552$

c 75. Find two values of t between 0 and 2π for which each statement is true, giving your answers correct to 6 decimal places.
(a) $\sin t = .62879$ (b) $\cos t = .34176$ (c) $\tan t = -3.14159$
Note: On many calculators, you would press $.62879$ $\boxed{\text{INV}}$ $\boxed{\sin}$ to get one answer to (a).

76. If $\pi/2 < t < \pi$, then $t_0 = \pi - t$. In a similar manner, express t_0 in terms of t in each case.
(a) $3\pi/2 < t < 2\pi$ (b) $5\pi < t < 11\pi/2$ (c) $-2\pi < t < -3\pi/2$

77. If $0° < \phi < 90°$, express the reference angle θ_0 in terms of ϕ in each case.
(a) $\theta = 180° + \phi$ (b) $\theta = 270° - \phi$ (c) $\theta = \phi - 90°$

78. Without using tables or a calculator, round to the nearest degree the smallest positive angle θ satisfying $\tan \theta = -40,000$.

79. If θ is a fourth quadrant angle whose terminal side coincides with the line $3x + 5y = 0$, find $\sin \theta$.

c 80. In calculus, you will learn that

$$\sin t = t - \frac{t^3}{3!} + \frac{t^5}{5!} - \frac{t^7}{7!} + \cdots$$

and

$$\cos t = 1 - \frac{t^2}{2!} + \frac{t^4}{4!} - \frac{t^6}{6!} + \cdots$$

Here, $n! = 1 \cdot 2 \cdot 3 \cdots n$ (for example, $2! = 1 \cdot 2 = 2$ and $3! = 1 \cdot 2 \cdot 3 = 6$). These series are used to construct Tables A and B. If we use just the first three terms in the sine series, we obtain

$$\sin t \approx t - \frac{t^3}{6} + \frac{t^5}{120} = \left[\left(\frac{t^2}{120} - \frac{1}{6}\right)t^2 + 1\right]t$$

Use the first three terms of these series to approximate each of the following and compare with the corresponding value in Table B.
(a) $\sin(.1)$ (b) $\sin(.4)$ (c) $\cos(.2)$

81. Determine ϕ in Figure 50 so that the path ACB has minimum length.

82. **TEASER**   Let α, β, and γ be acute angles such that $\tan \alpha = 1$, $\tan \beta = 2$, and $\tan \gamma = 3$.
(a) Use your calculator to approximate $\alpha + \beta + \gamma$.
(b) Make a conjecture about the exact value of $\alpha + \beta + \gamma$.
(c) Construct a clever geometric diagram to prove your conjecture.

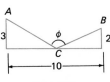

Figure 50

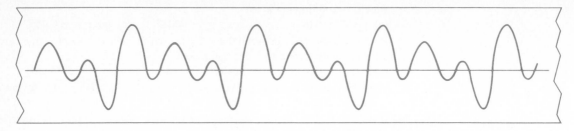

When heart beats, brain activity, or sound waves from a musical instrument are changed into visual images by means of an oscilloscope, they give a regular repetitive pattern which may look something like the diagram above. This repetitive behavior is a characteristic feature of the graphs of the trigonometric functions. In fact, almost any repetitive pattern can be approximated by appropriate combinations of the trigonometric functions.

2-6 Graphs of the Trigonometric Functions

Recall that to graph $y = f(x)$, we first construct a table of values of ordered pairs (x, y), then plot the corresponding points, and finally connect those points with a smooth curve. Here we want to graph $y = \sin t$, $y = \cos t$, and so on, and we will follow a similar procedure. Notice that we use t rather than x as the independent variable because we used t as the variable (radian measure of an angle) in our definition of the trigonometric functions.

We begin with the graphs of the sine and cosine functions. You should become so well acquainted with these two graphs that you can sketch them quickly whenever you need them. This will aid you in two ways. First, these graphs will help you remember many of the important properties of the sine and cosine functions. Second, knowing them will help you graph other more complicated trigonometric functions.

THE GRAPH OF $y = \sin t$

We begin with a table of values (Figure 51).

t	0	$\dfrac{\pi}{6}$	$\dfrac{\pi}{4}$	$\dfrac{\pi}{3}$	$\dfrac{\pi}{2}$	$\dfrac{3\pi}{4}$	π	$\dfrac{5\pi}{4}$	$\dfrac{3\pi}{2}$	$\dfrac{7\pi}{4}$	2π
$y = \sin t$	0	$\dfrac{1}{2}$	$\dfrac{\sqrt{2}}{2}$	$\dfrac{\sqrt{3}}{2}$	1	$\dfrac{\sqrt{2}}{2}$	0	$-\dfrac{\sqrt{2}}{2}$	-1	$-\dfrac{\sqrt{2}}{2}$	0

Figure 51

We have listed values of t between 0 and 2π. That is sufficient to graph one period (shown in Figure 52 as a heavy curve). From there on we can continue the curve indefinitely in either direction in a repetitive fashion, for we learned earlier that $\sin(t + 2\pi) = \sin t$.

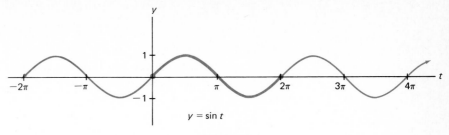

$y = \sin t$

Figure 52

THE GRAPH OF $y = \cos t$

The cosine function is a copycat; its graph is just like that of the sine function but pushed $\pi/2$ units to the left (Figure 53).

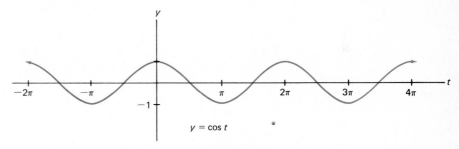

$y = \cos t$

Figure 53

To see that the graph of the cosine function is correct, we might make a table of values and proceed as we did for the sine function. In fact, we ask you to do just that in Problem 1. Alternatively, we can show that

$$\cos t = \sin\left(t + \frac{\pi}{2}\right)$$

This follows directly from identities we have observed earlier.

$$\sin\left(t + \frac{\pi}{2}\right) = \sin\left(\frac{\pi}{2} - (-t)\right)$$
$$= \cos(-t) \qquad \text{(cofunction identity)}$$
$$= \cos t \qquad \text{(cosine is even)}$$

PROPERTIES EASILY OBSERVED FROM THESE GRAPHS

1. Both sine and cosine are periodic with 2π as period.
2. $-1 \le \sin t \le 1$ and $-1 \le \cos t \le 1$.
3. $\sin t = 0$ if $t = -\pi, 0, \pi, 2\pi$, and so on.
   $\cos t = 0$ if $t = -\pi/2, \pi/2, 3\pi/2$, and so on.

4. $\sin t > 0$ in quadrants I and II.
   $\cos t > 0$ in quadrants I and IV.
5. $\sin(-t) = -\sin t$ and $\cos(-t) = \cos t$.
   The sine is an odd function; its graph is symmetric with respect to the origin. The cosine is an even function; its graph is symmetric with respect to the y-axis.
6. We can see immediately where the sine and cosine functions are increasing and where they are decreasing. For example, the sine function decreases for $\pi/2 \le t \le 3\pi/2$.

THE GRAPH OF $y = \tan t$

Since the tangent function is defined by

$$\tan t = \frac{\sin t}{\cos t}$$

we need to beware of values of t for which $\cos t = 0$: $-\pi/2, \pi/2, 3\pi/2$, and so forth. In fact, from Section 1-5, we know that we should expect vertical asymptotes at these places. Notice also that

$$\tan(-t) = \frac{\sin(-t)}{\cos(-t)} = \frac{-\sin t}{\cos t} = -\tan t$$

which means that the graph of the tangent will be symmetric with respect to the origin. Using these two pieces of information, a small table of values, and the fact that the tangent is periodic, we obtain the graph in Figure 54.

To confirm that the graph is correct near $t = \pi/2$, we suggest looking at Table B. Notice that the $\tan t$ steadily increases until at $t = 1.57$, we read $\tan t = 1255.8$. But as t takes the short step to 1.58, $\tan t$ takes a tremendous plunge to -108.65. In that short space, t has passed through $\pi/2 \approx 1.5708$ and $\tan t$ has shot up to celestial heights only to fall to a bottomless pit, from which, however, it manages to escape as t moves to the right.

While we knew the tangent would have to repeat itself every 2π units since the sine and cosine do this, we now notice that it actually repeats itself on intervals of length π. Since the word *period* denotes the length of the shortest interval after which a function repeats itself, the tangent function has period π. For an algebraic demonstration, see Problem 45 of Section 2-4.

THE GRAPH OF $y = \sec t$

Since $\sec t = 1/\cos t$, one way of getting the graph of the secant is by graphing the cosine and then taking reciprocals of the y-coordinates (Figure 55). Note that since $\cos t = 0$ at $t = -\pi/2, \pi/2, 3\pi/2$, and so on, the graph of $\sec t$ must have vertical asymptotes at these points.

Just like the cosine, the secant is an even function; that is, $\sec(-t) = \sec t$. And, like the cosine, secant has period 2π. However, notice that if $\cos t$ increases or decreases throughout an interval, $\sec t$ does just the opposite. For example, $\cos t$ decreases for $0 < t < \pi/2$, whereas $\sec t$ increases there.

t	0	$\frac{\pi}{4}$	$\frac{\pi}{3}$	$\frac{\pi}{2}$	$\frac{2\pi}{3}$	$\frac{3\pi}{4}$	π	$\frac{5\pi}{4}$	$\frac{3\pi}{2}$	$\frac{7\pi}{4}$	2π
$y = \tan t$	0	1	$\sqrt{3}$	undefined	$-\sqrt{3}$	-1	0	1	undefined	-1	0

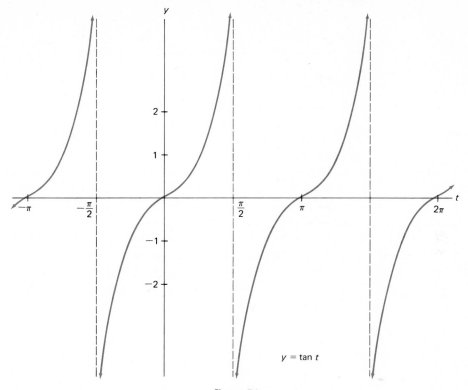

Figure 54

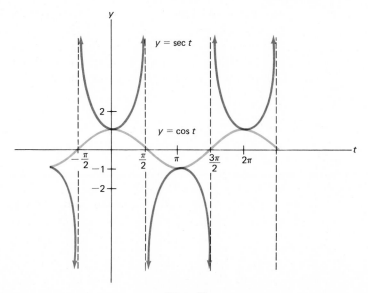

Figure 55

Be sure to study Example A below. It introduces the graph of $y = A \sin Bt$. This important topic is explored more fully in Section 4-5 in connection with simple harmonic motion.

Problem Set 2-6

1. Make a table of values and then sketch the graph of $y = \cos t$.
2. What real numbers constitute the domain of the cosine? The range?
3. Sketch the graph of $y = \cot t$ for $-2\pi \le t \le 2\pi$, being sure to show the asymptotes.
4. What real numbers constitute the entire domain of the cotangent? The range?
5. Using the corresponding fact about the cosine, demonstrate algebraically that $\sec(t + 2\pi) = \sec t$.
6. Sketch the graph of $y = \csc t$.
7. What is the domain of the secant? The range?
8. What is the domain of the cosecant? The range?
9. What is the period of the cotangent? The secant?
10. On the interval $-2\pi \le t \le 2\pi$, where is the cotangent increasing?
11. Which is true: $\cot(-t) = \cot t$ or $\cot(-t) = -\cot t$?
12. Which is true: $\csc(-t) = \csc t$ or $\csc(-t) = -\csc t$?

EXAMPLE A (Some Sine-Related Graphs) Sketch the graph of each of the following for $-2\pi \le t \le 4\pi$.
(a) $y = 2 \sin t$ (b) $y = \sin 2t$ (c) $y = 3 \sin 4t$

Solution.
(a) We could graph $y = 2 \sin t$ from a table of values. It is easier, though, to graph $\sin t$ (dotted graph below) and then multiply the ordinates by 2 (Figure 56). Since the graph bobs up and down between $y = -2$ and $y = 2$, we say that it has an **amplitude** of 2. The period is 2π, the same as for $\sin t$.

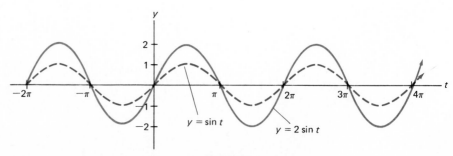

Figure 56

t	$-\pi$	$-\dfrac{3\pi}{4}$	$-\dfrac{\pi}{2}$	$-\dfrac{\pi}{4}$	$-\dfrac{\pi}{12}$	0	$\dfrac{\pi}{12}$	$\dfrac{\pi}{4}$	$\dfrac{\pi}{2}$	$\dfrac{3\pi}{4}$	π
$2t$	-2π	$-\dfrac{3\pi}{2}$	$-\pi$	$-\dfrac{\pi}{2}$	$-\dfrac{\pi}{6}$	0	$\dfrac{\pi}{6}$	$\dfrac{\pi}{2}$	π	$\dfrac{3\pi}{2}$	2π
$\sin 2t$	0	1	0	-1	$-\dfrac{1}{2}$	0	$\dfrac{1}{2}$	1	0	-1	0

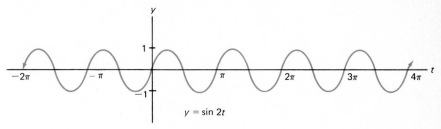

$y = \sin 2t$

Figure 57

(b) Here a table of values is advisable, since this is our first example of this type (Figure 57). This graph goes through a complete cycle as t increases from 0 to π; that is, the period of $\sin 2t$ is π instead of 2π as it was for $\sin t$. The amplitude is 1, just as for $\sin t$.

(c) We can save a lot of work once we recognize how the character of the graph of $A \sin Bt$ (and $A \cos Bt$) is determined by the numbers A and B $(B > 0)$. The amplitude (which tells how far the graph rises and falls from its median position) is given by $|A|$. The period is given by $2\pi/B$. Thus for $y = 3 \sin 4t$, the amplitude is 3 and the period is $2\pi/4 = \pi/2$. For a quick sketch, we use these two numbers to determine the high and low points and the t-intercepts, connecting these points with a smooth, wavelike curve (Figure 58).

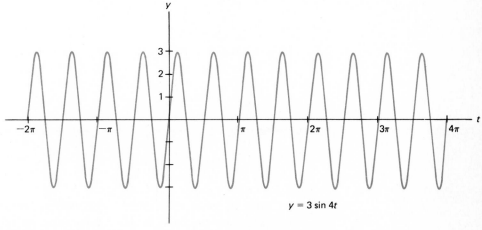

$y = 3 \sin 4t$

Figure 58

In Problems 13–22, determine the amplitude and the period. Then sketch the graph on the indicated interval.

13. $y = 3 \cos t, \ -\pi \leq t \leq \pi$

14. $y = \frac{1}{2} \cos t, \ -\pi \leq t \leq \pi$

15. $y = -\sin t, \ -\pi \leq t \leq \pi$

16. $y = -2 \cos t, \ -\pi \leq t \leq \pi$

17. $y = \cos 4t, \ -\pi \leq t \leq \pi$

18. $y = \cos 3t, \ -\pi/2 \leq t \leq \pi/2$

19. $y = 2 \sin \frac{1}{2}t, \ -2\pi \leq t \leq 2\pi$

20. $y = 3 \sin \frac{1}{3}t, \ -3\pi \leq t \leq 3\pi$

21. $y = 2 \cos 3t, \ -\pi \leq t \leq \pi$

22. $y = 4 \sin 3t, \ -\pi \leq t \leq \pi$

EXAMPLE B (Graphing Sums of Trigonometric Functions) Sketch the graph of the equation $y = 2 \sin t + \cos 2t$.

Solution. We graph $y = 2 \sin t$ and $y = \cos 2t$ on the same coordinate plane (these appear as dotted-line curves in Figure 59) and then add ordinates. Notice that for any t, the ordinates (y-values) of the dotted curves are added to obtain the desired ordinate. The graph of $y = 2 \sin t + \cos 2t$ is quite different from the separate (dotted) graphs but it does repeat itself; it has period 2π.

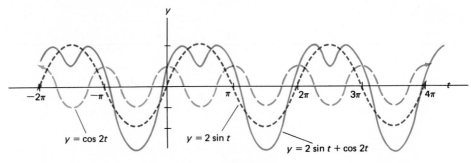

Figure 59

Sketch each graph by the method of adding ordinates. Show at least one complete period.

23. $y = 2 \sin t + \cos t$

24. $y = \sin t + 2 \cos t$

25. $y = \sin 2t + \cos t$

26. $y = \sin t + \cos 2t$

27. $y = \sin \frac{1}{2}t + \frac{1}{2} \sin t$

28. $y = \cos \frac{1}{2}t + \frac{1}{2} \cos t$

MISCELLANEOUS PROBLEMS

In Problems 29–32, sketch each graph on the indicated interval.

29. $y = -\cos t, \ -\pi \leq t \leq \pi$

30. $y = 3 \sin t, \ -\pi \leq t \leq \pi$

31. $y = \sin 4t, \ 0 \leq t \leq \pi$

32. $y = 3 \cos \frac{1}{2}t, \ -2\pi \leq t \leq 2\pi$

33. What are the amplitudes and periods for the graphs in Problems 29 and 31?

34. What are the amplitudes and periods for the graphs in Problems 30 and 32?

35. Determine the period and sketch the graph of each of the following, showing at least three periods.

(a) $y = \tan 2t$

(b) $y = 3 \tan(t/2)$

36. Follow the directions of Problem 35.
    (a) $y = 2 \cot 2t$ (b) $y = \sec 3t$

37. Sketch, using the same axes, the graphs of
    (a) $f(t) = \sin t$;   (b) $g(t) = 3 + \sin t$;   (c) $h(t) = \sin(t - \pi/4)$.

38. Sketch, using the same axes, the graphs of
    (a) $f(t) = \cos t$;   (b) $g(t) = -2 + \cos t$;   (c) $h(t) = \cos(t + \pi/3)$.

39. Sketch the graph of $y = \cos 3t + 2 \sin t$ for $-\pi \le t \le \pi$. Use the method of adding ordinates.

40. Sketch the graph of $y = t + \sin t$ on $-4\pi \le t \le 4\pi$. Use the method of adding ordinates.

© 41. Sketch the graph of $y = t - \cos t$ for $0 \le t \le 6$ by actually calculating the y values corresponding to $t = 0, .5, 1, 1.5, 2, 2.5, \ldots, 6$.

42. By sketching the graphs of $y = t$ and $y = 3 \sin t$ on the same coordinate axes, determine approximately all solutions of $t = 3 \sin t$.

43. The strength I of current (in amperes) in a wire of an alternating current circuit might satisfy

$$I = 30 \sin(120\pi t)$$

where time t is measured in seconds.
    (a) What is the period?
    (b) How many cycles (periods) are there in one second?
    (c) What is the maximum strength of the current?

© 44. Sketch the graph of $y = (\sin t)/t$ on $-3\pi \le t \le 3\pi$. Be sure to plot several points for t near 0 (for example, $t = -.5, -.2, -.1, .1, .2, .5$). What value does y seem to approach as t approaches 0?

45. Consider $y = \sin(1/t)$ on the interval $0 < t \le 1$.
    (a) Where does its graph cross the t-axis?
    (b) Evaluate y for $t = 2/\pi, 2/3\pi, 2/5\pi, 2/7\pi, \ldots$.
    (c) Sketch the graph as best you can, using a large unit on the t-axis.

46. **TEASER** How many solutions does each equation have on the indicated interval for t?
    (a) $\sin t = t/60$, $t \ge 0$
    (b) $\sin(1/t) = t/60$, $t \ge .06$
    (c) $\sin(1/t) = t/60$, $t > 0$

Chapter Summary

The word **trigonometry** means triangle measurement. In its elementary historical form, it is the study of how to **solve** right triangles when appropriate information is given. The main tools are the three trigonometric ratios sin θ, cos θ, and tan θ, which were first defined only for acute angles θ.

In order to give the subject its modern general form, we first generalized the notion of an angle θ, allowing θ to have arbitrary size and measuring it either in **degrees** or **radians.** Such an angle θ can be placed in **standard position** in a coordinate system, where it will cut off an arc of directed length t (the radian measure of θ) stretching from $(1, 0)$ to (x, y) on the unit circle. This allowed us to make the key definitions

$$\sin \theta = \sin t = y \qquad \cos \theta = \cos t = x$$

on which all of modern trigonometry rests.

From the above definitions, we derived several identities, of which the most important is

$$\sin^2 t + \cos^2 t = 1$$

We also defined four additional functions

$$\tan t = \frac{\sin t}{\cos t} \qquad \cot t = \frac{\cos t}{\sin t}$$

$$\sec t = \frac{1}{\cos t} \qquad \csc t = \frac{1}{\sin t}$$

To evaluate the trigonometric functions, we may use either a scientific calculator or Tables A and B in the Appendix. If the tables are used, the notions of **reference angle** and **reference number** become important. Finally we graphed several of the trigonometric functions, noting especially their **periodic** behavior.

Chapter Review Problem Set

1. Solve the following right triangles ($\gamma = 90°$).
   (a) $\alpha = 47.1°$, $c = 36.9$ (b) $a = 417$, $c = 573$

2. At a distance of 10 feet from a wall, the angle of elevation of the top of a mural with respect to eye level is 18° and the corresponding angle of depression of the bottom is 10°. How high is the mural?

3. Change 33° to radians. Change $9\pi/4$ radians to degrees.

4. How far does a wheel of radius 30 centimeters roll along level ground in making 100 revolutions?

5. Calculate each of the following without use of tables or a calculator.
   (a) $\sin(7\pi/6)$ (b) $\cos(11\pi/6)$
   (c) $\tan(13\pi/4)$ (d) $\sin(41\pi/6)$

6 Evaluate.
   (a) $\sin 411°$ (b) $\cos 1312°$
   (c) $\tan 5.77$ (d) $\sin 13.12$

7. Write in terms of $\sin t$.
   (a) $\sin(-t)$ (b) $\sin(t + 4\pi)$

   (c) $\sin(\pi + t)$ (d) $\cos\left(\dfrac{\pi}{2} - t\right)$

8. For what values of t between 0 and 2π is
   (a) $\cos t > 0$ (b) $\cos 2t > 0$?

9. If $(-5, -12)$ is on the terminal side of an angle θ in standard position, find
   (a) $\cot \theta$; \hspace{2cm} (b) $\sec \theta$.

10. If $\sin \theta = \frac{2}{3}$ and θ is a second quadrant angle, find $\tan \theta$.

11. Sketch the graph of $y = 3 \cos 2t$ for $-\pi \leq t \leq 2\pi$.

12. Sketch the graph of $y = \sin t + \sin 2t$ using the method of adding ordinates.

13. What is the range of the sine function? Of the cosecant function?

14. Using the facts that the sine function is odd and the cosine function is even, show that cotangent is an odd function.

15. Give the general definition of $\cos t$ based on the unit circle.

For just as in nature itself there is no middle ground between truth and falsehood, so in rigorous proofs one must either establish his point beyond doubt, or else beg the question inexcusably. There is no chance of keeping one's feet by invoking limitations, distinctions, verbal distortions, or other mental acrobatics. One must with a few words and at the first assault become Caesar or nothing at all.

Galileo

CHAPTER 3

Trigonometric Identities and Equations

3-1   Identities

Complicated combinations of the six trigonometric functions occur often in mathematics. It is important that we, like the professor above, be able to write a complicated trigonometric expression in a simpler or more convenient form. To do this requires two things. We must be good at algebra and we must know the fundamental identities of trigonometry.

THE FUNDAMENTAL IDENTITIES

We list eleven fundamental identities, which should be memorized.

1. $\tan t = \dfrac{\sin t}{\cos t}$

2. $\cot t = \dfrac{\cos t}{\sin t} = \dfrac{1}{\tan t}$

3. $\sec t = \dfrac{1}{\cos t}$

4. $\csc t = \dfrac{1}{\sin t}$

5. $\sin^2 t + \cos^2 t = 1$

6. $1 + \tan^2 t = \sec^2 t$

7. $1 + \cot^2 t = \csc^2 t$

8. $\sin\left(\dfrac{\pi}{2} - t\right) = \cos t$

9. $\cos\left(\dfrac{\pi}{2} - t\right) = \sin t$

10. $\sin(-t) = -\sin t$

11. $\cos(-t) = \cos t$

We have seen all these identities before. The first four are actually definitions; the others were established either in the text or the problem sets of Sections 2-3 and 2-4.

PROVING NEW IDENTITIES

The professor's work in our opening cartoon can be viewed in two ways. The more likely way of looking at it is that she wanted to simplify the complicated expression

$$(\sec t + \tan t)(1 - \sin t)$$

But it could be that someone had conjectured that

$$(\sec t + \tan t)(1 - \sin t) = \cos t$$

is an identity and that the professor was trying to prove it. It is this second concept we want to discuss now.

Suppose someone claims that a certain equation is an identity—that is, true for all values of the variable for which both sides make sense. How can you check on such a claim? The procedure used by the professor is one we urge you to follow. Start with the more complicated looking side and try to use a chain of equalities to produce the other side.

Suppose we wish to prove that

$$\sin t + \cos t \cot t = \csc t$$

is an identity. We begin with the left side and rewrite it step by step, using algebra and the fundamental identities, until we get the right side.

$$\sin t + \cos t \cot t = \sin t + \cos t \left(\frac{\cos t}{\sin t} \right)$$

$$= \frac{\sin^2 t + \cos^2 t}{\sin t}$$

$$= \frac{1}{\sin t}$$

$$= \csc t$$

When proving that an equation is an identity, it pays to look before you leap. Changing the more complicated side to sines and cosines, as in the above example, is often the best thing to do. But not always. Sometimes the simpler side gives us a clue as to how we should reshape the other side. For example, the left side of

$$\tan t = \frac{(\sec t - 1)(\sec t + 1)}{\tan t}$$

suggests that we try to rewrite the right side in terms of $\tan t$. This can be done by multiplying out the numerator and making use of the fundamental identity $\sec^2 t = 1 + \tan^2 t$.

$$\frac{(\sec t - 1)(\sec t + 1)}{\tan t} = \frac{\sec^2 t - 1}{\tan t} = \frac{\tan^2 t}{\tan t} = \tan t$$

Proving an identity is something like a game in that it requires a strategy. If one strategy does not work, try another, and still another, until you succeed.

A POINT OF LOGIC

Why all the fuss about working with just one side of a conjectured identity? First of all, it offers good practice in manipulating trigonometric expressions. But there is also a point of logic. If you operate on both sides simultaneously, you are in effect assuming that you already have an identity. That is bad logic and it can be corrected only by carefully checking that each step is reversible. To make this point clear, consider the equation

$$1 - x = x - 1$$

which is certainly not an identity. Yet when we square both sides we get

$$1 - 2x + x^2 = x^2 - 2x + 1$$

which is an identity. The trouble here is that squaring both sides is not a reversible operation.

The situation contrasts sharply with our procedure for solving conditional equations, in which we often perform an operation on both sides. For example, in the case of the equation

$$\sqrt{2x + 1} = 1 - x$$

we even square both sides. We are protected from error here by checking our solutions in the original equation.

Problem Set 3-1

1. Express entirely in terms of $\sin t$.
   (a) $\cos^2 t$ (b) $\tan t \cos t$
   (c) $\dfrac{3}{\csc^2 t} + 2\cos^2 t - 2$ (d) $\cot^2 t$

2. Express entirely in terms of $\cos t$.
   (a) $\sin^2 t$ (b) $\tan^2 t$
   (c) $\csc^2 t$ (d) $(1 + \sin t)^2 - 2\sin t$

3. Express entirely in terms of $\tan t$.
   (a) $\cot^2 t$ (b) $\sec^2 t$
   (c) $\sin t \sec t$ (d) $2\sec^2 t - 2\tan^2 t + 1$

4. Express entirely in terms of $\sec t$.
   (a) $\cos^4 t$ (b) $\tan^2 t$
   (c) $\tan t \csc t$ (d) $\tan^2 t - 2\sec^2 t + 5$

EXAMPLE A (Proving Identities)   Prove that the following is an identity.

$$\csc \theta - \sin \theta = \cot \theta \cos \theta$$

Solution. The left side looks inviting, as $\csc \theta = 1/\sin \theta$. We rewrite it a step at a time.

$$\csc \theta - \sin \theta = \frac{1}{\sin \theta} - \sin \theta$$

$$= \frac{1 - \sin^2 \theta}{\sin \theta}$$

$$= \frac{\cos^2 \theta}{\sin \theta}$$

$$= \frac{\cos \theta}{\sin \theta} \cdot \cos \theta$$

$$= \cot \theta \cos \theta$$

Prove that each of the following is an identity.

5. $\cos t \sec t = 1$
6. $\sin t \csc t = 1$
7. $\tan x \cot x = 1$
8. $\sin x \sec x = \tan x$
9. $\cos y \csc y = \cot y$
10. $\tan y \cos y = \sin y$
11. $\cot \theta \sin \theta = \cos \theta$
12. $\dfrac{\sec \theta}{\csc \theta} = \tan \theta$
13. $\dfrac{\tan u}{\sin u} = \dfrac{1}{\cos u}$
14. $\dfrac{\sin u}{\csc u} + \dfrac{\cos u}{\sec u} = 1$
15. $(1 + \sin z)(1 - \sin z) = \dfrac{1}{\sec^2 z}$
16. $(\sec z - 1)(\sec z + 1) = \tan^2 z$
17. $(1 - \sin^2 x)(1 + \tan^2 x) = 1$
18. $(1 - \cos^2 x)(1 + \cot^2 x) = 1$
19. $\sec t - \sin t \tan t = \cos t$
20. $\sin t(\csc t - \sin t) = \cos^2 t$
21. $\dfrac{\sec^2 t - 1}{\sec^2 t} = \sin^2 t$
22. $\dfrac{1 - \csc^2 t}{\csc^2 t} = \dfrac{-1}{\sec^2 t}$
23. $\cos t(\tan t + \cot t) = \csc t$
24. $\dfrac{1}{\sin t \cos t} - \dfrac{\cos t}{\sin t} = \tan t$

EXAMPLE B (Expressing All Trigonometric Functions in Terms of One of Them) If $\pi/2 < t < \pi$, express $\cos t$, $\tan t$, $\cot t$, $\sec t$, and $\csc t$ in terms of $\sin t$.

Solution. Since $\cos^2 t = 1 - \sin^2 t$ and cosine is negative in quadrant II,

$$\cos t = -\sqrt{1 - \sin^2 t}$$

Also

$$\tan t = \frac{\sin t}{\cos t} = -\frac{\sin t}{\sqrt{1 - \sin^2 t}}$$

$$\cot t = \frac{1}{\tan t} = -\frac{\sqrt{1 - \sin^2 t}}{\sin t}$$

$$\sec t = \frac{1}{\cos t} = -\frac{1}{\sqrt{1 - \sin^2 t}}$$

$$\csc t = \frac{1}{\sin t}$$

25. If $\pi/2 < t < \pi$, express $\sin t$, $\tan t$, $\cot t$, $\sec t$, and $\csc t$ in terms of $\cos t$.
26. If $\pi < t < 3\pi/2$, express $\sin t$, $\cos t$, $\cot t$, $\sec t$, and $\csc t$ in terms of $\tan t$.
27. If $\pi/2 < t < \pi$ and $\sin t = \frac{4}{5}$, find the values of the other five functions for the same value of t. *Hint:* Use the results of Example B.
28. If $\pi < t < 3\pi/2$ and $\tan t = 2$, find $\sin t$, $\cos t$, $\cot t$, $\sec t$, and $\csc t$.

EXAMPLE C (How to Proceed When Neither Side Is Simple)   Prove that

$$\frac{\sin t}{1 - \cos t} = \frac{1 + \cos t}{\sin t}$$

is an identity.

Solution.   Since both sides are equally complicated, it would seem to make no difference which side we choose to manipulate. We will try to transform the left side into the right side. Seeing $1 + \cos t$ in the numerator of the right side suggests multiplying the left side by $(1 + \cos t)/(1 + \cos t)$.

$$\frac{\sin t}{1 - \cos t} = \frac{\sin t}{1 - \cos t} \cdot \frac{1 + \cos t}{1 + \cos t} = \frac{\sin t(1 + \cos t)}{1 - \cos^2 t}$$

$$= \frac{\sin t(1 + \cos t)}{\sin^2 t}$$

$$= \frac{1 + \cos t}{\sin t}$$

Prove that each of the following is an identity.

29. $\dfrac{\sec t - 1}{\tan t} = \dfrac{\tan t}{\sec t + 1}$

30. $\dfrac{1 - \tan \theta}{1 + \tan \theta} = \dfrac{\cot \theta - 1}{\cot \theta + 1}$

Hint: In Problem 30, multiply numerator and denominator of the left side by $\cot \theta$.

31. $\dfrac{\tan^2 x}{\sec x + 1} = \dfrac{1 - \cos x}{\cos x}$

32. $\dfrac{\cot x}{\csc x + 1} = \dfrac{\csc x - 1}{\cot x}$

33. $\dfrac{\sin t + \cos t}{\tan^2 t - 1} = \dfrac{\cos^2 t}{\sin t - \cos t}$

34. $\dfrac{\sec t - \cos t}{1 + \cos t} = \sec t - 1$

MISCELLANEOUS PROBLEMS

35. Express $[(\sin x + \cos x)^2 - 1]\sec x \csc^3 x$ as follows.
    (a) Entirely in terms of $\sin x$.
    (b) Entirely in terms of $\tan x$.

36. If $\sec t = 8$, find the values of (a) $\cos t$; (b) $\cot^2 t$; (c) $\csc^2 t$.

In Problems 37–56, prove that each equation is an identity. Do this by taking one side and showing by a chain of equalities that it is equal to the other side.

37. $(1 + \tan^2 t)(\cos t + \sin t) = (1 + \tan t)\sec t$

38. $1 - (\cos t + \sin t)(\cos t - \sin t) = 2 \sin^2 t$

39. $2 \sec^2 y - 1 = \dfrac{1 + \sin^2 y}{\cos^2 y}$

40. $(\sin x + \cos x)(\sec x + \csc x) = 2 + \tan x + \cot x$

41. $\dfrac{\cos z}{1 + \cos z} = \dfrac{\sin z}{\sin z + \tan z}$

42. $2 \sin^2 t + 3 \cos^2 t + \sec^2 t = (\sec t + \cos t)^2$

43. $(\csc t + \cot t)^2 = \dfrac{1 + \cos t}{1 - \cos t}$

44. $\sec^4 y - \tan^4 y = \dfrac{1 + \sin^2 y}{\cos^2 y}$

45. $\dfrac{\cos x + \sin x}{\cos x - \sin x} = \dfrac{1 + \tan x}{1 - \tan x}$

46. $\dfrac{1 + \cos x}{1 - \cos x} - \dfrac{1 - \cos x}{1 + \cos x} = 4 \cot x \csc x$

47. $(\sec t + \tan t)(\csc t - 1) = \cot t$

48. $\sec t + \cos t = \sin t \tan t + 2 \cos t$

49. $\dfrac{\cos^3 t + \sin^3 t}{\cos t + \sin t} = 1 - \sin t \cos t$

50. $\dfrac{\tan x}{1 + \tan x} + \dfrac{\cot x}{1 - \cot x} = \dfrac{\tan x + \cot x}{\tan x - \cot x}$

51. $\dfrac{1 - \cos \theta}{1 + \cos \theta} = \left(\dfrac{1 - \cos \theta}{\sin \theta}\right)^2$

52. $\dfrac{(\sec^2 \theta + \tan^2 \theta)^2}{\sec^4 \theta - \tan^4 \theta} = \sec^2 \theta + \tan^2 \theta$

53. $(\csc t - \cot t)^4 (\csc t + \cot t)^4 = 1$

54. $(\sec t + \tan t)^5 (\sec t - \tan t)^6 = \dfrac{1 - \sin t}{\cos t}$

55. $\sin^6 u + \cos^6 u = 1 - 3 \sin^2 u \cos^2 u$

56. $\dfrac{\cos^2 x - \cos^2 y}{\cot^2 x - \cot^2 y} = \sin^2 x \sin^2 y$

57. In a later section, we will learn that

$$\tan 3x = \frac{3 \tan x - \tan^3 x}{1 - 3 \tan^2 x}$$

Taking this for granted, show that

$$\cot 3x = \frac{3 \cot x - \cot^3 x}{1 - 3 \cot^2 x}$$

Note the similarity in form of these two identities.

58. **TEASER** Generalize Problem 57 by showing that if $\tan kx = f(\tan x)$ and if k is an odd number, then $\cot kx = f(\cot x)$. *Hint:* Let $x = \pi/2 - y$.

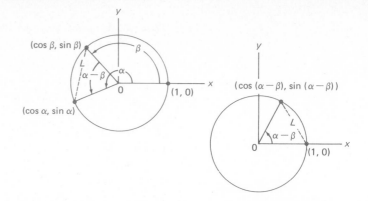

Equal Chords

In the two unit circles at the right, the two dotted chords have the same length, being chords for angles of the same size, namely, $\alpha - \beta$. Out of this simple observation, we can extract several of the most important identities of trigonometry.

3-2   Addition Laws

When you study calculus, you will meet expressions like $\cos(\alpha + \beta)$ and $\sin(\alpha - \beta)$. It will be very important to rewrite these expressions directly in terms of $\sin \alpha$, $\cos \alpha$, $\sin \beta$, and $\cos \beta$. It might be tempting to replace $\cos(\alpha + \beta)$ by $\cos \alpha + \cos \beta$ and $\sin(\alpha - \beta)$ by $\sin \alpha - \sin \beta$, but that would be terribly wrong. To see this, let's try $\alpha = \pi/3$ and $\beta = \pi/6$.

$$\cos\left(\frac{\pi}{3} + \frac{\pi}{6}\right) = \cos\frac{\pi}{2} = 0 \qquad \cos\frac{\pi}{3} + \cos\frac{\pi}{6} = \frac{1}{2} + \frac{\sqrt{3}}{2} \approx 1.4$$

$$\sin\left(\frac{\pi}{3} - \frac{\pi}{6}\right) = \sin\frac{\pi}{6} = .5 \qquad \sin\frac{\pi}{3} - \sin\frac{\pi}{6} = \frac{\sqrt{3}}{2} - \frac{1}{2} \approx .4$$

To obtain correct expressions is the goal of this section.

A KEY IDENTITY

> **Distance Formula**
>
> The distance between (x_1, y_1) and (x_2, y_2) is
> $$\sqrt{(x_2 - x_1)^2 + (y_2 - y_1)^2}$$

Figure 1

The opening display shows two chords of equal length L. Using the formula for the distance between two points (Figure 1) and the identity $\sin^2 \theta + \cos^2 \theta = 1$, we have the following expression for the square of the chord on the right.

$$L^2 = [\cos(\alpha - \beta) - 1]^2 + \sin^2(\alpha - \beta)$$
$$= \cos^2(\alpha - \beta) - 2\cos(\alpha - \beta) + 1 + \sin^2(\alpha - \beta)$$
$$= [\cos^2(\alpha - \beta) + \sin^2(\alpha - \beta)] + 1 - 2\cos(\alpha - \beta)$$
$$= 2 - 2\cos(\alpha - \beta)$$

A similar calculation for the square of the chord on the left gives

$$L^2 = (\cos \alpha - \cos \beta)^2 + (\sin \alpha - \sin \beta)^2$$

$$= \cos^2 \alpha - 2 \cos \alpha \cos \beta + \cos^2 \beta + \sin^2 \alpha - 2 \sin \alpha \sin \beta + \sin^2 \beta$$

$$= 1 - 2 \cos \alpha \cos \beta - 2 \sin \alpha \sin \beta + 1$$

$$= 2 - 2(\cos \alpha \cos \beta + \sin \alpha \sin \beta)$$

When we equate these two expressions for L^2, we get our key identity

$$\cos(\alpha - \beta) = \cos \alpha \cos \beta + \sin \alpha \sin \beta$$

Our derivation is based on a picture in which α and β are positive angles with $\alpha > \beta$. Minor modifications would establish the identity for arbitrary angles α and β and hence also for their radian measures s and t. Thus for all real numbers s and t,

$$\boxed{\cos(s - t) = \cos s \cos t + \sin s \sin t}$$

We can use this identity to calculate $\cos(\pi/12)$ by thinking of $\pi/12$ as $\pi/3 - \pi/4$.

$$\cos \frac{\pi}{12} = \cos\left(\frac{\pi}{3} - \frac{\pi}{4}\right) = \cos \frac{\pi}{3} \cos \frac{\pi}{4} + \sin \frac{\pi}{3} \sin \frac{\pi}{4}$$

$$= \frac{1}{2} \cdot \frac{\sqrt{2}}{2} + \frac{\sqrt{3}}{2} \cdot \frac{\sqrt{2}}{2} = \frac{\sqrt{2} + \sqrt{6}}{4} \approx .9659$$

In words, this identity says: *The cosine of a difference is the cosine of the first times the cosine of the second plus the sine of the first times the sine of the second.* It is important to memorize this identity in words so you can easily apply it to $\cos(3u - v)$, $\cos[s - (-t)]$, or even $\cos[(\pi/2 - s) - t]$, as we shall have to do soon.

RELATED IDENTITIES

In the boxed identity above, we replace t by $-t$ and use the fundamental identities $\cos(-t) = \cos t$ and $\sin(-t) = -\sin t$ to get

$$\cos[s - (-t)] = \cos s \cos(-t) + \sin s \sin(-t)$$

$$= \cos s \cos t + (\sin s)(-\sin t)$$

This gives us the **addition law for cosines.**

$$\boxed{\cos(s + t) = \cos s \cos t - \sin s \sin t}$$

We illustrate this law by calculating $\cos(13\pi/12)$.

$$\cos \frac{13\pi}{12} = \cos\left(\frac{3\pi}{4} + \frac{\pi}{3}\right) = \cos \frac{3\pi}{4} \cos \frac{\pi}{3} - \sin \frac{3\pi}{4} \sin \frac{\pi}{3}$$

$$= \frac{-\sqrt{2}}{2} \cdot \frac{1}{2} - \frac{\sqrt{2}}{2} \cdot \frac{\sqrt{3}}{2} = -\frac{\sqrt{2} + \sqrt{6}}{4} \approx -.9659$$

There is also an identity involving $\sin(s + t)$. To derive this identity, we use the cofunction identity $\sin u = \cos(\pi/2 - u)$ to write

$$\sin(s + t) = \cos\left[\frac{\pi}{2} - (s + t)\right] = \cos\left[\left(\frac{\pi}{2} - s\right) - t\right]$$

Then we use our key identity for the cosine of a difference to obtain

$$\cos\left(\frac{\pi}{2} - s\right)\cos t + \sin\left(\frac{\pi}{2} - s\right)\sin t$$

Two applications of cofunction identities give us the result we want, the **addition law for sines.**

$$\boxed{\sin(s + t) = \sin s \cos t + \cos s \sin t}$$

Finally, replacing t by $-t$ in this last result leads to

$$\boxed{\sin(s - t) = \sin s \cos t - \cos s \sin t}$$

If we let $s = \pi/2$ in the latter, we get another important identity—but it is one we already know.

$$\sin\left(\frac{\pi}{2} - t\right) = \sin\frac{\pi}{2}\cos t - \cos\frac{\pi}{2}\sin t = 1 \cdot \cos t - 0 \cdot \sin t = \cos t$$

CAUTION

$$\sin\left(\theta + \frac{\pi}{2}\right) = \sin\theta + \sin\frac{\pi}{2}$$
$$= \sin\theta + 1$$

$$\sin\left(\theta + \frac{\pi}{2}\right)$$
$$= \sin\theta\cos\frac{\pi}{2} + \cos\theta\sin\frac{\pi}{2}$$
$$= \cos\theta$$

Problem Set 3-2

Find the value of each expression. Note that in each case, the answers to parts (a) and (b) are different.

1. (a) $\sin\dfrac{\pi}{4} + \sin\dfrac{\pi}{6}$ (b) $\sin\left(\dfrac{\pi}{4} + \dfrac{\pi}{6}\right)$

2. (a) $\cos\dfrac{\pi}{4} + \cos\dfrac{\pi}{6}$ (b) $\cos\left(\dfrac{\pi}{4} + \dfrac{\pi}{6}\right)$

3. (a) $\cos\dfrac{\pi}{4} - \cos\dfrac{\pi}{6}$ (b) $\cos\left(\dfrac{\pi}{4} - \dfrac{\pi}{6}\right)$

4. (a) $\sin\dfrac{\pi}{4} - \sin\dfrac{\pi}{6}$ (b) $\sin\left(\dfrac{\pi}{4} - \dfrac{\pi}{6}\right)$

Use the identities derived in this section to show that the equalities in Problems 5–12 are identities.

5. $\sin(t + \pi) = -\sin t$

6. $\cos(t + \pi) = -\cos t$

7. $\sin\left(t + \dfrac{3\pi}{2}\right) = -\cos t$

8. $\cos\left(t + \dfrac{3\pi}{2}\right) = \sin t$

9. $\sin\left(t - \dfrac{\pi}{2}\right) = -\cos t$

10. $\cos\left(t - \dfrac{\pi}{2}\right) = \sin t$

11. $\cos\left(t + \dfrac{\pi}{3}\right) = \dfrac{1}{2}\cos t - \dfrac{\sqrt{3}}{2}\sin t$

12. $\sin\left(t + \dfrac{\pi}{3}\right) = \dfrac{1}{2}\sin t + \dfrac{\sqrt{3}}{2}\cos t$

EXAMPLE A (Recognizing Expressions as Single Sines or Cosines) Write as a single sine or cosine.

(a) $\sin\frac{7}{6}\cos\frac{1}{6} + \cos\frac{7}{6}\sin\frac{1}{6}$

(b) $\cos(x + h)\cos h + \sin(x + h)\sin h$

Solution.

(a) By the addition law for sines,

$$\sin\tfrac{7}{6}\cos\tfrac{1}{6} + \cos\tfrac{7}{6}\sin\tfrac{1}{6} = \sin(\tfrac{7}{6} + \tfrac{1}{6}) = \sin\tfrac{4}{3}$$

(b) This we recognize as the cosine of a difference

$$\cos(x + h)\cos h + \sin(x + h)\sin h = \cos[(x + h) - h] = \cos x$$

Write each of the following as a single sine or cosine.

13. $\cos\frac{1}{2}\cos\frac{3}{2} - \sin\frac{1}{2}\sin\frac{3}{2}$

14. $\cos 2 \cos 3 + \sin 2 \sin 3$

15. $\sin\dfrac{7\pi}{8}\cos\dfrac{\pi}{8} + \cos\dfrac{7\pi}{8}\sin\dfrac{\pi}{8}$

16. $\sin\dfrac{5\pi}{16}\cos\dfrac{\pi}{16} - \cos\dfrac{5\pi}{16}\sin\dfrac{\pi}{16}$

17. $\cos 33° \cos 27° - \sin 33° \sin 27°$

18. $\sin 49° \cos 41° + \cos 49° \sin 41°$

19. $\sin(\alpha + \beta)\cos\beta - \cos(\alpha + \beta)\sin\beta$

20. $\cos(\alpha + \beta)\cos(\alpha - \beta) - \sin(\alpha + \beta)\sin(\alpha - \beta)$

EXAMPLE B (Using the Addition Laws) Suppose that α is a first quadrant angle with $\cos\alpha = \frac{4}{5}$ and β is a second quadrant angle with $\sin\beta = \frac{12}{13}$. Evaluate $\sin(\alpha + \beta)$ and $\cos(\alpha + \beta)$ and then determine the quadrant for $\alpha + \beta$.

Solution. We are going to need $\sin\alpha$ and $\cos\beta$. We can find them by using the identity $\sin^2\theta + \cos^2\theta = 1$, but we have to be careful about signs.

$$\sin\alpha = \sqrt{1 - \cos^2\alpha} = \sqrt{1 - \tfrac{16}{25}} = \tfrac{3}{5}$$

$$\cos\beta = -\sqrt{1 - \sin^2\beta} = -\sqrt{1 - \tfrac{144}{169}} = -\tfrac{5}{13}$$

We chose the plus sign in the first case because α is a first quadrant angle and the minus sign in the second because β is a second quadrant angle, where the cosine is negative. Then

$$\sin(\alpha + \beta) = \sin \alpha \cos \beta + \cos \alpha \sin \beta$$

$$= \left(\frac{3}{5}\right)\left(\frac{-5}{13}\right) + \left(\frac{4}{5}\right)\left(\frac{12}{13}\right) = \frac{33}{65}$$

$$\cos(\alpha + \beta) = \cos \alpha \cos \beta - \sin \alpha \sin \beta$$

$$= \left(\frac{4}{5}\right)\left(\frac{-5}{13}\right) - \left(\frac{3}{5}\right)\left(\frac{12}{13}\right) = \frac{-56}{65}$$

Since $\sin(\alpha + \beta)$ is positive and $\cos(\alpha + \beta)$ is negative, $\alpha + \beta$ is a second quadrant angle.

21. If α and β are third quadrant angles with $\sin \alpha = -\frac{4}{5}$ and $\cos \beta = -\frac{5}{13}$, find $\sin(\alpha + \beta)$ and $\cos(\alpha + \beta)$. In what quadrant does the terminal side of $\alpha + \beta$ lie?

22. Let α and β be second quadrant angles with $\sin \alpha = \frac{2}{3}$ and $\sin \beta = \frac{3}{4}$. Find $\sin(\alpha + \beta)$ and $\cos(\alpha + \beta)$ and determine the quadrant for $\alpha + \beta$.

23. Let α be a first quadrant angle with $\sin \alpha = 1/\sqrt{10}$ and β be a second quadrant angle with $\cos \beta = -\frac{1}{2}$. Find $\sin(\alpha - \beta)$ and $\cos(\alpha - \beta)$ and determine the quadrant for $\alpha - \beta$.

24. Let α and β be second and third quadrant angles, respectively, with $\cos \alpha = \cos \beta = -\frac{3}{7}$. Find $\sin(\alpha - \beta)$ and $\cos(\alpha - \beta)$ and determine the quadrant for $\alpha - \beta$.

EXAMPLE C (Tangent Identities)   Verify the **addition law for tangents**.

$$\tan(s + t) = \frac{\tan s + \tan t}{1 - \tan s \tan t}$$

Solution.

$$\tan(s + t) = \frac{\sin(s + t)}{\cos(s + t)}$$

$$= \frac{\sin s \cos t + \cos s \sin t}{\cos s \cos t - \sin s \sin t}$$

$$= \frac{\dfrac{\sin s \cos t}{\cos s \cos t} + \dfrac{\cos s \sin t}{\cos s \cos t}}{\dfrac{\cos s \cos t}{\cos s \cos t} - \dfrac{\sin s \sin t}{\cos s \cos t}}$$

$$= \frac{\tan s + \tan t}{1 - \tan s \tan t}$$

The key step was the third one, in which we divided both the numerator and the denominator by $\cos s \cos t$.

Establish that each equation in Problems 25–28 is an identity.

25. $\tan(s - t) = \dfrac{\tan s - \tan t}{1 + \tan s \tan t}$

26. $\tan(s + \pi) = \tan s$

27. $\tan\left(t + \dfrac{\pi}{4}\right) = \dfrac{1 + \tan t}{1 - \tan t}$

28. $\tan\left(t - \dfrac{\pi}{3}\right) = \dfrac{\tan t - \sqrt{3}}{1 + \sqrt{3} \tan t}$

MISCELLANEOUS PROBLEMS

29. Express in terms of $\sin t$ and $\cos t$.
   (a) $\sin(t - \tfrac{5}{6}\pi)$ (b) $\cos(\tfrac{\pi}{6} - t)$

30. Express $\tan(\theta + \tfrac{3}{4}\pi)$ in terms of $\tan \theta$.

31. Let α and β be first and third quadrant angles, respectively, with $\sin \alpha = \tfrac{2}{3}$ and $\cos \beta = -\tfrac{1}{3}$. Evaluate each of the following exactly.
   (a) $\cos \alpha$ (b) $\sin \beta$ (c) $\cos(\alpha + \beta)$
   (d) $\sin(\alpha - \beta)$ (e) $\tan(\alpha + \beta)$ (f) $\sin(2\beta)$

32. If $0 \le t \le \pi/2$ and $\cos(t + \pi/6) = .8$, find the exact value of $\sin t$ and $\cos t$.
   Hint: $t = (t + \pi/6) - \pi/6$.

33. Evaluate each of the following (the easy way).
   (a) $\sin(t + \pi/3)\cos t - \cos(t + \pi/3)\sin t$
   (b) $\cos 175° \cos 25° + \sin 175° \sin 25°$
   (c) $\sin t \cos(1 - t) + \cos t \sin(1 - t)$

34. Find the exact value of $\cos 85° \cos 40° + \cos 5° \cos 50°$.

35. Show that each of the following is an identity.
   (a) $\sin(x + y)\sin(x - y) = \sin^2 x - \sin^2 y$
   (b) $\dfrac{\sin(x + y)}{\cos(x - y)} = \dfrac{\tan x + \tan y}{1 + \tan x \tan y}$
   (c) $\dfrac{\cos 5t}{\sin t} - \dfrac{\sin 5t}{\cos t} = \dfrac{\cos 6t}{\sin t \cos t}$

36. Show that the following are identities.
   (a) $\cot(u + v) = \dfrac{\cot u \cot v - 1}{\cot u + \cot v}$
   (b) $\dfrac{\sin(u + v)}{\sin(u - v)} = \dfrac{\tan u + \tan v}{\tan u - \tan v}$
   (c) $\dfrac{\cos 2t}{\sin t} + \dfrac{\sin 2t}{\cos t} = \csc t$

37. Let θ be the smallest counterclockwise angle from the line $y = m_1 x + b_1$ to the line $y = m_2 x + b_2$, where $m_1 m_2 \ne -1$. Show that

$$\tan \theta = \dfrac{m_2 - m_1}{1 + m_1 m_2}$$

38. Find the counterclockwise angle θ from the line $3x - 4y = 1$ to the line $2x + 6y = 3$. (See Problem 37.)

39. Use the addition and subtraction laws (the four boxed formulas of this section) to prove the following **product identities**.
   (a) $\cos s \cos t = \tfrac{1}{2}[\cos(s + t) + \cos(s - t)]$
   (b) $\sin s \sin t = -\tfrac{1}{2}[\cos(s + t) - \cos(s - t)]$
   (c) $\sin s \cos t = \tfrac{1}{2}[\sin(s + t) + \sin(s - t)]$
   (d) $\cos s \sin t = \tfrac{1}{2}[\sin(s + t) - \sin(s - t)]$

40. Use the identities of Problem 39 to prove the following **factoring identities**.
   Hint: Let $u = s + t$ and $v = s - t$.

(a) $\cos u + \cos v = 2 \cos \dfrac{u+v}{2} \cos \dfrac{u-v}{2}$

(b) $\cos u - \cos v = -2 \sin \dfrac{u+v}{2} \sin \dfrac{u-v}{2}$

(c) $\sin u + \sin v = 2 \sin \dfrac{u+v}{2} \cos \dfrac{u-v}{2}$

(d) $\sin u - \sin v = 2 \cos \dfrac{u+v}{2} \sin \dfrac{u-v}{2}$

41. Evaluate each of the following exactly.
    (a) $\cos 105° \cos 45°$ (b) $\sin 15° - \sin 75°$
    (c) $\cos 15° + \cos 30° + \cos 45° + \cos 60° + \cos 75°$

42. Show that each of the following is an identity.

    (a) $\dfrac{\cos 9t + \cos 3t}{\sin 9t - \sin 3t} = \cot 3t$ (b) $\dfrac{\sin 3u + \sin 7u}{\cos 3u + \cos 7u} = \tan 5u$

    (c) $\cos 10\beta + \cos 2\beta + 2 \cos 8\beta \cos 6\beta = 4 \cos^2 6\beta \cos 2\beta$

43. Stack three identical squares and consider angles α, β, and γ as shown in Figure 2. Prove that $\alpha + \beta = \gamma$.

44. **TEASER** Consider an oblique triangle (no right angles) with angles α, β, and γ. Prove that

$$\tan \alpha + \tan \beta + \tan \gamma = \tan \alpha \tan \beta \tan \gamma$$

Figure 2

I could do this problem if $\sin (2t) = 2 \sin t$

Wishful Thinking

"Wishful thinking is imagining good things you don't have . . . . [It] may be bad as too much salt is bad in the soup and even a little garlic is bad in the chocolate pudding. I mean, wishful thinking may be bad if there is too much of it or in the wrong place, but it is good in itself and may be a great help in life and in problem solving."

George Polya
in *Mathematical Discovery*

3-3 Double-Angle and Half-Angle Formulas

George Polya would agree that the student in our opening panel is wishing for too much. And there is a better way than wishing to get formulas for $\sin 2t$ and $\cos 2t$. All we have to do is to think of $2t$ as $t + t$ and apply the addition laws of the previous section.

$$\sin(t + t) = \sin t \cos t + \cos t \sin t = 2 \sin t \cos t$$
$$\cos(t + t) = \cos t \cos t - \sin t \sin t = \cos^2 t - \sin^2 t$$

DOUBLE-ANGLE FORMULAS

We have just derived two very important results. They are called *double-angle formulas*, though double-number formulas would perhaps be more appropriate.

$$\sin 2t = 2 \sin t \cos t$$
$$\cos 2t = \cos^2 t - \sin^2 t$$

Suppose $\sin t = \frac{2}{3}$ and $\pi/2 < t < \pi$. Then we can calculate both $\sin 2t$ and $\cos 2t$, but we must first find $\cos t$. Since $\pi/2 < t < \pi$, the cosine is negative, and therefore

$$\cos t = -\sqrt{1 - \sin^2 t} = -\sqrt{1 - \frac{4}{25}} = -\frac{\sqrt{21}}{5}$$

The double-angle formulas now give

$$\sin 2t = 2 \sin t \cos t = 2\left(\frac{2}{5}\right)\left(\frac{-\sqrt{21}}{5}\right) = \frac{-4\sqrt{21}}{25} \approx -.73$$

$$\cos 2t = \cos^2 t - \sin^2 t = \left(\frac{-\sqrt{21}}{5}\right)^2 - \left(\frac{2}{5}\right)^2 = \frac{17}{25} \approx .68$$

CAUTION

$$\cos 6\theta = 6 \cos \theta$$

$$\cos 6\theta = 2 \cos^2 3\theta - 1$$

There are two other forms of the cosine double-angle formula that are often useful. If, in the expression $\cos^2 t - \sin^2 t$, we replace $\cos^2 t$ by $1 - \sin^2 t$, we obtain

$$\cos 2t = 1 - 2 \sin^2 t$$

and, alternatively, if we replace $\sin^2 t$ by $1 - \cos^2 t$, we have

$$\cos 2t = 2 \cos^2 t - 1$$

Of course, in all that we have done, we may replace the number t by the angle θ; hence the name double-angle formulas.

Once we grasp the generality of the four boxed formulas, we can write numerous others that follow from them. For example,

$$\sin 6\theta = 2 \sin 3\theta \cos 3\theta$$

$$\cos 4u = \cos^2 2u - \sin^2 2u$$

$$\cos t = 1 - 2 \sin^2\left(\frac{t}{2}\right)$$

$$\cos t = 2 \cos^2\left(\frac{t}{2}\right) - 1$$

The last two of these identities lead us directly to the half-angle formulas.

HALF-ANGLE FORMULAS

In the identity $\cos t = 1 - 2 \sin^2(t/2)$, we solve for $\sin(t/2)$.

$$2 \sin^2\left(\frac{t}{2}\right) = 1 - \cos t$$

$$\sin^2\left(\frac{t}{2}\right) = \frac{1 - \cos t}{2}$$

$$\sin\left(\frac{t}{2}\right) = \pm\sqrt{\frac{1 - \cos t}{2}}$$

Similarly, if we solve $\cos t = 2 \cos^2(t/2) - 1$ for $\cos(t/2)$, the result is

$$\cos\left(\frac{t}{2}\right) = \pm\sqrt{\frac{1 + \cos t}{2}}$$

In both of these formulas, the choice of the plus or minus sign is determined by the interval on which $t/2$ lies. For example,

$$\cos\left(\frac{5\pi}{8}\right) = \cos\left(\frac{5\pi/4}{2}\right) = -\sqrt{\frac{1 + \cos(5\pi/4)}{2}}$$

$$= -\sqrt{\frac{1 - \sqrt{2}/2}{2}} = -\frac{\sqrt{2 - \sqrt{2}}}{2}$$

We chose the minus sign because $5\pi/8$ corresponds to an angle in quadrant II, where the cosine is negative.

As a second example, suppose that $\cos \theta = .4$, where θ is a fourth quadrant angle. Then we can calculate $\sin(\theta/2)$ and $\cos(\theta/2)$, observing first that $\theta/2$ is necessarily a second quadrant angle.

$$\sin\left(\frac{\theta}{2}\right) = \sqrt{\frac{1 - \cos\theta}{2}} = \sqrt{\frac{1 - .4}{2}} \approx .548$$

$$\cos\left(\frac{\theta}{2}\right) = -\sqrt{\frac{1 + \cos\theta}{2}} = -\sqrt{\frac{1 + .4}{2}} \approx -.837$$

Problem Set 3-3

1. If $\cos t = \frac{4}{5}$ with $0 < t < \pi/2$, show that $\sin t = \frac{3}{5}$. Then use formulas from this section to calculate
   (a) $\sin 2t$;   (b) $\cos 2t$;
   (c) $\cos(t/2)$;   (d) $\sin(t/2)$.
2. If $\sin t = -\frac{2}{3}$ with $3\pi/2 < t < 2\pi$, show that $\cos t = \sqrt{5}/3$. Then calculate
   (a) $\sin 2t$;   (b) $\cos 2t$;
   (c) $\cos(t/2)$;   (d) $\sin(t/2)$.

Use formulas from this section to simplify the expressions in Problems 3–16. For example, $2\sin(.5)\cos(.5) = \sin 1$.

3. $2\sin 5t \cos 5t$ 4. $2\sin 3\theta \cos 3\theta$
5. $\cos^2(3t/2) - \sin^2(3t/2)$ 6. $\cos^2(7\pi/8) - \sin^2(7\pi/8)$
7. $2\cos^2(y/4) - 1$ 8. $2\cos^2(\alpha/3) - 1$
9. $1 - 2\sin^2(.6t)$ 10. $2\sin^2(\pi/8) - 1$
11. $\sin^2(\pi/8) - \cos^2(\pi/8)$ 12. $2\sin(.3)\cos(.3)$
13. $\dfrac{1 + \cos x}{2}$ 14. $\dfrac{1 - \cos y}{2}$
15. $\dfrac{1 - \cos 4\theta}{2}$ 16. $\dfrac{1 + \cos 8u}{2}$

17. Use the half-angle formulas to calculate
    (a) $\sin(\pi/8)$;   (b) $\cos(112.5°)$.
18. Calculate, using half-angle formulas,
    (a) $\cos 67.5°$;   (b) $\sin(\pi/12)$.
19. Use the addition law for tangents (Example C of Section 3-2) to show that
$$\tan 2t = \frac{2\tan t}{1 - \tan^2 t}$$
20. Use the identity of Problem 19 to evaluate $\tan 2t$ given that
    (a) $\tan t = 3$;
    (b) $\cos t = \frac{4}{5}$ and $0 < t < \pi/2$.
21. Use the half-angle formulas for sine and cosine to show that
$$\tan\left(\frac{t}{2}\right) = \pm\sqrt{\frac{1 - \cos t}{1 + \cos t}}$$
22. Use the identity of Problem 21 to evaluate
    (a) $\tan(\pi/8)$;   (b) $\tan 112.5°$.

EXAMPLE (Using Double-Angle and Half-Angle Formulas to Prove New Identities) Prove that the following are identities.

(a) $\sin 3t = 3 \sin t - 4 \sin^3 t$

(b) $\tan \dfrac{t}{2} = \dfrac{\sin t}{1 + \cos t}$

Solution.

(a) We think of $3t$ as $2t + t$ and use the addition law for sines and then double-angle formulas.

$$\begin{aligned}
\sin 3t &= \sin(2t + t) \\
&= \sin 2t \cos t + \cos 2t \sin t \\
&= (2 \sin t \cos t) \cos t + (1 - 2 \sin^2 t) \sin t \\
&= 2 \sin t (1 - \sin^2 t) + \sin t - 2 \sin^3 t \\
&= 2 \sin t - 2 \sin^3 t + \sin t - 2 \sin^3 t \\
&= 3 \sin t - 4 \sin^3 t
\end{aligned}$$

(b) This is the unambiguous form for $\tan(t/2)$ (see Problem 21). To prove it, think of t as $2(t/2)$ and apply double-angle formulas to the right side.

$$\begin{aligned}
\frac{\sin t}{1 + \cos t} &= \frac{\sin(2(t/2))}{1 + \cos(2(t/2))} \\
&= \frac{2 \sin(t/2) \cos(t/2)}{1 + 2 \cos^2(t/2) - 1} \\
&= \frac{\sin(t/2)}{\cos(t/2)} \\
&= \tan \frac{t}{2}
\end{aligned}$$

Now prove that each of the following is an identity.

23. $\cos 3t = 4 \cos^3 t - 3 \cos t$

24. $(\sin t + \cos t)^2 = 1 + \sin 2t$

25. $\csc 2t + \cot 2t = \cot t$

26. $\sin^2 t \cos^2 t = \frac{1}{8}(1 - \cos 4t)$

27. $\dfrac{\sin \theta}{1 - \cos \theta} = \cot \dfrac{\theta}{2}$

28. $1 - 2 \sin^2 \theta = 2 \cot 2\theta \sin \theta \cos \theta$

29. $\dfrac{2 \tan \alpha}{1 + \tan^2 \alpha} = \sin 2\alpha$

30. $\dfrac{1 - \tan^2 \alpha}{1 + \tan^2 \alpha} = \cos 2\alpha$

31. $\sin 4\theta = 4 \sin \theta (2 \cos^3 \theta - \cos \theta)$

 Hint: $4\theta = 2(2\theta)$.

32. $\cos 4\theta = 8 \cos^4 \theta - 8 \cos^2 \theta + 1$

MISCELLANEOUS PROBLEMS

33. Write a simple expression for each of the following.
    (a) $2 \sin(x/2) \cos(x/2)$
    (b) $\cos^2 3t - \sin^2 3t$
    (c) $2 \sin^2(y/4) - 1$
    (d) $(\cos 4t - 1)/2$
    (e) $(1 - \cos 4t)/(1 + \cos 4t)$
    (f) $(\sin 6y)/(1 + \cos 6y)$

34. Find the exact value of each of the following.
    (a) $\sin 15° \cos 15°$
    (b) $\cos^2 105° - \sin^2 105°$
    (c) $\sin 15°$
    (d) $\cos 105°$

35. If $\pi < t < 3\pi/2$ and $\cos t = -5/13$, find each value.
    (a) $\sin 2t$
    (b) $\cos(t/2)$
    (c) $\tan(t/2)$

36. If the trigonometric point $P(t)$ on the unit circle has coordinates $(-\frac{3}{5}, \frac{4}{5})$, find the coordinates for each of the following points.
    (a) $P(2t)$
    (b) $P(t/2)$
    (c) $P(4t)$

In Problems 37–52, prove that each equation is an identity.

37. $\cos^4 z - \sin^4 z = \cos 2z$

38. $(1 - \cos 4x)/\tan^2 2x = 2 \cos^2 2x$

39. $1 + (1 - \cos 8t)/(1 + \cos 8t) = \sec^2 4t$

40. $\sec 2t = (\sec^2 t)/(2 - \sec^2 t)$

41. $\tan(\theta/2) - \sin \theta = (-\sin \theta)/(1 + \sec \theta)$

42. $(2 - \sec^2 2\theta) \tan 4\theta = 2 \tan 2\theta$

43. $3 \cos 2t + 4 \sin 2t = (3 \cos t - \sin t)(\cos t + 3 \sin t)$

44. $\csc 2x - \cot 2x = \tan x$

45. $2(\cos 3x \cos x + \sin 3x \sin x)^2 = 1 + \cos 4x$

46. $\dfrac{1 + \sin 2x + \cos 2x}{1 + \sin 2x - \cos 2x} = \cot x$

47. $\tan 3t = \dfrac{3 \tan t - \tan^3 t}{1 - 3 \tan^2 t}$

48. $\cos^4 u = \frac{3}{8} + \frac{1}{2} \cos 2u + \frac{1}{8} \cos 4u$

49. $\sin^4 u + \cos^4 u = \frac{3}{4} + \frac{1}{4} \cos 4u$

50. $\cos^6 u - \sin^6 u = \cos 2u - \frac{1}{4} \sin^2 2u \cos 2u$

51. Prove that $\cos^2 x + \cos^2 2x + \cos^2 3x = 1 + 2 \cos x \cos 2x \cos 3x$ is an identity. *Hint:* Use half-angle formulas and factoring identities.

52. Calculate $(\sin 2t)[3 + (16 \sin^2 t - 16)\sin^2 t] - \sin 6t$ for $t = 1, 2,$ and 3. Guess at an identity and then prove it.

53. If $\alpha, \beta,$ and γ are the three angles of a triangle, prove that

$$\sin 2\alpha + \sin 2\beta + \sin 2\gamma = 4 \sin \alpha \sin \beta \sin \gamma$$

54. Show that $\cos x \cos 2x \cos 4x \cos 8x \cos 16x = (\sin 32x)/(32 \sin x)$.

55. Figure 3 shows two abutting circles of radius 1, one centered at the origin, the other at $(2, 0)$. Find the exact coordinates of P, the point where the line through $(-1, 0)$ and tangent to the second circle meets the first circle.

56. **TEASER** Determine the exact value of

$$\sin 1° \sin 3° \sin 5° \sin 7° \cdots \sin 175° \sin 177° \sin 179°$$

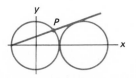

Figure 3

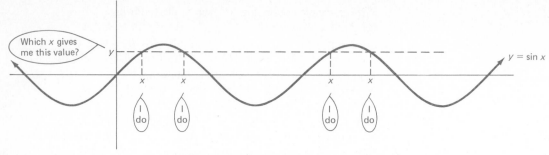

The sine function is not one-to-one.

3-4 Inverse Trigonometric Functions

The diagram above shows why the ordinary sine function does not have an inverse. We learned in Section 1-5 (a section worth reviewing now) that only one-to-one functions have inverses. To be one-to-one means that for each y there is at most one x that corresponds to it. The sine function is about as far from being one-to-one as possible. For each y between -1 and 1, there are infinitely many x's giving that y-value. To make the sine function have an inverse, we will have to restrict its domain drastically.

THE INVERSE SINE

Consider the graph of the sine function again (Figure 4). We want to restrict its domain in such a way that the sine assumes its full range of values but takes on each value only once. There are many possible choices, but the one commonly used is $-\pi/2 \le x \le \pi/2$. Notice the corresponding part of the sine graph below. From now on, whenever we need an inverse sine function, we always assume the domain of the sine has been restricted to $-\pi/2 \le x \le \pi/2$.

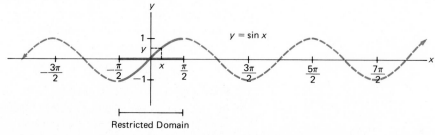

Figure 4

Having done this, we see that each y corresponds to exactly one x. We write $x = \sin^{-1} y$ (x is the inverse sine of y). Thus

$$\sin^{-1}\!\left(\frac{1}{2}\right) = \frac{\pi}{6}$$

$$\sin^{-1}(1) = \frac{\pi}{2}$$

$$\sin^{-1}(-1) = -\frac{\pi}{2}$$

$$\sin^{-1}\!\left(\frac{-\sqrt{2}}{2}\right) = -\frac{\pi}{4}$$

Please note that $\sin^{-1} y$ does not mean $1/(\sin y)$; you should not think of -1 as an exponent when used as a superscript on a function.

An alternate notation for $x = \sin^{-1} y$ is $x = \arcsin y$ (x is the arcsine of y). This is appropriate notation, since $\pi/6 = \arcsin \frac{1}{2}$ could be interpreted as saying that $\pi/6$ is the arc (on the unit circle) whose sine is $\frac{1}{2}$.

Recall from Section 1-5 that if f is a one-to-one function, then

$$x = f^{-1}(y) \quad \text{if and only if} \quad y = f(x)$$

Here the corresponding statement is

$$x = \sin^{-1} y \quad \text{if and only if} \quad y = \sin x \quad \text{and} \quad -\frac{\pi}{2} \le x \le \frac{\pi}{2}$$

Moreover

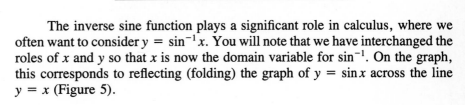

$$\sin(\sin^{-1} y) = y \quad \text{for} \quad -1 \le y \le 1$$

$$\sin^{-1}(\sin x) = x \quad \text{for} \quad -\frac{\pi}{2} \le x \le \frac{\pi}{2}$$

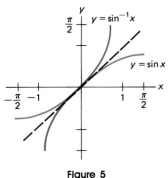

Figure 5

The inverse sine function plays a significant role in calculus, where we often want to consider $y = \sin^{-1} x$. You will note that we have interchanged the roles of x and y so that x is now the domain variable for $\sin^{-1}$. On the graph, this corresponds to reflecting (folding) the graph of $y = \sin x$ across the line $y = x$ (Figure 5).

THE INVERSE COSINE

One look at the graph of $y = \cos x$ should convince you that we cannot restrict the domain of the cosine to the same interval as that for the sine (Figure 6). We choose rather to use the interval $0 \le x \le \pi$, in which the cosine is one-to-one.

Having made the needed restriction, we may reasonably talk about $\cos^{-1}$. Moreover,

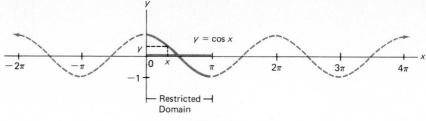

Figure 6

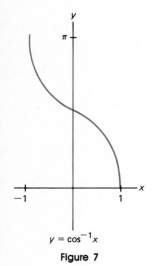

$$x = \cos^{-1} y \quad \text{if and only if} \quad y = \cos x \quad \text{and} \quad 0 \le x \le \pi$$

In particular,

$$\cos^{-1} 1 = 0$$

$$\cos^{-1} \frac{\sqrt{3}}{2} = \frac{\pi}{6}$$

$$\cos^{-1} 0 = \frac{\pi}{2}$$

$$\cos^{-1}(-1) = \pi$$

The graph of $y = \cos^{-1} x$ is shown in Figure 7. It is the graph of $y = \cos x$ reflected across the line $y = x$.

$y = \cos^{-1} x$

Figure 7

THE INVERSE TANGENT

To make $y = \tan x$ have an inverse, we restrict x to $-\pi/2 < x < \pi/2$. Thus

$$x = \tan^{-1} y \quad \text{if and only if} \quad y = \tan x \quad \text{and} \quad -\frac{\pi}{2} < x < \frac{\pi}{2}$$

The graphs of the tangent function and its inverse are shown in Figure 8.
    Notice that the graph of $y = \tan^{-1} x$ has horizontal asymptotes at $y = \pi/2$ and $y = -\pi/2$.

THE INVERSE SECANT

The secant function has an inverse, provided we restrict its domain to $0 \le x \le \pi$, excluding $\pi/2$. Thus

$$x = \sec^{-1} y \quad \text{if and only if} \quad y = \sec x \quad \text{and} \quad 0 \le x \le \pi, x \ne \frac{\pi}{2}$$

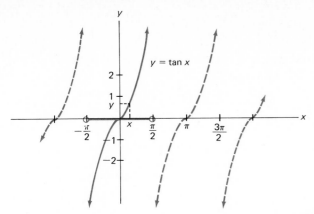

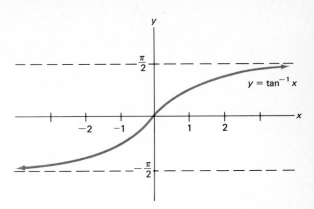

Figure 8

(Some authors choose to restrict the domain of the secant to $\{x : 0 \leq x < \pi/2$ or $\pi \leq x < 3\pi/2\}$. For this reason, check an author's definition before using any stated fact about the inverse secant.) The graphs of $y = \sec x$ and $y = \sec^{-1} x$ are shown in Figure 9.

Since the secant and cosine are reciprocals of each other, it is not surprising that $\sec^{-1} x$ is related to $\cos^{-1} x$. In fact, for every x in the domain of $\sec^{-1} x$, we have

$$\sec^{-1} x = \cos^{-1}\left(\frac{1}{x}\right)$$

This follows from the fact that each side is limited in value to the interval 0 to π and has the same cosine.

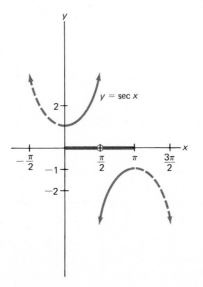

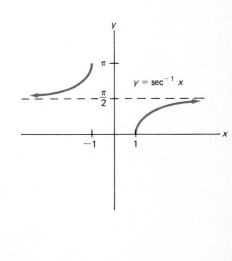

Figure 9

$$\cos(\sec^{-1} x) = \frac{1}{\sec(\sec^{-1} x)} = \frac{1}{x}$$

The two other inverse trigonometric functions, $\cot^{-1} x$ and $\csc^{-1} x$, are of less importance. They are introduced in Problem 66.

INVERSE TRIGONOMETRIC FUNCTIONS AND CALCULATORS

Most scientific calculators have been programmed to give values of $\sin^{-1}$, $\cos^{-1}$, and $\tan^{-1}$ that are consistent with the definitions we have given. For example, to obtain $\sin^{-1}(.32)$ on many calculators, press the buttons .32 $\boxed{\text{INV}}$ $\boxed{\sin}$; on other calculators, there is a button marked $\sin^{-1}$ and you simply press .32 $\boxed{\sin^{-1}}$. Normally, you cannot get $\sec^{-1}$ directly; instead you must use the identity $\sec^{-1} x = \cos^{-1}(1/x)$. Of course, in every case you must put your calculator in the appropriate mode, depending on whether you want the answer in degrees or radians.

Here is a problem that requires thinking in addition to use of a calculator. Find all values of t between 0 and 2π for which $\tan t = 2.12345$. A calculator (in radian mode) immediately gives the solution

$$t_0 = \tan^{-1}(2.12345) = 1.13067$$

However, there is another solution having t_0 as reference number, namely,

$$t = \pi + t_0 = 4.27227$$

THREE IDENTITIES

Here are three identities connecting sines, cosines, and their inverses.

(i) $\qquad\qquad\qquad \cos(\sin^{-1} x) = \sqrt{1 - x^2}$

(ii) $\qquad\qquad\qquad \sin(\cos^{-1} x) = \sqrt{1 - x^2}$

(iii) $\qquad\qquad \sin^{-1} x + \cos^{-1} x = \dfrac{\pi}{2}$

To prove the first identity, we let $\theta = \sin^{-1} x$. Remember that this means that $x = \sin \theta$, with $-\pi/2 \le \theta \le \pi/2$. Then

$$\cos(\sin^{-1} x) = \cos \theta = \pm\sqrt{1 - \sin^2 \theta} = \pm\sqrt{1 - x^2}$$

Finally, choose the plus sign because $\cos \theta$ is positive for $-\pi/2 \le \theta \le \pi/2$. The second identity is proved in a similar fashion.

To prove the third identity, let

$$\alpha = \sin^{-1} x \qquad \beta = \cos^{-1} x$$

and note that we must show that $\alpha + \beta = \pi/2$. Now

$$\sin(\alpha + \beta) = \sin \alpha \cos \beta + \cos \alpha \sin \beta$$

$$= \sin(\sin^{-1} x) \cos(\cos^{-1} x) + \cos(\sin^{-1} x) \sin(\cos^{-1} x)$$

$$= x \cdot x + \sqrt{1 - x^2} \cdot \sqrt{1 - x^2}$$
$$= x^2 + 1 - x^2$$
$$= 1$$

From this, we conclude that $\alpha + \beta$ is either $\pi/2$ or some number that differs from $\pi/2$ by a multiple of 2π. But since $-\pi/2 \le \alpha \le \pi/2$ and $0 \le \beta \le \pi$, it follows that

$$-\frac{\pi}{2} \le \alpha + \beta \le \frac{3\pi}{2}$$

The only possibility on this interval is $\alpha + \beta = \pi/2$.

Problem Set 3-4

Find the exact value of each of the following (without using a calculator).

1. $\sin^{-1}(\sqrt{3}/2)$
2. $\cos^{-1} \frac{1}{2}$
3. $\arcsin(\sqrt{2}/2)$
4. $\arccos(\sqrt{2}/2)$
5. $\tan^{-1} 0$
6. $\tan^{-1} 1$
7. $\tan^{-1} \sqrt{3}$
8. $\tan^{-1}(\sqrt{3}/3)$
9. $\arccos(-\frac{1}{2})$
10. $\arcsin(-\frac{1}{2})$
11. $\sec^{-1} \sqrt{2}$
12. $\sec^{-1}(-2/\sqrt{3})$

Use a calculator or Table B to find each value (in radians) in Problems 13–18.

13. $\sin^{-1} .21823$
14. $\cos^{-1} .30582$
15. $\sin^{-1}(-0.21823)$
16. $\cos^{-1}(-0.30582)$
17. $\tan^{-1} .20660$
18. $\tan^{-1}(1.2602)$

c 19. Calculate, using $\sec^{-1} x = \cos^{-1}(1/x)$.
  (a) $\sec^{-1} 1.4263$ (b) $\sec^{-1}(-2.6715)$

c 20. Calculate.
  (a) $\sec^{-1}(\pi + 1)$ (b) $\sec^{-1}(-\sqrt{5}/2)$

Solve for t, where $0 \le t < 2\pi$. Use a calculator if you have one.

21. $\sin t = .3416$
22. $\cos t = .9812$
23. $\tan t = 3.345$
24. $\sec t = 1.342$

Find the following without the use of tables or a calculator.

25. $\sin(\sin^{-1} \frac{2}{3})$
26. $\cos(\cos^{-1}(-\frac{1}{4}))$
27. $\tan(\tan^{-1} 10)$
28. $\cos^{-1}(\cos(\pi/2))$
29. $\sin^{-1}(\sin(\pi/3))$
30. $\tan^{-1}(\tan(\pi/4))$
31. $\sin^{-1}(\cos(\pi/4))$
32. $\cos^{-1}(\sin(-\pi/6))$
33. $\cos(\sin^{-1} \frac{4}{5})$
34. $\sin(\cos^{-1} \frac{3}{5})$
  Hint: Use the identities established on page 106.
35. $\cos(\tan^{-1} \frac{1}{2})$
36. $\cos(\tan^{-1}(-\frac{3}{4}))$
37. $\cos(\sec^{-1} 3)$
38. $\sec(\cos^{-1}(-.4))$
39. $\sec^{-1}(\sec(2\pi/3))$
40. $\sec(\sec^{-1} 2.56)$

41. $\cos(\sin^{-1}(-.2564))$ 　　　　　42. $\tan^{-1}(\sin 14.1)$

43. $\sin^{-1}(\cos 1.12)$ 　　　　　　　44. $\cos^{-1}(\cos^{-1} .91)$

45. $\tan(\sec^{-1} 2.5)$ 　　　　　　　　46. $\sec^{-1}(\sin 1.67)$

EXAMPLE A (Complicated Evaluations Involving Inverses) Evaluate
(a) $\sin(2 \cos^{-1} \frac{2}{3})$; 　　(b) $\tan(\tan^{-1} 2 + \sin^{-1} \frac{4}{5})$.

Solution.

(a) Let $\theta = \cos^{-1}(\frac{2}{3})$ so that $\cos \theta = \frac{2}{3}$ and

$$\sin \theta = \sqrt{1 - \cos^2 \theta} = \sqrt{1 - \frac{4}{9}} = \frac{\sqrt{5}}{3}$$

Then apply the double-angle formula for $\sin 2\theta$ as indicated below.

$$\sin\left(2 \cos^{-1} \frac{2}{3}\right) = \sin 2\theta$$

$$= 2 \sin \theta \cos \theta$$

$$= 2\frac{\sqrt{5}}{3} \cdot \frac{2}{3}$$

$$= \frac{4}{9}\sqrt{5}$$

(b) Let $\alpha = \tan^{-1} 2$ and $\beta = \sin^{-1}(\frac{4}{5})$ and apply the identity

$$\tan(\alpha + \beta) = \frac{\tan \alpha + \tan \beta}{1 - \tan \alpha \tan \beta}$$

Now $\tan \alpha = 2$ and

$$\tan \beta = \frac{\sin \beta}{\cos \beta} = \frac{\frac{4}{5}}{\sqrt{1 - \left(\frac{4}{5}\right)^2}} = \frac{\frac{4}{5}}{\frac{3}{5}} = \frac{4}{3}$$

Therefore

$$\tan(\alpha + \beta) = \frac{2 + \frac{4}{3}}{1 - 2 \cdot \frac{4}{3}} = -2$$

Evaluate by using the method of Example A, not by using a calculator.

47. $\sin(2 \cos^{-1} \frac{3}{5})$ 　　　　　　48. $\sin(2 \cos^{-1} \frac{1}{2})$

49. $\cos(2 \sin^{-1}(-\frac{3}{5}))$ 　　　　　50. $\tan(2 \tan^{-1} \frac{1}{3})$

51. $\sin(\cos^{-1} \frac{3}{5} + \cos^{-1} \frac{5}{13})$ 　　　52. $\tan(\tan^{-1} \frac{1}{2} + \tan^{-1}(-3))$

53. $\cos(\sec^{-1} \frac{3}{2} - \sec^{-1} \frac{4}{3})$ 　　　54. $\sin(\sin^{-1} \frac{4}{5} + \sec^{-1} 3)$

EXAMPLE B (More Identities)   Show that

$$\cos(2\ \tan^{-1} x) = \frac{1 - x^2}{1 + x^2}$$

Solution.   We will apply the double-angle formula

$$\cos 2\theta = 2 \cos^2 \theta - 1$$

Here $\theta = \tan^{-1} x$, so that $x = \tan \theta$. Then

$$\cos(2\ \tan^{-1} x) = \cos(2\theta)$$

$$= 2 \cos^2 \theta - 1$$

$$= \frac{2}{\sec^2 \theta} - 1$$

$$= \frac{2}{1 + \tan^2 \theta} - 1$$

$$= \frac{2}{1 + x^2} - 1$$

$$= \frac{1 - x^2}{1 + x^2}$$

Show that each of the following is an identity.

55. $\tan(\sin^{-1} x) = \dfrac{x}{\sqrt{1 - x^2}}$

56. $\sin(\tan^{-1} x) = \dfrac{x}{\sqrt{1 + x^2}}$

57. $\tan(2\ \tan^{-1} x) = \dfrac{2x}{1 - x^2}$

58. $\cos(2\ \sin^{-1} x) = 1 - 2x^2$

59. $\cos(2\ \sec^{-1} x) = \dfrac{2}{x^2} - 1$

60. $\sec(2\ \tan^{-1} x) = \dfrac{1 + x^2}{1 - x^2}$

MISCELLANEOUS PROBLEMS

61. Without using tables or a calculator, find each value (in radians).
    (a) $\arcsin(-\sqrt{3}/2)$ (b) $\tan^{-1}(-\sqrt{3})$ (c) $\sec^{-1}(-2)$

62. Calculate each of the following (radian mode).
    (a) $\dfrac{2 \arccos(.956)}{3 \arcsin(-.846)}$ (b) $.3624\ \sec^{-1}(4.193)$
    (c) $\cos^{-1}(2 \sin .1234)$ (d) $\sin[\arctan(4.62) - \arccos(-.48)]$

63. Without using tables or a calculator, find each value. Then check using your calculator.
    (a) $\tan[\tan^{-1}(43)]$ (b) $\cos[\sin^{-1}(\frac{5}{13})]$
    (c) $\sin[\frac{\pi}{4} + \sin^{-1}(.8)]$ (d) $\cos[\sin^{-1}(.6) + \sec^{-1}(3)]$

64. Try to calculate each of the following and then explain why your calculator gives you an error message.
    (a) $\cos[\sin^{-1}(2)]$ (b) $\cos^{-1}(\tan 2)$ (c) $\tan[\arctan 3 + \arctan(\frac{1}{3})]$

65. Solve for x.
    (a) $\cos(\sin^{-1}x) = \frac{3}{4}$ (b) $\sin(\cos^{-1}x) = \sqrt{.19}$
    (c) $\sin^{-1}(3x - 5) = \frac{\pi}{6}$ (d) $\tan^{-1}(x^2 - 3x + 3) = \frac{\pi}{4}$

66. To determine inverses for cotangent and cosecant, we restrict their domains to $0 < x < \pi$ and $-\pi/2 \le x \le \pi/2$, $x \ne 0$, respectively. With these restrictions understood, find each value.
    (a) $\cot^{-1}(\sqrt{3})$ (b) $\cot^{-1}(-1/\sqrt{3})$ (c) $\cot^{-1}(0)$
    (d) $\csc^{-1}(2)$ (e) $\csc^{-1}(-1)$ (f) $\csc^{-1}(-2/\sqrt{3})$

67. It is always true that $\sin(\sin^{-1}x) = x$, but it is not always true that $\sin^{-1}(\sin x) = x$. For example, $\sin^{-1}(\sin \pi) \ne \pi$. Instead,

$$\sin^{-1}(\sin \pi) = \sin^{-1}(0) = 0$$

Find each value.
    (a) $\sin^{-1}[\sin(\pi/2)]$ (b) $\sin^{-1}[\sin(3\pi/4)]$
    (c) $\sin^{-1}[\sin(5\pi/4)]$ (d) $\sin^{-1}[\sin(3\pi/2)]$
    (e) $\cos^{-1}[\cos(3\pi)]$ (f) $\tan^{-1}[\tan(13\pi/4)]$

68. Sketch the graph of $y = \sin^{-1}(\sin x)$ for $-2\pi \le x \le 4\pi$. *Hint:* See Problem 67.

69. For each of the following right triangles, write θ explicitly in terms of x.

(a)
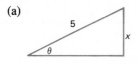

(b)
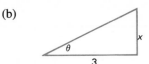

(c)
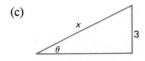

(d)

70. In some computer languages (for example, BASIC and FORTRAN), the only built-in inverse trigonometric function is $\tan^{-1}$. Establish the following identities, which show why this is sufficient.
    (a) $\sin^{-1} x = \tan^{-1}\left(\dfrac{x}{\sqrt{1 - x^2}}\right)$

    (b) $\cos^{-1} x = \dfrac{\pi}{2} - \sin^{-1} x = \dfrac{\pi}{2} - \tan^{-1}\left(\dfrac{x}{\sqrt{1 - x^2}}\right)$

71. Assume that your calculator's $\tan^{-1}$ button is working but not the $\sin^{-1}$ or $\cos^{-1}$ buttons. Use the results in Problem 70 to calculate each of the following.
    (a) $\sin^{-1}(.6)$ (b) $\sin^{-1}(-.3)$
    (c) $\cos^{-1}(.8)$ (d) $\cos^{-1}(-.9)$

72. Show that $\arctan(\frac{1}{4}) + \arctan(\frac{3}{5}) = \pi/4$. *Hint:* Show that both sides have the same tangent, using the formula for $\tan(\alpha + \beta)$.

73. In 1706, John Machin used the following formula to calculate π to 100 decimal places, a tremendous feat for its day. Establish this formula. *Hint:* Apply the addition formula for the tangent to the left side. Think of the right side as $4\theta = 2(2\theta)$ and use the tangent double angle formula twice.

$$\frac{\pi}{4} + \arctan\left(\frac{1}{239}\right) = 4 \arctan\left(\frac{1}{5}\right)$$

74. Show that

$$\arctan\left(\frac{1}{3}\right) + \arctan\left(\frac{1}{5}\right) + \arctan\left(\frac{1}{7}\right) + \arctan\left(\frac{1}{8}\right) = \frac{\pi}{4}$$

75. A picture 4 feet high is hung on a museum wall so that its bottom is 7 feet above the floor. A viewer whose eye level is 5 feet above the floor stands b feet from the wall.
   (a) Express θ, the vertical angle subtended by the picture at her eye, explicitly in terms of b.
   (b) Calculate θ when $b = 8$.
   (c) Determine b so $\theta = 30°$.

76. **TEASER** A goat is tethered to a stake at the edge of a circular pond of radius r by means of a rope of length kr, $0 < k \le 2$. Find an explicit formula for its grazing area in terms of r and k.

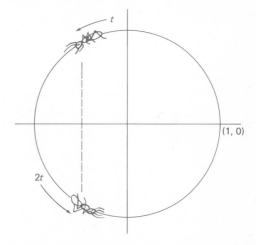

Two Bugs on a Circle

Two bugs crawl around the unit circle starting together at $(1, 0)$, one moving at one unit per second, the other moving twice as fast. When will one bug be directly above the other bug?

3-5 Trigonometric Equations

What does the bug problem have to do with trigonometric equations? Well, you should agree that after t seconds the slow bug, having traveled t units along the unit circle, is at $(\cos t, \sin t)$. The fast bug is at $(\cos 2t, \sin 2t)$. One bug will be directly above the other bug when their two x-coordinates are equal. This means we must solve the equation

$$\cos 2t = \cos t$$

Specifically, we must find the first $t > 0$ that makes this equality true. We shall solve this equation in due time, but first we ought to solve some simpler trigonometric equations.

SIMPLE EQUATIONS

Suppose we are asked to solve the equation

$$\sin t = \frac{1}{2}$$

for t. The number $t = \pi/6$ occurs to us right away. But that is not the only answer. All numbers that measure angles in the first or second quadrant and have $\pi/6$ as their reference number are solutions. Thus,

$$\ldots, -\frac{11\pi}{6}, -\frac{7\pi}{6}, \frac{\pi}{6}, \frac{5\pi}{6}, \frac{13\pi}{6}, \ldots$$

all work. In fact, one characteristic of trigonometric equations is that, if they have one solution, they have infinitely many solutions.

Let us alter the problem. Suppose we wish to solve $\sin t = \frac{1}{2}$ for $0 \le t < 2\pi$. Then the answers are $\pi/6$ and $5\pi/6$. In the following pages, we shall assume that, unless otherwise specified, we are to find only those solutions on the interval $0 \le t < 2\pi$.

For a second example, let us solve $\sin t = -.5234$ for $0 \le t < 2\pi$. Our unthinking action is to use a calculator to find $\sin^{-1}(-.5234)$, but note that this gives us the value $-.55084$, which is not on the required interval. The better way to proceed is to recognize that the given equation has two solutions but that they both have t_0 as their reference number, where

$$t_0 = \sin^{-1}(.5234) = .55084$$

The two solutions we seek are $\pi + t_0$ and $2\pi - t_0$, that is, 3.69243 and 5.73235.

Do you remember how we solved the equation $x^2 = 4x$? We rewrote it with 0 on one side, factored the other side, and then set each factor equal to 0 (see Figure 10). We follow exactly the same procedure with our next trigonometric equation. To solve

$$\cos t \cot t = -\cos t$$

use the following steps:

$$\cos t \cot t + \cos t = 0$$

$$\cos t (\cot t + 1) = 0$$

$$\cos t = 0, \cot t + 1 = 0$$

$$\cos t = 0, \qquad \cot t = -1$$

$$x^2 = 4x$$
$$x^2 - 4x = 0$$
$$x(x - 4) = 0$$
$$x = 0, x - 4 = 0$$
$$x = 0, 4$$

Figure 10

Thus our problem is reduced to solving two simple equations. The first has the two solutions, $\pi/2$ and $3\pi/2$; the second has solutions $3\pi/4$ and $7\pi/4$. Thus the set of all solutions of

$$\cos t \cot t = -\cos t$$

on the interval $0 \le t < 2\pi$ is

$$\left\{ \frac{\pi}{2}, \frac{3\pi}{4}, \frac{3\pi}{2}, \frac{7\pi}{4} \right\}$$

EQUATIONS OF QUADRATIC FORM

In Problem Set 1-1, we solved quadratic equations by a number of techniques (factoring, taking square roots, and using the quadratic formula). We use the same techniques here. For example,

$$\cos^2 t = \frac{3}{4}$$

is analogous to $x^2 = \frac{3}{4}$. We solve such an equation by taking square roots.

$$\cos t = \pm \frac{\sqrt{3}}{2}$$

The set of solutions on $0 \le t < 2\pi$ is

$$\left\{ \frac{\pi}{6}, \frac{5\pi}{6}, \frac{7\pi}{6}, \frac{11\pi}{6} \right\}$$

As a second example, consider the equation

$$2 \sin^2 t - \sin t - 1 = 0$$

Think of it as being like

$$2x^2 - x - 1 = 0$$

Now

$$2x^2 - x - 1 = (2x + 1)(x - 1)$$

and so

$$2 \sin^2 t - \sin t - 1 = (2 \sin t + 1)(\sin t - 1)$$

When we set each factor equal to zero and solve, we get

$$2 \sin t + 1 = 0 \qquad\qquad \sin t - 1 = 0$$
$$\sin t = -\tfrac{1}{2} \qquad\qquad \sin t = 1$$
$$t = \frac{7\pi}{6}, \frac{11\pi}{6} \qquad\qquad t = \frac{\pi}{2}$$

The set of all solutions on $0 \le t < 2\pi$ is

$$\left\{ \frac{\pi}{2}, \frac{7\pi}{6}, \frac{11\pi}{6} \right\}$$

USING IDENTITIES TO SOLVE EQUATIONS

Consider the equation

$$\tan^2 x = \sec x + 1$$

The identity $\sec^2 x = \tan^2 x + 1$ suggests writing everything in terms of $\sec x$.

$$\sec^2 x - 1 = \sec x + 1$$

$$\sec^2 x - \sec x - 2 = 0$$

$$(\sec x + 1)(\sec x - 2) = 0$$

$$\sec x + 1 = 0 \qquad \sec x - 2 = 0$$

$$\sec x = -1 \qquad \sec x = 2$$

$$x = \pi \qquad x = \frac{\pi}{3}, \frac{5\pi}{3}$$

Thus, the set of solutions on $0 \le t < 2\pi$ is $\{\pi/3, \pi, 5\pi/3\}$. Unfamiliarity with the secant may hinder you at the last step. If so, use $\sec x = 1/\cos x$ to write the equations in terms of cosines and solve the equations

$$\cos x = -1 \qquad \cos x = \frac{1}{2}$$

SOLUTION TO THE TWO-BUG PROBLEM

Our opening display asked when one bug would first be directly above the other. We reduced that problem to solving

$$\cos 2t = \cos t$$

for t. Using a double-angle formula, we may write

$$2 \cos^2 t - 1 = \cos t$$

$$2 \cos^2 t - \cos t - 1 = 0$$

$$(2 \cos t + 1)(\cos t - 1) = 0$$

$$\cos t = -\frac{1}{2} \qquad \cos t = 1$$

$$t = \frac{2\pi}{3}, \frac{4\pi}{3} \qquad t = 0$$

The smallest positive solution is $t = 2\pi/3$. After a little over 2 seconds, the slow bug will be directly above the fast bug.

Problem Set 3-5

Solve each of the following, finding all solutions on the interval 0 to 2π, excluding 2π.

1. $\sin t = 0$
2. $\cos t = 1$
3. $\sin t = -1$
4. $\tan t = -\sqrt{3}$
5. $\sin t = 2$
6. $\sec t = \frac{1}{2}$
7. $2 \cos x + \sqrt{3} = 0$
8. $2 \sin x + 1 = 0$
9. $\tan^2 x = 1$
10. $4 \sin^2 \theta - 3 = 0$
11. $(2 \cos \theta + 1)(2 \sin \theta - \sqrt{2}) = 0$
12. $(\sin \theta - 1)(\tan \theta + 1) = 0$
13. $\sin^2 x + \sin x = 0$
14. $2 \cos^2 x - \cos x = 0$
15. $\tan^2 \theta = \sqrt{3} \tan \theta$
16. $\cot^2 \theta = -\cot \theta$
17. $2 \sin^2 x = 1 + \cos x$
18. $\sec^2 x = 1 + \tan x$

ⓒ 19. $\tan^2 x - 3 \tan x + 1 = 0$ ⓒ 20. $\cos 2t = 3 \sin t$

CAUTION

$\boxed{\begin{array}{c} \tan^2 \theta = \sqrt{3} \tan \theta \\ \tan \theta = \sqrt{3} \end{array}}$ ✗

$\begin{array}{c} \tan^2 \theta = \sqrt{3} \tan \theta \\ \tan \theta (\tan \theta - \sqrt{3}) = 0 \\ \tan \theta = 0 \qquad \tan \theta = \sqrt{3} \end{array}$

EXAMPLE A (Solving by Squaring Both Sides) Solve

$$1 - \cos t = \sqrt{3} \sin t$$

Solution. Since the identity relating sines and cosines involves their squares, we begin by squaring both sides. Then we express everything in terms of $\cos t$ and solve.

$$(1 - \cos t)^2 = 3 \sin^2 t$$
$$1 - 2 \cos t + \cos^2 t = 3(1 - \cos^2 t)$$
$$\cos^2 t - 2 \cos t + 1 = 3 - 3 \cos^2 t$$
$$4 \cos^2 t - 2 \cos t - 2 = 0$$
$$2 \cos^2 t - \cos t - 1 = 0$$
$$(2 \cos t + 1)(\cos t - 1) = 0$$
$$\cos t = -\frac{1}{2} \qquad \cos t = 1$$
$$t = \frac{2\pi}{3}, \frac{4\pi}{3} \qquad t = 0$$

Since squaring may introduce extraneous solutions, it is important to check our answers. We find that $4\pi/3$ is extraneous, since substituting $4\pi/3$ for t in the original equation gives us $1 + \frac{1}{2} = -\frac{3}{2}$. However, 0 and $2\pi/3$ are solutions, as you should verify.

Solve each of the following equations on the interval $0 \le t < 2\pi$; check your answers.

21. $\sin t + \cos t = 1$
22. $\sin t - \cos t = 1$
23. $\sqrt{3}(1 - \sin t) = \cos t$
24. $1 + \sin t = \sqrt{3} \cos t$
25. $\sec t + \tan t = 1$
26. $\tan t - \sec t = 1$

EXAMPLE B (Finding All of the Solutions)   Find the entire set of solutions of the equation $\cos 2t = \cos t$.

Solution.   In the text, we found 0, $2\pi/3$, and $4\pi/3$ to be the solutions for $0 \le t < 2\pi$. Clearly we get new solutions by adding 2π again and again to any of these numbers. The same holds true for subtracting 2π. In fact, the entire solution set consists of all those numbers of the form $2\pi k$, $2\pi/3 + 2\pi k$, or $4\pi/3 + 2\pi k$, where k is any integer.

Find the entire solution set of each of the following equations.

27. $\sin t = \frac{1}{2}$ 28. $\cos t = -\frac{1}{2}$ 29. $\tan t = 0$

30. $\tan t = -\sqrt{3}$ 31. $\sin^2 t = \frac{1}{4}$ 32. $\cos^2 t = 1$

EXAMPLE C (Multiple-Angle Equations)   Find all solutions of $\cos 4t = \frac{1}{2}$ on the interval $0 \le t < 2\pi$.

Solution.   There will be more answers than you think. We know that $\cos 4t$ equals $\frac{1}{2}$ when

$$4t = \frac{\pi}{3}, \frac{5\pi}{3}, \frac{7\pi}{3}, \frac{11\pi}{3}, \frac{13\pi}{3}, \frac{17\pi}{3}, \frac{19\pi}{3}, \frac{23\pi}{3}$$

that is, when

$$t = \frac{\pi}{12}, \frac{5\pi}{12}, \frac{7\pi}{12}, \frac{11\pi}{12}, \frac{13\pi}{12}, \frac{17\pi}{12}, \frac{19\pi}{12}, \frac{23\pi}{12}$$

The reason that there are 8 solutions instead of 2 is that $\cos 4t$ completes 4 periods on the interval $0 \le t < 2\pi$.

Solve each of the following equations, finding all solutions on the interval $0 \le t < 2\pi$.

33. $\sin 2t = 0$ 34. $\cos 2t = 0$ 35. $\sin 4t = 1$

36. $\cos 4t = 1$ 37. $\tan 2t = -1$ 38. $\tan 3t = 0$

MISCELLANEOUS PROBLEMS

In Problems 39–56, find all solutions to the given equation on $0 \le x < 2\pi$.

39. $2 \sin^2 x = \sin x$ 40. $2 \cos x \sin x + \cos x = 0$

C 41. $\cos^2 x = \frac{1}{3}$ C 42. $\tan^2 x + 2 \tan x = 0$

43. $2 \tan x - \sec^2 x = 0$ 44. $\tan^2 x = 1 + \sec x$

45. $\tan 2x = 3 \tan x$ 46. $\cos(x/2) - \cos x = 1$

C 47. $\sin^2 x + 3 \sin x - 1 = 0$ C 48. $\tan^2 x - 2 \tan x - 10 = 0$

49. $\sin 2x + \sin x + 4 \cos x = -2$

50. $\cos x + \sin x = \sec x + \sec x \tan x$

51. $\sin x \cos x = -\sqrt{3}/4$ 52. $4 \sin x - 4 \sin^3 x + \cos x = 0$

C 53. $\cos x - 2 \sin x = 2$ C 54. $\sin x + \cos x = \frac{1}{3}$

55. $\cos^8 x - \sin^8 x = 0$ 56. $\cos^6 x + \sin^6 x = \frac{13}{16}$

57. A ray of light from the lamp L in Figure 11 reflects off a mirror to the object O.

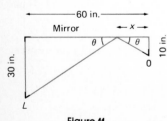

Figure 11

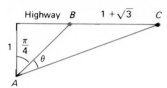

Highway B $1 + \sqrt{3}$ C

1 $\dfrac{\pi}{4}$ θ

A

Figure 12

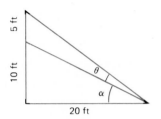

5 ft

10 ft

θ

α

20 ft

Figure 13

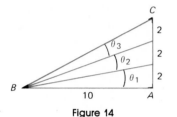

C

θ_3

θ_2

θ_1

2

2

2

B 10 A

Figure 14

(a) Find the distance x.

(b) Write an equation for θ.

(c) Solve this equation.

58. Tom and John are lost in a desert 1 mile from a highway, at point A in Figure 12. Each strikes out in a different direction to get to the highway. Tom gets to the highway at point B and John arrives at point C, $1 + \sqrt{3}$ miles farther down the road. Write an equation for θ and solve it.

59. Mr. Quincy built a slide with a 10-foot rise and 20-foot base (Figure 13). (a) Find the angle α in degrees. (b) By how much (θ in Figure 13) would the angle of the slide increase if he made the rise 15 feet, keeping the base at 20 feet?

60. Find the angles θ_1, θ_2, and θ_3 shown in Figure 14. Your answers should convince you that the angle ABC is not trisected.

61. Solve the equation
$$\sin 4t + \sin 3t + \sin 2t = 0$$
Hint: Use the identity $\sin u + \sin v = 2 \sin((u + v)/2) \cos((u - v)/2)$.

62. Solve the equation
$$\cos 5t + \cos 3t - 2 \cos t = 0$$
Hint: Use the identity $\cos u + \cos v = 2 \cos((u + v)/2) \cos((u - v)/2)$.

63. Solve $\cos^8 u + \sin^8 u = \frac{41}{128}$ for $0 \le u \le \pi$. *Hint:* Begin by using half-angle formulas.

64. **TEASER** Show that $t = \pi/4$ is the only solution on $0 \le t \le \pi$ to the equation

$$\frac{a + b \cos t}{b + a \sin t} = \frac{a + b \sin t}{b + a \cos t}$$

Chapter Summary

An **identity** is an equality that is true for all values of the unknown for which both sides of the equality make sense. Our first task was to establish the fundamental identities of trigonometry, here arranged by category.

Basic Identities

1. $\tan t = \dfrac{\sin t}{\cos t}$ 2. $\cot t = \dfrac{\cos t}{\sin t} = \dfrac{1}{\tan t}$

3. $\sec t = \dfrac{1}{\cos t}$ 4. $\csc t = \dfrac{1}{\sin t}$

5. $\sin^2 t + \cos^2 t = 1$ 6. $1 + \tan^2 t = \sec^2 t$

7. $1 + \cot^2 t = \csc^2 t$

Cofunction Identities

8. $\sin\left(\dfrac{\pi}{2} - t\right) = \cos t$ 9. $\cos\left(\dfrac{\pi}{2} - t\right) = \sin t$

Odd-Even Identities

10. $\sin(-t) = -\sin t$ 11. $\cos(-t) = \cos t$

Addition Formulas

12. $\sin(s + t) = \sin s \cos t + \cos s \sin t$

13. $\sin(s - t) = \sin s \cos t - \cos s \sin t$

14. $\cos(s + t) = \cos s \cos t - \sin s \sin t$

15. $\cos(s - t) = \cos s \cos t + \sin s \sin t$

Double-Angle Formulas

16. $\sin 2t = 2 \sin t \cos t$

17. $\cos 2t = \cos^2 t - \sin^2 t = 1 - 2 \sin^2 t = 2 \cos^2 t - 1$

Half-Angle Formulas

18. $\sin \dfrac{t}{2} = \pm \sqrt{\dfrac{1 - \cos t}{2}}$ 19. $\cos \dfrac{t}{2} = \pm \sqrt{\dfrac{1 + \cos t}{2}}$

Once we have memorized the fundamental identities, we can use them to prove thousands of other identities. The suggested technique is to take one side of a proposed identity and show by a chain of equalities that it is equal to the other.

A **trigonometric equation** is an equality involving trigonometric functions that is true only for some values of the unknown (for example, $\sin 2t = \frac{1}{2}$). Here our job is to solve the equation, that is, to find the values of the unknown that make it true.

With their natural domains, the trigonometric functions are not one-to-one and therefore do not have inverses. However, there are standard ways to restrict the domains so that inverses exist. Here are the results.

$$x = \sin^{-1} y \quad \text{means} \quad y = \sin x \quad \text{and} \quad \frac{-\pi}{2} \le x \le \frac{\pi}{2}$$

$$x = \cos^{-1} y \quad \text{means} \quad y = \cos x \quad \text{and} \quad 0 \le x \le \pi$$

$$x = \tan^{-1} y \quad \text{means} \quad y = \tan x \quad \text{and} \quad \frac{-\pi}{2} < x < \frac{\pi}{2}$$

$$x = \sec^{-1} y \quad \text{means} \quad y = \sec x \quad \text{and} \quad 0 \le x \le \pi, x \ne \frac{\pi}{2}$$

Chapter Review Problem Set

1. Prove that the following are identities.

   (a) $\cot \theta \cos \theta = \csc \theta - \sin \theta$

   (b) $\dfrac{\cos x \tan^2 x}{\sec x + 1} = 1 - \cos x$

2. Express each of the following in terms of $\sin x$ and simplify.

(a) $\dfrac{(\cos^2 x - 1)(1 + \tan^2 x)}{\csc x}$ (b) $\dfrac{\cos^2 x \csc x}{1 + \csc x}$

3. Use appropriate identities to simplify and then calculate each of the following.

(a) $2 \cos^2 22.5° - 1$

(b) $\sin 37° \cos 53° + \cos 37° \sin 53°$

(c) $\cos 108° \cos 63° + \sin 108° \sin 63°$

4. If $\cos t = -\frac{4}{5}$ and $\pi < t < 3\pi/2$, calculate

(a) $\sin 2t$;   (b) $\sin(t/2)$.

5. Prove that the following are identities.

(a) $\sin 2t \cos t - \cos 2t \sin t = \sin t$

(b) $\sec 2t + \tan 2t = \dfrac{\cos t + \sin t}{\cos t - \sin t}$

(c) $\dfrac{\cos(\alpha + \beta)}{\cos \alpha \cos \beta} = \tan \alpha (\cot \alpha - \tan \beta)$

6. Solve the following trigonometric equations for t, $0 \le t < 2\pi$.

(a) $\cos t = -\sqrt{3}/2$

(b) $(2 \sin t + 1) \tan t = 0$

(c) $\cos^2 t + 2 \cos t - 3 = 0$

(d) $\sin t - \cos t = 1$

(e) $\sin 3t = 1$

7. What is the standard way to restrict the domain of sine, cosine, and tangent so that they have inverses?

8. Calculate each of the following without the help of a calculator.

(a) $\sin^{-1}(-\sqrt{3}/2)$ (b) $\cos^{-1}(-\sqrt{3}/2)$

(c) $\tan^{-1}(-\sqrt{3})$ (d) $\tan(\tan^{-1} 6)$

(e) $\cos^{-1}(\cos 3\pi)$ (f) $\sin(\cos^{-1} \frac{2}{3})$

(g) $\cos(2 \cos^{-1} .7)$ (h) $\sin(2 \cos^{-1} \frac{5}{13})$

9. Sketch the graph of $y = \tan^{-1} x$.

10. Find an approximate value for $\tan^{-1}(-1000)$.

11. Show that

$$\frac{\pi}{4} = \arctan \frac{1}{2} + \arctan \frac{1}{3}$$

*Thus one sees in the sciences
many brilliant theories which
have remained unapplied for
a long time suddenly
becoming the foundation of
most important applications,
and likewise applications very
simple in appearance giving
birth to ideas of the most
abstract theories.*

—*Marquis de Condorcet*

CHAPTER 4

Applications of Trigonometry

Consider an arbitrary triangle with angles α, β, γ, and corresponding opposite sides a, b, c, respectively. Then

$$\frac{\sin \alpha}{a} = \frac{\sin \beta}{b} = \frac{\sin \gamma}{c}$$

Equivalently,

$$\frac{a}{\sin \alpha} = \frac{b}{\sin \beta} = \frac{c}{\sin \gamma}$$

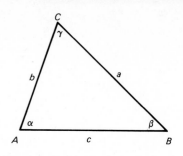

4-1   Oblique Triangles: Law of Sines

We learned in Section 2-1 how to solve a right triangle. But can we solve an oblique triangle—that is, one without a 90° angle? One valuable tool is the **law of sines,** stated above. It is valid for any triangle whatever, but we initially establish it for the case where all angles are acute.

PROOF OF THE LAW OF SINES

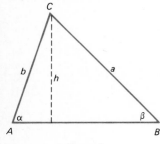

Figure 1

Consider a triangle with all acute angles, labeled as in Figure 1. Drop a perpendicular of length h from vertex C to the opposite side. Then by right-triangle trigonometry,

$$\sin \alpha = \frac{h}{b} \qquad \sin \beta = \frac{h}{a}$$

If we solve for h in these two equations and equate the results, we obtain

$$b \sin \alpha = a \sin \beta$$

Finally, dividing both sides by ab yields

$$\frac{\sin \alpha}{a} = \frac{\sin \beta}{b}$$

Since the roles of β and γ can be interchanged, the same reasoning gives

$$\frac{\sin \alpha}{a} = \frac{\sin \gamma}{c}$$

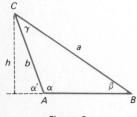

Figure 2

Next consider a triangle with an obtuse angle α (90° < α < 180°). Drop a perpendicular of length h from vertex C to the extension of AB (see Figure 2). Notice that angle α' is the reference angle for α and so $\sin \alpha = \sin \alpha'$. It follows from right-triangle trigonometry that

$$\sin \alpha = \sin \alpha' = \frac{h}{b} \qquad \sin \beta = \frac{h}{a}$$

just as in the acute case. The rest of the argument is identical with that case.

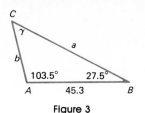

Figure 3

SOLVING A TRIANGLE (AAS)

Suppose that we know two angles and any side of a triangle. For example, suppose that in triangle ABC, $\alpha = 103.5°$, $\beta = 27.5°$, and $c = 45.3$ (Figure 3). Our task is to find γ, a, and b.

1. Since $\alpha + \beta + \gamma = 180°$, $\gamma = 180° - (103.5° + 27.5°) = 49°$.
2. By the law of sines,

$$\frac{a}{\sin 103.5°} = \frac{45.3}{\sin 49°}$$

$$a = \frac{(45.3)(\sin 103.5°)}{\sin 49°}$$

$$= \frac{(45.3)(\sin 76.5°)}{\sin 49°}$$

$$= \frac{(45.3)(.9724)}{.7547}$$

$$\approx 58.4$$

3. Also by the law of sines,

$$\frac{b}{\sin 27.5°} = \frac{45.3}{\sin 49°}$$

$$b = \frac{(45.3)(\sin 27.5°)}{\sin 49°}$$

$$= \frac{(45.3)(.4617)}{.7547}$$

$$\approx 27.7$$

SOLVING A TRIANGLE (SSA)

Suppose that two sides and the angle opposite one of them are given. This is called the **ambiguous case** because the given information may not determine a unique triangle.

If α, a, and b are given, we consider trying to construct a triangle fitting these data by first drawing angle α, then marking off b on one of its sides thus determining vertex C. Finally, we attempt to locate vertex B by striking off a circular arc of radius a with center at C. If $a \geq b$, this can always be done in a unique way. Figure 4 illustrates this for both α acute and α obtuse. However if $a < b$, there are several possibilities (Figure 5).

Fortunately, we are able to decide which of these possibilities is the case if we draw an approximate picture and then attempt to apply the law of sines. First, note that if $a \geq b$ there is one triangle corresponding to the data and for it β is an acute angle. Application of the law of sines will give $\sin \beta$, which allows determination of β.

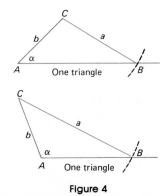

Figure 4

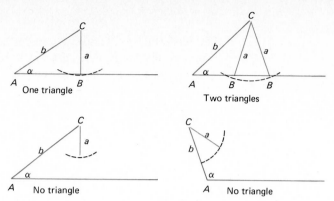

Figure 5

If $a < b$, we may attempt to apply the law of sines. If it yields $\sin \beta = 1$, we have a unique right triangle. If it yields $\sin \beta < 1$, we have two triangles corresponding to the two angles β_1 and β_2 (one acute, the other obtuse) with this sine. If it yields $\sin \beta > 1$, we have an inconsistency in the data; no triangle satisfying the data exists.

Suppose, for example, that we are given $\alpha = 36°$, $a = 9.4$, and $b = 13.1$ (Figure 6). We must find β, γ, and c. Since $a < b$, there may be zero, one, or two triangles. We proceed to compute $\sin \beta$.

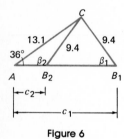

Figure 6

1. $\dfrac{\sin \beta}{13.1} = \dfrac{\sin 36°}{9.4}$

$\sin \beta = \dfrac{(13.1) \sin 36°}{9.4} = \dfrac{(13.1)(.5878)}{9.4} \approx .8191$

Since $\sin \beta < 1$, there are two triangles.

$$\beta_1 = 55° \qquad \beta_2 = 125°$$

2. $\gamma_1 = 180° - (36° + 55°) = 89°$

$\gamma_2 = 180° - (36° + 125°) = 19°$

3. $\dfrac{c_1}{\sin 89°} = \dfrac{9.4}{\sin 36°}$

$c_1 = \dfrac{9.4}{\sin 36°} (\sin 89°)$

$= \dfrac{9.4}{.5878} (.9998) \approx 16.0$

$\dfrac{c_2}{\sin 19°} = \dfrac{9.4}{\sin 36°}$

$c_2 = \dfrac{9.4}{\sin 36°} (\sin 19°)$

$= \dfrac{9.4}{.5878} (.3256) \approx 5.2$

Problem Set 4-1

Solve the triangles of Problems 1–10 using either Table A or a calculator.

1. $\alpha = 42.6°$, $\beta = 81.9°$, $a = 14.3$
2. $\beta = 123°$, $\gamma = 14.2°$, $a = 295$
3. $\alpha = \gamma = 62°$, $b = 50$
4. $\alpha = \beta = 14°$, $c = 30$
5. $\alpha = 115°$, $a = 46$, $b = 34$
6. $\beta = 143°$, $a = 46$, $b = 84$
7. $\alpha = 30°$, $a = 8$, $b = 5$
8. $\beta = 60°$, $a = 11$, $b = 12$
9. $\alpha = 30°$, $a = 5$, $b = 8$
10. $\beta = 60°$, $a = 12$, $b = 11$

11. Two observers stationed 110 meters apart at A and B on the bank of a river are looking at a tower situated at a point C on the opposite bank. They measure angles CAB and CBA to be 43° and 57°, respectively. How far is the first observer from the tower?

12. A telegraph pole leans away from the sun at an angle of 11° to the vertical. The pole casts a shadow 96 feet long on horizontal ground when the angle of elevation of the sun is 23°. Find the length of the pole (see Figure 7).

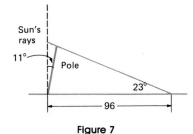

Figure 7

13. A vertical pole 60 feet long is standing by the side of an inclined road. It casts a shadow 138 feet long directly downhill along the road when the angle of elevation of the sun is 58°. Find the angle of inclination θ of the road (see Figure 8).

14. Two forest rangers 15 miles apart at points A and B observe a fire at a point C. The ranger at A measures angle CAB as 43.6° and the one at B measures angle CBA as 79.3°. How far is the fire from each ranger? How far is the fire from a straight road that goes from A to B?

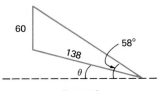

Figure 8

EXAMPLE (An Important Area Formula) Consider a triangle with two sides b and c and included angle α. Show that the area A of the triangle is

$$A = \frac{1}{2}bc \sin \alpha$$

Solution. Let h denote the altitude of the triangle as shown in the diagrams of Figure 9. Whether α is acute or obtuse, we have $\sin \alpha = h/c$, that is, $h = c \sin \alpha$. We conclude that

$$A = \tfrac{1}{2}bh = \tfrac{1}{2}bc \sin \alpha$$

15. Find the area of the triangle with sides $b = 20$ and $c = 30$ and included angle $\alpha = 40°$.

16. Find the area of the triangle with $a = 14.6$, $b = 31.7$, and $\gamma = 130.2°$.

17. Find the area of the triangle with $c = 30.1$, $\alpha = 25.3°$, and $\beta = 112.2°$.

18. Find the area of the triangle with $a = 20$, $\alpha = 29°$, and $\gamma = 46°$.

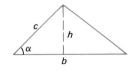
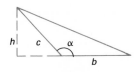

Figure 9

MISCELLANEOUS PROBLEMS

19. The children's slide at the park is 30 feet long and inclines 36° from the horizontal. The ladder to the top is 18 feet long. How steep is the ladder, that is, what

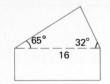

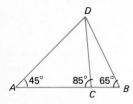

Figure 10

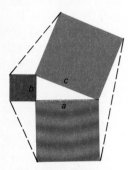

Figure 11

Figure 12

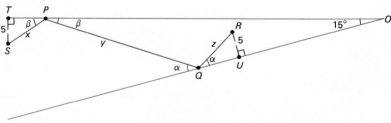

Figure 13

angle does it make with the horizontal? Assume the slide is straight and that the bottom end of the slide is at the same level as the bottom end of the ladder.

20. Prevailing winds have caused an old tree to incline 11° eastward from the vertical. The sun in the west is 32° above the horizontal. How long a shadow is cast by the tree if the tree measures 114 feet from top to bottom?

21. A rectangular room, 16 feet by 30 feet, has an open beam ceiling. The two parts of the ceiling make angles of 65° and 32° with the horizontal (an end view is shown in Figure 10). Find the total area of the ceiling.

22. Sheila Sather, traveling north on a straight road at a constant rate of 60 miles per hour, sighted flames shooting up into the air at a point 20° west of north. Exactly 1 hour later, the fire was 59° west of south. Determine the shortest distance from the road to the fire.

23. A lighthouse stands at a certain distance out from a straight shoreline. It throws a beam of light that revolves at a constant rate of one revolution per minute. A short time after shining on the nearest point on the shore, the beam reaches a point on the shore that is 2640 feet from the lighthouse, and 3 seconds later it reaches a point 2000 feet farther along the shore. How far is the lighthouse from the shore?

24. In Figure 11, AC is 10 meters longer than CB. Determine the length of CD.

25. Four line segments of lengths 3, 4, 5, and 6 radiate like spokes from a common point. Their outer ends are the vertices of a quadrilateral Q. Determine the maximum possible area of Q.

26. Figure 12 illustrates the Pythagorean theorem ($a^2 + b^2 = c^2$). A rubber band is stretched around this figure. Show that the area of the region enclosed by the rubber band is $2(ab + c^2)$.

27. Let 2ϕ denote the angle at a point of the regular 6-pointed star shown in Figure 13. Express the area A of this star in terms of ϕ and the edge length r.

28. **TEASER** Figure 14 shows two mirrors intersecting at an angle of 15°. A light ray from S is reflected at P and again at Q and then is absorbed at R. It is given that $ST = RU = 5$, $OT = 50$, and $OU = 20$. Find the length $x + y + z$ of the path of the light ray. As indicated, the angle of incidence equals the angle of reflection.

Figure 14

Consider an arbitrary triangle with angles α, β, γ and corresponding opposite sides a, b, c, respectively. Then

$$a^2 = b^2 + c^2 - 2bc \cos \alpha$$

$$b^2 = a^2 + c^2 - 2ac \cos \beta$$

$$c^2 = a^2 + b^2 - 2ab \cos \gamma$$

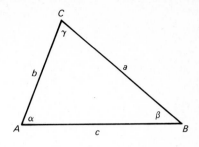

4-2 Oblique Triangles: Law of Cosines

When two sides and the included angle (SAS) or three sides (SSS) of a triangle are given, we cannot apply the law of sines to solve the triangle. Rather, we need the law of cosines, stated above in symbols. Actually it is wise to learn the law in words.

The square of any side is equal to the sum of the squares of the other two sides minus twice the product of those sides times the cosine of the angle between them.

Notice what happens when $\gamma = 90°$ so that $\cos \gamma = 0$. The law of cosines

$$c^2 = a^2 + b^2 - 2ab \cos \gamma$$

becomes

$$c^2 = a^2 + b^2$$

which is just the Pythagorean theorem. In fact, you should think of the law of cosines as a generalization of the Pythagorean theorem, with the term $-2ab \cos \gamma$ acting as a correction term when γ is not 90°.

PROOF OF THE LAW OF COSINES

Assume first that angle α is acute. Drop a perpendicular CD from vertex C to side AB as shown in Figure 15. Label the lengths of CD, AD, and DB by h, x, and $c - x$, respectively.

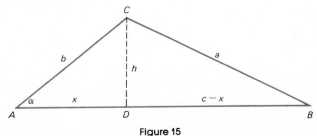

Figure 15

Consider the two right triangles ADC and BDC. By the Pythagorean theorem

$$h^2 = b^2 - x^2 \quad \text{and} \quad h^2 = a^2 - (c - x)^2$$

Equating these two expressions for h^2 gives

$$a^2 - (c - x)^2 = b^2 - x^2$$
$$a^2 = b^2 - x^2 + (c - x)^2$$
$$a^2 = b^2 - x^2 + c^2 - 2cx + x^2$$
$$a^2 = b^2 + c^2 - 2cx$$

Now $\cos \alpha = x/b$, and so $x = b \cos \alpha$. Thus

$$a^2 = b^2 + c^2 - 2cb \cos \alpha$$

which is the result we wanted.

Next we give the proof of the law of cosines for the obtuse angle case. Again drop a perpendicular from vertex C to side AB extended and label the resulting diagram as shown in Figure 16. From consideration of triangles ADC and BDC and the Pythagorean theorem, we obtain

$$h^2 = b^2 - x^2 \quad \text{and} \quad h^2 = a^2 - (c + x)^2$$

Algebra analogous to that used in the acute angle case yields

$$a^2 = b^2 + c^2 + 2cx$$

Now α' is the reference angle for α, and so $\cos \alpha = -\cos \alpha'$. Also $\cos \alpha' = x/b$. Therefore,

$$x = b \cos \alpha' = -b \cos \alpha$$

When we substitute this expression for x in the equation above, we get

$$a^2 = b^2 + c^2 - 2cb \cos \alpha$$

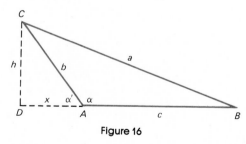

Figure 16

SOLVING A TRIANGLE (SAS)

Consider a triangle with $b = 18.1$, $c = 12.3$, and $\alpha = 115°$ (Figure 17). We want to determine a, β, and γ.

1. By the law of cosines,

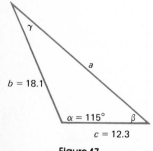

$b = 18.1$

$\alpha = 115°$

$c = 12.3$

Figure 17

$$a^2 = (18.1)^2 + (12.3)^2 - 2(18.1)(12.3) \cos 115°$$
$$= 327.61 + 151.29 - (445.26)(-\cos 65°)$$
$$= 327.61 + 151.29 + (445.26)(.4226)$$
$$= 667.08$$
$$a \approx 25.8$$

2. Now we can use the law of sines.

$$\frac{\sin \beta}{18.1} = \frac{\sin 115°}{25.8}$$

$$\sin \beta = \frac{(18.1) \sin 115°}{25.8} = \frac{(1.81)(\sin 65°)}{25.8}$$

$$= \frac{(18.1)(.9063)}{25.8} = .6358$$

$$\beta \approx 39.5°$$

3. $\gamma \approx 180° - (115° + 39.5°) = 25.5°$.

SOLVING A TRIANGLE (SSS)

If $a = 13.1$, $b = 15.5$, and $c = 17.2$, then we must determine the three angles.

1. By the law of cosines,

$$a^2 = b^2 + c^2 - 2bc \cos \alpha$$

Thus

$$\cos \alpha = \frac{b^2 + c^2 - a^2}{2bc}$$

$$= \frac{(15.5)^2 + (17.2)^2 - (13.1)^2}{2(15.5)(17.2)} = .6836$$

$$\alpha \approx 46.9°$$

2. By the law of sines,

$$\frac{\sin \beta}{15.5} = \frac{\sin 46.9°}{13.1}$$

$$\sin \beta = \frac{(15.5)(\sin 46.9°)}{13.1} = \frac{(15.5)(.7302)}{13.1} = .8640$$

$$\beta \approx 59.8°$$

3. $\gamma = 180° - (46.9° + 59.8°) = 73.3°$.

Problem Set 4-2

In Problems 1–8, solve the triangles satisfying the given data. Use either Table C or a calculator.

1. $\alpha = 60°$, $b = 14$, $c = 10$
2. $\beta = 60°$, $a = c = 8$
3. $\gamma = 120°$, $a = 8$, $b = 10$
4. $\alpha = 150°$, $b = 35$, $c = 40$
5. $a = 5$, $b = 6$, $c = 7$
6. $a = 10$, $b = 20$, $c = 25$
7. $a = 12.2$, $b = 19.1$, $c = 23.8$
8. $a = .11$, $b = .21$, $c = .31$

9. At one corner of a triangular field, the angle measures 52.4°. The sides that meet at this corner are 100 meters and 120 meters long. How long is the third side?

10. To approximate the distance between two points A and B on opposite sides of a swamp, a surveyor selects a point C and measures it to be 140 meters from A and 260 meters from B. Then she measures the angle ACB, which turns out to be 49°. What is the calculated distance from A to B?

11. Two runners start from the same point at 12:00 noon, one of them heading north at 6 miles per hour and the other heading 68° east of north at 8 miles per hour (Figure 18). What is the distance between them at 3:00 that afternoon?

12. A 50-foot pole stands on top of a hill which slants 20° from the horizontal. How long must a rope be to reach from the top of the pole to a point 88 feet directly downhill (that is, on the slant) from the base of the pole?

13. A triangular garden plot has sides of length 35 meters, 40 meters, and 60 meters. Find the largest angle of the triangle.

14. A piece of wire 60 inches long is bent into the shape of a triangle. Find the angles of the triangle if two of the sides have lengths 24 inches and 20 inches.

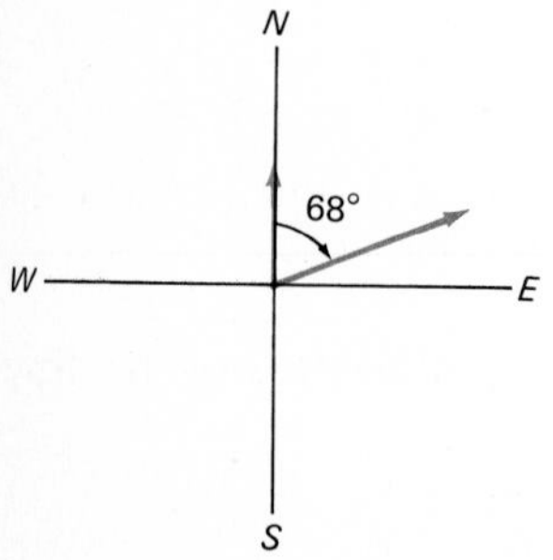

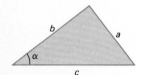

Figure 18

EXAMPLE (Heron's Area Formula)   Show that a triangle with sides a, b, and c (Figure 19) and semiperimeter $s = (a + b + c)/2$ has area A given by

$$A = \sqrt{s(s - a)(s - b)(s - c)}$$

Figure 19

Solution.   The proof is subtle, depending on the clever matching of the area formula from the last section with the law of cosines. Begin by writing the law of cosines in the form

$$2bc \cos \alpha = b^2 + c^2 - a^2$$

a formula we will use shortly. Next, take the area formula $A = \frac{1}{2} bc \sin \alpha$ in its squared form and manipulate it very carefully.

$$A^2 = \frac{1}{4}b^2c^2 \sin^2 \alpha = \frac{1}{4}b^2c^2(1 - \cos^2 \alpha)$$

$$= \frac{1}{16}(2bc)(1 + \cos \alpha)(2bc)(1 - \cos \alpha)$$

$$= \frac{1}{16}(2bc + 2bc \cos \alpha)(2bc - 2bc \cos \alpha)$$

$$= \frac{1}{16}(2bc + b^2 + c^2 - a^2)(2bc - b^2 - c^2 + a^2)$$

$$= \frac{1}{16}[(b + c)^2 - a^2][a^2 - (b - c)^2]$$

$$= \frac{(b + c + a)(b + c - a)(a - b + c)(a + b - c)}{2 \quad\quad 2 \quad\quad 2 \quad\quad 2}$$

$$= \left[\frac{a + b + c}{2}\right]\left[\frac{a + b + c}{2} - a\right]\left[\frac{a + b + c}{2} - b\right]\left[\frac{a + b + c}{2} - c\right]$$

$$= s(s - a)(s - b)(s - c)$$

15. The area of the right triangle with sides 3, 4, and 5 is 6. Confirm that Heron's formula gives the same answer.

16. Find the area of the triangle with sides 31, 42, and 53.

17. Find the area of the triangle with sides 5.9, 6.7, and 10.3.

18. Use the answer you got to Problem 16 to find the length h of the shortest altitude of the triangle with sides 31, 42, and 53.

MISCELLANEOUS PROBLEMS

19. A triangular garden plot has sides measuring 42 meters, 50 meters, and 63 meters. Find the measure of the smallest angle.

20. A diagonal and a side of a parallelogram measure 80 centimeters and 25 centimeters, respectively, and the angle between them measures 47°. Find the length of the other diagonal. Recall that the diagonals of a parallelogram bisect each other.

21. Two cars, starting from the intersection of two straight highways, travel along the highways at speeds of 55 miles per hour and 65 miles per hour, respectively. If the angle of intersection of the highways measures 72°, how far apart are the cars after 36 minutes?

22. Buoys A, B, and C mark the vertices of a triangular racing course on a lake. Buoys A and B are 4200 feet apart, buoys A and C are 3800 feet apart, and angle CAB measures 100°. If the winning boat in a race covered the course in 6.4 minutes, what was its average speed in miles per hour?

23. A quadrilateral Q has sides of length 1, 2, 3, and 4, respectively. The angle between the first pair of sides is 120°. Find the angle between the other pair of sides and also the exact area of Q.

24. For the triangle ABC in Figure 20, let r be the radius of the inscribed circle and let $s = (a + b + c)/2$ be its semiperimeter.
    (a) Show that the area of the triangle is rs.
    (b) Show that $r = \sqrt{(s - a)(s - b)(s - c)/s}$.
    (c) Find r for a triangle with sides 5, 6, and 7.

25. Consider a triangle with sides of length 4, 5, and 6. Show that one of its angles is twice another. *Hint:* Show that the cosine of twice one angle is equal to the cosine of another angle.

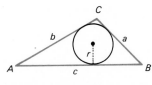

Figure 20

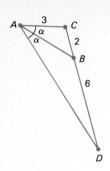

26. In the triangle with sides of length a, b, and c, let a_1, b_1, and c_1 denote the lengths of the corresponding medians from these sides to the opposite vertices. Show that

$$a_1^2 + b_1^2 + c_1^2 = \frac{3}{4}(a^2 + b^2 + c^2)$$

27. Determine the length of AB in Figure 21.

28. **TEASER** The two hands of a clock are 4 and 5 inches long, respectively. At some time between 1:45 and 2, the tips of the hands are 8 inches apart. What time is it then?

Figure 21

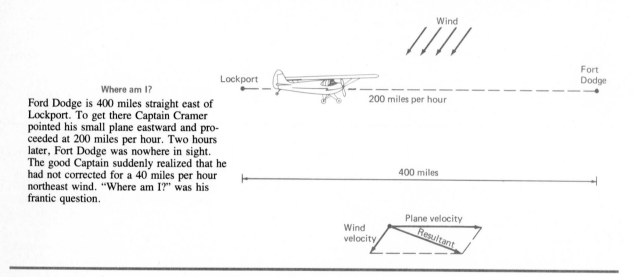

Where am I?

Ford Dodge is 400 miles straight east of Lockport. To get there Captain Cramer pointed his small plane eastward and proceeded at 200 miles per hour. Two hours later, Fort Dodge was nowhere in sight. The good Captain suddenly realized that he had not corrected for a 40 miles per hour northeast wind. "Where am I?" was his frantic question.

4-3   Vectors and Scalars

Many quantities that occur in science (for example, length, mass, volume, and electric charge) can be specified by giving a single number. We call these quantities *scalar quantities;* the numbers that measure their magnitudes are called **scalars.** Other quantities, such as velocity, force, torque, and displacement, must be specified by giving both a magnitude and a direction. We call such quantities **vectors** and represent them by arrows (directed line segments). The length of the arrow is the **magnitude** of the vector; its direction is the **direction** of the vector. Thus, in our opening diagram, the plane's velocity appears as an arrow 200 units long pointing eastward, while the wind velocity is shown as an arrow 40 units long pointing southwest. But how shall we put these two vectors together, that is, find their resultant? Before we try to answer, we introduce more terminology.

Tail   Head

Figure 22

Equivalent
vectors

Figure 23

Arrows that we draw, like those shot from a bow, have two ends. There is the initial or feather end, which we shall call the **tail**, and the pointed or terminal end, which we shall call the **head** (Figure 22). Two vectors are considered to be **equivalent** if they have the same magnitude and direction (Figure 23). We shall symbolize vectors by boldface letters, such as **u** and **v**. (Since this is hard to accomplish in normal writing, you might use $\vec{u}$ and $\vec{v}$.)

ADDITION OF VECTORS

To find the **sum,** or resultant, of **u** and **v**, move **v** without changing its magnitude or direction until its tail coincides with the head of **u**. Then **u** + **v** is the vector connecting the tail of **u** to the head of **v** (see the left diagram in Figure 24.)

As an alternative way to find **u** + **v**, move **v** so that its tail coincides with that of **u**. Then **u** + **v** is the vector with this common tail that is the diagonal of the parallelogram with **u** and **v** as sides. This method (called the *parallelogram law*) is illustrated on the right in Figure 24.

You should convince yourself that addition is commutative and associative, that is, **u** + **v** = **v** + **u** and (**u** + **v**) + **w** = **u** + (**v** + **w**).

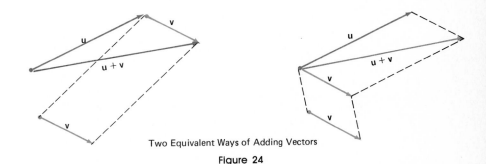

Two Equivalent Ways of Adding Vectors

Figure 24

SCALAR MULTIPLICATION AND SUBTRACTION

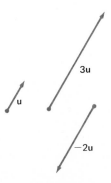

3u

u

−2u

Figure 25

If **u** is a vector, then 3**u** is the vector with the same direction as **u** but three times as long; −2**u** is twice as long as **u** and oppositely directed (Figure 25). More generally, $c\mathbf{u}$ has magnitude $|c|$ times that of **u** and is similarly or oppositely directed, depending on whether c is positive or negative. In particular, $(-1)\mathbf{u}$ (usually written −**u**) has the same length as **u** but the opposite direction. It is called the **negative** of **u** because when we add it to **u**, the result is a vector that has shriveled to a point. This special vector (the only vector without direction) is called the **zero vector** and is denoted by **0**. It is the identity element for addition; that is, **u** + **0** = **0** + **u** = **u**. Finally, subtraction is defined by

$$\mathbf{u} - \mathbf{v} = \mathbf{u} + (-\mathbf{v})$$

CAPTAIN CRAMER'S QUESTION

Consider again the problem posed in the display that opens this section. Physicists tell us that velocities add as vectors. Consequently, our first problem is to add **u**, which points east and is 200 units long, to a vector **v**, which points southwest and is 40 units long. Specifically, our aim is to find the length of **u** + **v**, denoted by ‖**u** + **v**‖, and the angle α that **u** + **v** makes with **u**. The situation is shown in Figure 26. Note that we interpret a northeast wind to mean that $\beta = 45°$.

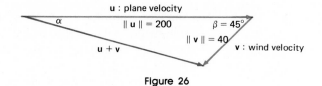

Figure 26

Now, by the law of cosines,

$$\|\mathbf{u} + \mathbf{v}\|^2 = (200)^2 + (40)^2 - 2(200)(40) \cos 45°$$

$$\approx 30{,}300$$

Thus

$$\|\mathbf{u} + \mathbf{v}\| \approx 174$$

Next, by the law of sines,

$$\frac{\sin \alpha}{40} = \frac{\sin 45°}{174}$$

or

$$\sin \alpha = (40 \sin 45°)/174 \approx .1626$$

$$\alpha \approx 9.4°$$

Where is Captain Cramer? Since his true velocity is the vector **u** + **v** and since he flew at this velocity for 2 hours, he is at a point 2(174) = 348 miles from Lockport along the line that makes an angle of 9.4° with the line between Lockport and Fort Dodge. Of course, he is also 80 miles southwest of Fort Dodge.

Problem Set 4-3

In Problems 1–4, draw the vector **w** *so that its tail is at the heavy dot.*

1. **w** = **u** + **v**

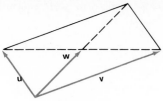

Figure 27

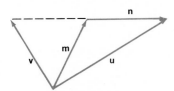

Figure 28

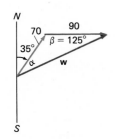

Figure 29

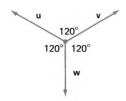

Figure 30

2. $\mathbf{w} = \mathbf{u} - \mathbf{v}$

3. $\mathbf{w} = -2\mathbf{u} + \frac{1}{2}\mathbf{v}$

4. $\mathbf{w} = \mathbf{u} - 3\mathbf{v}$

5. Figure 27 shows a parallelogram. Express $\mathbf{w}$ in terms of $\mathbf{u}$ and $\mathbf{v}$.
6. In the large triangle of Figure 28, $\mathbf{m}$ is a median (it bisects the side to which it is drawn). Express $\mathbf{m}$ and $\mathbf{n}$ in terms of $\mathbf{u}$ and $\mathbf{v}$.
7. In Figure 29, $\mathbf{w} = -(\mathbf{u} + \mathbf{v})$ and $\|\mathbf{u}\| = \|\mathbf{v}\| = 1$. Find $\|\mathbf{w}\|$.
8. Do Problem 7 if the top angle is 90° and the two side angles are 135°.

EXAMPLE A (Displacements Are Vectors) In navigation, directions are specified by giving an angle, called the **bearing,** with respect to a north-south line. Thus a bearing of N35°E denotes an angle whose initial side points north and whose terminal side is 35° east of north. If a ship sails 70 miles in the direction N35°E and then 90 miles straight east, what is its distance and bearing with respect to its starting point?

Solution. Our job is to determine the length and bearing of $\mathbf{w}$ (see Figure 30). We first use a little geometry to determine that $\beta = 125°$. Then, by the law of cosines,

$$\|\mathbf{w}\|^2 = (70)^2 + (90)^2 - 2(70)(90) \cos 125° \approx 20{,}227$$

$$\|\mathbf{w}\| \approx 142$$

By the law of sines,

$$\frac{\sin \alpha}{90} = \frac{\sin 125°}{142}$$

$$\sin \alpha = \frac{90 \sin 125°}{142} \approx .5192$$

$$\alpha \approx 31°$$

Thus the bearing of $\mathbf{w}$ is N66°E.

9. If I walk 10 miles N45°E and then 10 miles straight north, how far am I from my starting point?
10. In Problem 9, what is the bearing of my final position with respect to my starting point?
© 11. An airplane flew 100 kilometers in the direction S51°W and then 145 kilometers S39°W. What was the airplane's distance and bearing with respect to its starting point?

12. A ship sailed 11.2 miles straight north and then 48.3 miles N13.2°W. Find its distance and bearing with respect to the starting point.

EXAMPLE B (Velocities Are Vectors)   The river is flowing at 6 miles per hour and Jane's boat travels at 20 miles per hour in still water. In what direction should she head her boat if she wants to go straight across the river?

Solution.   Our job is simply to determine α in Figure 31.

$$\sin \alpha = \tfrac{6}{20} = .3000$$

$$\alpha \approx 17°$$

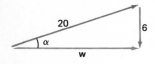

Figure 31

13. If the river (see Example B) is $\frac{1}{2}$ mile wide, how long will it take Jane to get across? *Hint:* First determine $\|\mathbf{w}\|$, which is her actual speed with respect to the shore.

14. If Jane (see Example B) had not corrected for the current (that is, if she had pointed her boat straight across), where would she have landed on the opposite shore?

15. A wind with velocity 58 miles per hour is blowing in the direction N20°W. An airplane which flies at 425 miles per hour in still air is supposed to fly straight north. How should the airplane be headed and how fast will it then be flying with respect to the ground?

16. A ship is sailing due south at 20 miles per hour. A man walks west (that is, at right angles to the side of the ship) across the deck at 3 miles per hour. What is the magnitude and direction of his velocity relative to the surface of the water?

EXAMPLE C (Forces Are Vectors)   A weight of 200 kilograms is supported by two wires, as shown in Figure 32. Find the magnitude of the tension in each wire.

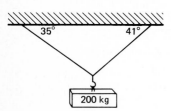

Figure 32

Solution.   The weight and the two tensions are forces which behave as vectors. These three forces must balance; that is, the two forces exerted by the wires must add together and cancel the downward force of the weight. This will happen if their sum is a vector of magnitude 200 pointing upward, as shown in Figure 33. This figure is a parallelogram composed of two congruent triangles. Using the given 35° and 41° angles and the fact that the angles of a triangle have a sum of 180°, we can find all the angles of the figure, as shown. By the law of sines,

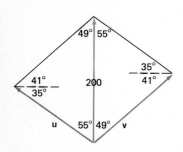

Figure 33

$$\frac{\|\mathbf{u}\|}{\sin 49°} = \frac{200}{\sin 76°}$$

$$\|\mathbf{u}\| = \frac{200 \sin 49°}{\sin 76°} \approx 156$$

Similarly,

$$\|\mathbf{v}\| = \frac{200 \sin 55°}{\sin 76°} \approx 169$$

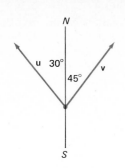

Figure 34

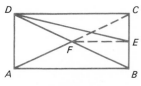

Figure 35

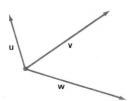

Figure 36

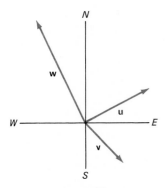

Figure 37

17. In Figure 34, $\|\mathbf{u}\| = \|\mathbf{v}\| = 10$. Find the magnitude and direction of a force $\mathbf{w}$ needed to counterbalance $\mathbf{u}$ and $\mathbf{v}$.

18. John pushes on a post from the direction S30°E with a force of 50 pounds. Wayne pushes on the same post from the direction S60°W with a force of 40 pounds. What is the magnitude and direction of the resultant force?

[c] 19. A body weighing 237.5 pounds is held in equilibrium by two ropes that make angles of 27.34° and 39.22°, respectively, with the vertical. Find the magnitude of the force exerted on the body by each rope.

20. A 250-kilogram weight rests on a smooth (friction negligible) inclined plane that makes an angle of 30° with the horizontal. What force parallel to the plane will just keep the weight from sliding down the plane? *Hint:* Consider the downward force of 250 kilograms to be the sum of two forces, one parallel to the plane and one perpendicular to it.

MISCELLANEOUS PROBLEMS

21. Draw the sum of the three vectors shown in Figure 35.

22. Draw $\mathbf{u} - \mathbf{v} + \frac{1}{2}\mathbf{w}$ for Figure 35.

23. Three vectors form the edges of a triangle and are oriented clockwise around it. What is their sum?

24. Four vectors each of length 1 point in the directions N, N30°E, N60°E, and E, respectively. Find the exact length and direction of their sum.

25. Refer to Figure 36. Express each of the following in terms of $\overrightarrow{AD}$ and $\overrightarrow{AB}$.
    (a) $\overrightarrow{BD}$ (b) $\overrightarrow{AF}$ (c) $\overrightarrow{DE}$ (d) $\overrightarrow{AF} - \overrightarrow{DE}$

26. Let $\mathbf{u}$, $\mathbf{v}$, and $\mathbf{w}$, be the vectors from the vertices of a triangle to the midpoints of the opposite edges (the medians). Show that $\mathbf{u} + \mathbf{v} + \mathbf{w} = \mathbf{0}$.

27. Alice and Bette left point P at the same time and met at point Q 2 hours later. To get there Alice walked a straight path but Bette first walked 1 mile south and then 2 miles in the direction S60°E. How fast did Alice walk, assuming that she walked at a constant rate?

[c] 28. Suppose that $\mathbf{u}$, $\mathbf{v}$, and $\mathbf{w}$ of Figure 37 point in the directions N60°E, S45°E, N25°W and have lengths 60, $30\sqrt{2}$, 100, respectively. Find the length of $\mathbf{u} + \mathbf{v} + \mathbf{w}$.

[c] 29. Two men are pushing an object along the ground. One is pushing with a force of 50 pounds in the direction N32°W and the other is pushing with a force of 100 pounds in the direction N30°E. In what direction is the object moving?

[c] 30. A pilot, flying in a wind which is blowing 80 miles per hour due south, discovers that she is heading due east when she points her plane in the direction N60°E. Find the air speed (speed in still air) of the plane.

[c] 31. What heading and air speed are required for a plane to fly 600 miles per hour due north if a wind of 56 miles per hour is blowing in the direction S12°E?

[c] 32. A spacecraft designed to softland on the moon has three legs whose feet form the vertices of an equilateral triangle on the ground. Each leg makes an angle of 35° with the vertical. If the impact force of 9000 pounds is evenly distributed, find the compression force on each leg.

[c] 33. What is the smallest force needed to keep a car weighing 3625 pounds from rolling down a hill that makes an angle of 10.35° with the horizontal?

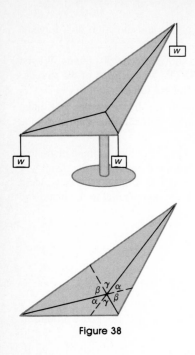

34. Work Example C a different way as follows. For equilibrium, the magnitude $\|\mathbf{u}\| \sin 55°$ of the leftward force must equal the magnitude $\|\mathbf{v}\| \sin 49°$ of the rightward force. Similarly, the downward force of 200 must just balance the upward force of $\|\mathbf{u}\| \cos 55° + \|\mathbf{v}\| \cos 49°$. Solve the resulting pair of equations for $\|\mathbf{u}\|$ and $\|\mathbf{v}\|$ and confirm that you get the same answers we got earlier.

35. Suppose as in Example C that a weight w is supported by two wires making angles α and $\beta = 60°$ with the ceiling and creating tensions of 90 pounds in the first wire and 75 pounds in the second wire. Determine α and w by reasoning as in Problem 34.

36. **TEASER** Consider a horizontal triangular table (Figure 38) with each vertex angle being less than 120°. Three strings are knotted together at P and pass over frictionless pulleys at the vertices. Identical weights w are attached to the free ends of the strings. Show that at equilibrium, the angles between the strings at P are equal, that is, show that $\alpha + \beta = \alpha + \gamma = \beta + \gamma$.

Figure 38

A New Look for an Old Vector

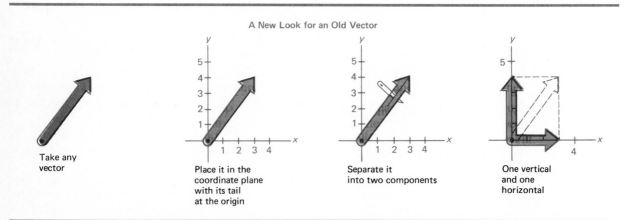

| Take any vector | Place it in the coordinate plane with its tail at the origin | Separate it into two components | One vertical and one horizontal |

4-4   The Algebra of Vectors

Our treatment of vectors in Section 4-3 was mainly geometric. To give the subject an algebraic appearance, we first suppose that all vectors have been placed in the ordinary cartesian coordinate plane with their tails attached to the origin, (0, 0). In this case, both the magnitude and the direction of a vector are completely determined by the position of its head.

Next we select two vectors to play a permanent and special role. The first, called $\mathbf{i}$, is the vector from (0, 0) to (1, 0); the second, called $\mathbf{j}$, is the vector from $(0, 0)$ to $(0, 1)$. Then, as Figure 39 makes clear, an arbitrary vector **u** with its head at (a, b) can be expressed uniquely in the form

$$\mathbf{u} = a\mathbf{i} + b\mathbf{j}$$

The vectors $a\mathbf{i}$ and $b\mathbf{j}$ are called the **horizontal** and **vertical vector components** of **u**, while a and b are called its **scalar components.** Notice that the length of **u** is easily expressed in terms of its scalar components: $\|\mathbf{u}\| = \sqrt{a^2 + b^2}$.

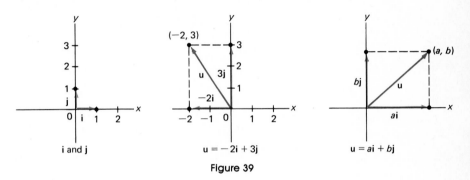

Figure 39

ALGEBRAIC OPERATIONS

To add the vectors $\mathbf{u} = a\mathbf{i} + b\mathbf{j}$ and $\mathbf{v} = c\mathbf{i} + d\mathbf{j}$, simply add the corresponding components; that is,

$$\mathbf{u} + \mathbf{v} = (a + c)\mathbf{i} + (b + d)\mathbf{j}$$

Similarly, to multiply **u** by the scalar k, multiply each component by k. Thus

$$k\mathbf{u} = (ka)\mathbf{i} + (kb)\mathbf{j}$$

To see that these new algebraic rules are equivalent to the old geometric ones, study Figure 40.

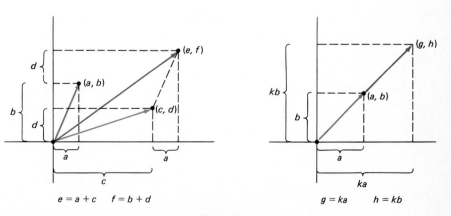

Figure 40

Once the rules for addition and scalar multiplication are established, the rule for subtraction follows easily.

$$\mathbf{u} - \mathbf{v} = \mathbf{u} + (-1)\mathbf{v} = a\mathbf{i} + b\mathbf{j} + (-1)(c\mathbf{i} + d\mathbf{j}) = (a - c)\mathbf{i} + (b - d)\mathbf{j}$$

Moreover, with vectors written in component form, it is a simple matter to establish all of the following properties. Keep in mind that $\mathbf{0} = 0\mathbf{i} + 0\mathbf{j}$.

ALGEBRAIC PROPERTIES OF VECTORS

1. $\mathbf{u} + \mathbf{v} = \mathbf{v} + \mathbf{u}$
2. $\mathbf{u} + (\mathbf{v} + \mathbf{w}) = (\mathbf{u} + \mathbf{v}) + \mathbf{w}$
3. $\mathbf{u} + \mathbf{0} = \mathbf{0} + \mathbf{u} = \mathbf{u}$
4. $\mathbf{u} + (-\mathbf{u}) = -\mathbf{u} + \mathbf{u} = \mathbf{0}$
5. $k(\mathbf{u} + \mathbf{v}) = k\mathbf{u} + k\mathbf{v}$
6. $(k + l)\mathbf{u} = k\mathbf{u} + l\mathbf{u}$
7. $(kl)\mathbf{u} = k(l\mathbf{u}) = l(k\mathbf{u})$
8. $1\mathbf{u} = \mathbf{u}$
9. $0\mathbf{u} = \mathbf{0} = k\mathbf{0}$

THE DOT PRODUCT

Is there a sensible way to multiply two vectors together? Yes; in fact, there are two kinds of products. One, called the vector product, requires three-dimensional space and therefore falls outside of the scope of this course. The other, called the **dot product** or **scalar product**, can be introduced now. If $\mathbf{u} = a\mathbf{i} + b\mathbf{j}$ and $\mathbf{v} = c\mathbf{i} + d\mathbf{j}$, then the dot product of $\mathbf{u}$ and $\mathbf{v}$ is the scalar given by

$$\mathbf{u} \cdot \mathbf{v} = ac + bd$$

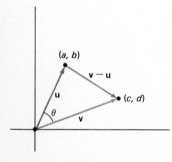

(a, b)

v − u

u

(c, d)

θ

v

Figure 41

Why would anyone be interested in the dot product? To answer this, we need its geometric interpretation. Suppose the heads of $\mathbf{u} = a\mathbf{i} + b\mathbf{j}$ and $\mathbf{v} = c\mathbf{i} + d\mathbf{j}$ are at (a, b) and (c, d), as shown in Figure 41. Then we may think of $\mathbf{v} - \mathbf{u}$ as the vector from (a, b) to (c, d). Let θ denote the smallest positive angle between $\mathbf{u}$ and $\mathbf{v}$. By the law of cosines,

$$\|\mathbf{v} - \mathbf{u}\|^2 = \|\mathbf{u}\|^2 + \|\mathbf{v}\|^2 - 2\|\mathbf{u}\|\,\|\mathbf{v}\| \cos \theta$$

$$(a - c)^2 + (b - d)^2 = a^2 + b^2 + c^2 + d^2 - 2\|\mathbf{u}\|\,\|\mathbf{v}\| \cos \theta$$

$$-2ac - 2bd = -2\|\mathbf{u}\|\,\|\mathbf{v}\| \cos \theta$$

$$ac + bd = \|\mathbf{u}\|\,\|\mathbf{v}\| \cos \theta$$

The last equality gives us a geometric formula for the dot product.

$$\mathbf{u} \cdot \mathbf{v} = \|\mathbf{u}\|\,\|\mathbf{v}\| \cos \theta$$

Of what use is this formula? For one thing, it gives us an easy way to tell when two vectors are perpendicular. Since $\cos \theta$ is zero if and only if θ is 90°, we see that:

Two vectors are perpendicular if and only if their dot product is zero.

More generally, we can use the formula to find the angle between any two vectors. For example, if $\mathbf{u} = 3\mathbf{i} + 4\mathbf{j}$ and $\mathbf{v} = -2\mathbf{i} + 3\mathbf{j}$, then

$$\cos \theta = \frac{\mathbf{u} \cdot \mathbf{v}}{\|\mathbf{u}\| \|\mathbf{v}\|} = \frac{(3)(-2) + (4)(3)}{\sqrt{9 + 16}\sqrt{4 + 9}} = \frac{6}{5\sqrt{13}} \approx .3328$$

We conclude that $\theta \approx 70.6°$.

Important applications of the dot product are to projections of one vector on another (see Example B) and to the concept of work from physics (see Example C).

Problem Set 4-4

In Problems 1–4, find $3\mathbf{u} - \mathbf{v}$, $\mathbf{u} \cdot \mathbf{v}$, and $\cos \theta$ for the given vectors $\mathbf{u}$ and $\mathbf{v}$.

1. $\mathbf{u} = 3\mathbf{i} - 4\mathbf{j}$, $\mathbf{v} = 5\mathbf{i} + 12\mathbf{j}$
2. $\mathbf{u} = \mathbf{i} + \sqrt{3}\mathbf{j}$, $\mathbf{v} = 6\mathbf{i} - 8\mathbf{j}$
3. $\mathbf{u} = 2\mathbf{i} - \mathbf{j}$, $\mathbf{v} = 3\mathbf{i} - 4\mathbf{j}$
4. $\mathbf{u} = \mathbf{i} + \mathbf{j}$, $\mathbf{v} = \mathbf{i} - \mathbf{j}$

5. If $\mathbf{u} = 14.1\mathbf{i} + 32.7\mathbf{j}$ and $\mathbf{v} = 19.2\mathbf{i} - 13.3\mathbf{j}$, find θ, the smallest positive angle between $\mathbf{u}$ and $\mathbf{v}$.

6. Determine the length of $2\mathbf{u} - 3\mathbf{v}$, where $\mathbf{u}$ and $\mathbf{v}$ are the vectors in Problem 5.

EXAMPLE A (Writing Vectors in the Form $a\mathbf{i} + b\mathbf{j}$)   Write the vector $\mathbf{w}$ from the point $P(1, 5)$ to the point $Q(6, 2)$ in the form $a\mathbf{i} + b\mathbf{j}$.

Solution.   We need the horizontal and vertical components of $\mathbf{w}$. They are obtained by subtracting the coordinates of the tail from those of the head. This gives $(6 - 1, 2 - 5)$ or $(5, -3)$, which are the coordinates of the head of $\mathbf{w}$ translated so that its tail is at the origin (Figure 42). Thus $\mathbf{w} = 5\mathbf{i} - 3\mathbf{j}$.

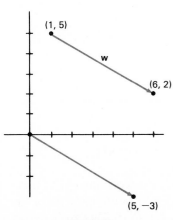

(1, 5)

w

(6, 2)

(5, −3)

Figure 42

In Problems 7–10, let $\mathbf{u}$ be the vector from P to Q and $\mathbf{v}$ be the vector from P to R. Write both vectors in the form $a\mathbf{i} + b\mathbf{j}$ and then find $\mathbf{u} \cdot \mathbf{v}$.

7. $P(1, 1)$, $Q(6, 3)$, $R(5, -2)$
8. $P(-1, 2)$, $Q(-3, 6)$, $R(0, -5)$
9. $P(1, 1)$, $Q(-3, -4)$, $R(-5, 6)$
10. $P(-1, -1)$, $Q(3, -5)$, $R(2, 4)$
11. If $\mathbf{u}$ is a vector 10 units long pointing in the direction N30°W, write $\mathbf{u}$ in the form $a\mathbf{i} + b\mathbf{j}$.

12. If **u** is a vector 9 units long pointing in the direction S21°W, write **u** in the form $a\mathbf{i} + b\mathbf{j}$.

13. Determine x so that $x\mathbf{i} + \mathbf{j}$ is perpendicular to $3\mathbf{i} - 4\mathbf{j}$.

14. Determine two vectors that are perpendicular to $2\mathbf{i} + 5\mathbf{j}$. *Hint:* Try $x\mathbf{i} + \mathbf{j}$ and $x\mathbf{i} - \mathbf{j}$.

15. Find a vector of unit length that has the same direction as $\mathbf{u} = 3\mathbf{i} - 4\mathbf{j}$. *Hint:* Try $\mathbf{u}/\|\mathbf{u}\|$.

16. Find two vectors of unit length that are perpendicular to $2\mathbf{i} + 3\mathbf{j}$. *Hint:* See Problems 14 and 15.

17. Find a vector twice as long as $\mathbf{u} = 2\mathbf{i} - 5\mathbf{j}$ and with opposite direction.

18. For any angle θ, show that $\mathbf{u} = (\cos\theta)\mathbf{i} + (\sin\theta)\mathbf{j}$ and $\mathbf{v} = (\sin\theta)\mathbf{i} - (\cos\theta)\mathbf{j}$ are perpendicular unit vectors.

19. Show that $\mathbf{u} \cdot \mathbf{u} = \|\mathbf{u}\|^2$ for any vector $\mathbf{u} = a\mathbf{i} + b\mathbf{j}$.

20. Show that $(k\mathbf{u}) \cdot \mathbf{v} = k(\mathbf{u} \cdot \mathbf{v})$.

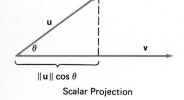

$\|\mathbf{u}\| \cos\theta$

Scalar Projection

Figure 43

EXAMPLE B (The Projection of One Vector on Another)   The scalar $\|\mathbf{u}\| \cos\theta$ is called the **scalar projection of u on v** for reasons that should be apparent from Figure 43. It is positive, zero, or negative, depending on whether θ is acute, right, or obtuse. If we multiply this scalar times $\mathbf{v}/\|\mathbf{v}\|$, we get the **vector projection of u on v.** Its geometric interpretation is shown in Figure 44. Derive formulas for these projections in terms of the dot product.

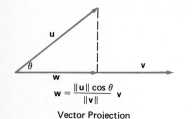

$\mathbf{w} = \dfrac{\|\mathbf{u}\| \cos\theta}{\|\mathbf{v}\|}\,\mathbf{v}$

Vector Projection

Figure 44

Solution.   Since

$$\|\mathbf{u}\| \cos\theta = \frac{\|\mathbf{u}\|\,\|\mathbf{v}\| \cos\theta}{\|\mathbf{v}\|}$$

it follows that

$$\text{scalar proj. } \mathbf{u} \text{ on } \mathbf{v} = \frac{\mathbf{u} \cdot \mathbf{v}}{\|\mathbf{v}\|}$$

$$\text{vector proj. } \mathbf{u} \text{ on } \mathbf{v} = \frac{\mathbf{u} \cdot \mathbf{v}}{\|\mathbf{v}\|^2}\,\mathbf{v} = \frac{\mathbf{u} \cdot \mathbf{v}}{\mathbf{v} \cdot \mathbf{v}}\,\mathbf{v}$$

In Problems 21–26, let $\mathbf{u} = 2\mathbf{i} + 9\mathbf{j}$, $\mathbf{v} = 4\mathbf{i} + 3\mathbf{j}$, *and* $\mathbf{w} = -5\mathbf{i} - 12\mathbf{j}$. *In each case, sketch the appropriate vectors and then find the indicated quantity.*

21. Scalar projection of **u** on **v**.

22. Vector projection of **u** on **v**.

23. Vector projection of **u** on **w**.

24. Scalar projection of **v** on **w**.

25. Scalar projection of **w** on **v**.

26. Vector projection of **v** on **w**.

EXAMPLE C (Work Done by a Force)   In physics, the **work** done by a force **F** in moving an object from P to Q is defined to be the product of the magnitude of that force times the distance from P to Q. This assumes that the force is in the direction of the motion. In the more general case where the force $\mathbf{F}$ is at an angle to the motion, we must replace the magnitude of $\mathbf{F}$ by its scalar projection in the direction of the motion. If both $\mathbf{F}$ and the displacement $\mathbf{D}$ are treated as vectors, the work done is

$$(\text{scalar proj. } \mathbf{F} \text{ on } \mathbf{D})\|\mathbf{D}\| = \frac{\mathbf{F} \cdot \mathbf{D}}{\|\mathbf{D}\|}\|\mathbf{D}\|$$

Thus

$$\boxed{\text{Work} = \mathbf{F} \cdot \mathbf{D}}$$

Use this to find the work done by a force of 80 pounds in the direction N60°E in moving an object from $(1, 0)$ to $(7, -2)$ as in Figure 45.

Solution.   It is simply a matter of writing $\mathbf{F}$ and $\mathbf{D}$ in the form $a\mathbf{i} + b\mathbf{j}$ and taking their dot product.

$$\mathbf{F} = 80 \cos 30°\mathbf{i} + 80 \sin 30°\mathbf{j}$$
$$= 40\sqrt{3}\mathbf{i} + 40\mathbf{j}$$
$$\mathbf{D} = 6\mathbf{i} - 2\mathbf{j}$$
$$\mathbf{F} \cdot \mathbf{D} = 240\sqrt{3} - 80 \approx 336$$

If the units of distance are feet, the work done is 336 foot-pounds.

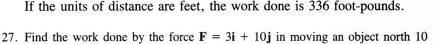

Figure 45

27. Find the work done by the force $\mathbf{F} = 3\mathbf{i} + 10\mathbf{j}$ in moving an object north 10 units.
28. Find the work done by a S70°E force of 100 dynes in moving an object 50 centimeters east.
29. Find the work done by a N45°E force of 50 dynes in moving an object from $(1, 1)$ to $(6, 9)$, with the distance measured in centimeters.
30. Find the work done by $\mathbf{F} = 3\mathbf{i} + 4\mathbf{j}$ in moving an object from $(0, 0)$ to $(-6, 0)$. Interpret the negative answer.

MISCELLANEOUS PROBLEMS

31. If $\mathbf{u} = 2\mathbf{i} + 3\mathbf{j}$ and $\mathbf{v} = 3\mathbf{i} + 4\mathbf{j}$, find $\|3\mathbf{u} - \mathbf{v}\|$.
32. For $\mathbf{u}$ and $\mathbf{v}$ as in Problem 31, find $\|\mathbf{u}\|$, $\|\mathbf{v}\|$, $\mathbf{u} \cdot \mathbf{v}$, and θ (the angle between $\mathbf{u}$ and $\mathbf{v}$).
33. Find two vectors of length 1 that are perpendicular to $3\mathbf{i} - 4\mathbf{j}$.
34. Find the vector $a\mathbf{i} + b\mathbf{j}$ that is 12 units long with the same direction as $3\mathbf{i} - 4\mathbf{j}$.
35. Find the vector projection of $5\mathbf{i} + 3\mathbf{j}$ on $3\mathbf{i} - 4\mathbf{j}$. Also find the angle between these two vectors.
36. Find the work done by a force of 100 pounds directed N45° E in moving an object along the line from $(1, 1)$ to $(7, 5)$, distances measured in feet.
37. Show that for any two vectors $\mathbf{u}$ and $\mathbf{v}$,

$$|\mathbf{u} \cdot \mathbf{v}| \leq \|\mathbf{u}\| \, \|\mathbf{v}\|$$

When will equality hold?

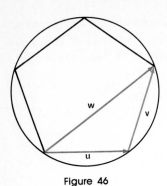

Figure 46

38. Prove that $\mathbf{u} \cdot \mathbf{v} = \mathbf{v} \cdot \mathbf{u}$ and that $\mathbf{u} \cdot (\mathbf{v} + \mathbf{w}) = \mathbf{u} \cdot \mathbf{v} + \mathbf{u} \cdot \mathbf{w}$.

39. If $\mathbf{u} + \mathbf{v}$ and $\mathbf{u} - \mathbf{v}$ are perpendicular, what can we conclude about $\|\mathbf{u}\|$ and $\|\mathbf{v}\|$?

40. Show that $\|\mathbf{u} + \mathbf{v}\|^2 + \|\mathbf{u} - \mathbf{v}\|^2 = 2(\|\mathbf{u}\|^2 + \|\mathbf{v}\|^2)$.

41. Find the exact value of $\sin 18°$, which may be needed to complete Problem 42. *Hint:* Let $\theta = 18°$ and note that $\cos 3\theta = \sin 2\theta = 2 \sin \theta \cos \theta$ and that $\cos 3\theta = \cos (2\theta + \theta) = (1 - 4 \sin^2 \theta) \cos \theta$.

42. **TEASER** Let $\mathbf{u}$ and $\mathbf{v}$ denote adjacent edges of a regular pentagon that is inscribed in a circle of radius 1 (Figure 46). Let $\mathbf{w} = \mathbf{u} + \mathbf{v}$.
    (a) Express $\mathbf{u} \cdot \mathbf{w}$ and $\|\mathbf{u}\| \|\mathbf{w}\|$ in terms of $\cos 36°$.
    (b) Use your calculator to guess the exact value of $(\|\mathbf{u}\| \|\mathbf{w}\|)^2$ and then prove that this is the correct value.

A Piston Problem

One end of an 8-foot shaft is attached to a piston that moves up and down. The other end is attached to a wheel by means of a horizontal slotted arm which fits over a peg P on the rim. Starting at an initial position of $\theta = \pi/4$, the wheel of radius 2 feet rotates at a rate of 3 radians per second. Find a formula for d, the vertical distance from the piston to the wheel center, after t seconds.

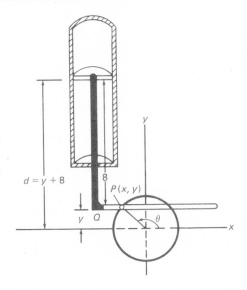

4-5   Simple Harmonic Motion

The up-and-down motion of the piston above is an example of what is called simple harmonic motion. Notice right away that the motion of the piston is essentially the same as that of the point Q. That means we want to find y; and y is just the y-coordinate of the peg P.

Perhaps the piston-wheel device seems complicated, so let's consider another version of the same problem. Imagine the wheel shown in Figure 47 to be turning at a uniform rate in the counterclockwise direction. Emanating from P, a point attached to the rim, is a horizontal beam of light, which projects a bright spot at Q on a nearby vertical wall. As the wheel turns, the spot at Q moves up and down. Our task is to express the y-coordinate of Q (which is also the y-coordinate of P) in terms of the elapsed time t.

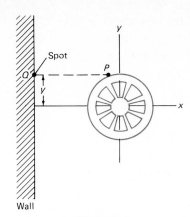

Figure 47

The solution to this problem depends on a number of factors (the rate at which the wheel turns, the radius of the wheel, and the location of P at $t = 0$). We think it wise to begin with a simple case and gradually extend to more general situations.

Case 1 Suppose the wheel has radius 1, that it turns at 1 radian per second, and that it starts at $\theta = 0$. Then at time t, θ will measure t radians and P will have y-coordinate

$$y = \sin t$$

(see Figure 48). Keep in mind that this equation describes the up-and-down motion of Q.

Case 2 Let everything be as in the first case, but now let the wheel turn at 3 radians per second (Figure 49). Then at time t, θ will measure $3t$ radians and both P and Q will have y-coordinate.

$$y = \sin 3t$$

Case 3 Next increase the radius of the wheel to 2 feet, but leave the other information as in Case 2 (Figure 50). Now the coordinates of P are $(2 \cos 3t, 2 \sin 3t)$ and

$$y = 2 \sin 3t$$

Case 4 Finally, let the wheel start at $\theta = \pi/4$ rather than $\theta = 0$. With the help of Figure 51, we see that

$$y = 2 \sin\left(3t + \frac{\pi}{4}\right)$$

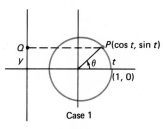

Case 1

Figure 48

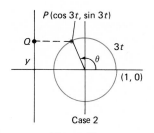

Case 2

Figure 49

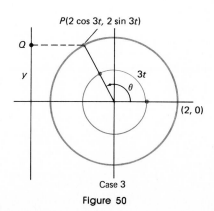

Case 3

Figure 50

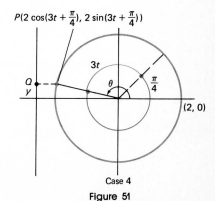

Case 4

Figure 51

Case 4 describes the wheel of the original piston-wheel problem. The number y measures the distance between Q and the x-axis, and $d = y + 8$ is the distance from the piston to the x-axis. Thus the answer to the question first posed is

$$d = 8 + 2 \sin\left(3t + \frac{\pi}{4}\right)$$

The number 8 does not interest us; it is the sine expression that is significant. As a matter of fact, equations of the form

$$y = A \sin(Bt + C) \quad \text{and} \quad y = A \cos(Bt + C)$$

with $B > 0$ arise often in physics. Any straight-line motion which can be described by one of these formulas is called **simple harmonic motion.** Cases 1–4 are examples of this motion. Other examples from physics occur in connection with the motion of a weight attached to a vibrating spring (Figure 52) and the motion of a water molecule in an ocean wave. Voltage in an alternating current, although it does not involve motion, is given by the same kind of sine (or cosine) equation.

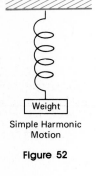

Weight

Simple Harmonic Motion

Figure 52

GRAPHS

The graphs of the four boxed equations given on page 145 are worthy of study. They are shown in Figure 53. Note how the graph of $y = \sin t$ is progressively modified as we move from Case 1 to Case 4.

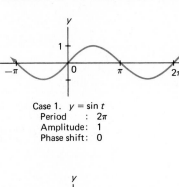

Case 1. $y = \sin t$
Period   : 2π
Amplitude: 1
Phase shift: 0

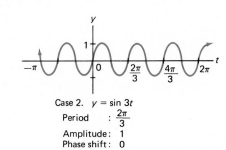

Case 2. $y = \sin 3t$
Period   : $\frac{2\pi}{3}$
Amplitude: 1
Phase shift: 0

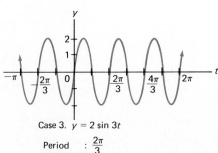

Case 3. $y = 2 \sin 3t$

Period   : $\frac{2\pi}{3}$

Amplitude : 2
Phase shift : 0

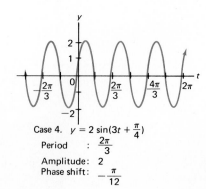

Case 4. $y = 2 \sin(3t + \frac{\pi}{4})$
Period   : $\frac{2\pi}{3}$
Amplitude: 2
Phase shift: $-\frac{\pi}{12}$

Figure 53

Under each graph are listed three important numbers, numbers that identify the critical features of the graph. The **period** is the length of the shortest interval after which the graph repeats itself. The **amplitude** is the maximum distance of the graph from its median position (the t-axis). The **phase shift** measures the distance the graph is shifted horizontally from its normal position.

You might have expected a phase shift of $-\pi/4$ in Case 4, since the initial angle of the wheel measured $\pi/4$ radians. But, note that factoring 3 from $3t + \pi/4$ gives

$$y = 2 \sin\left(3t + \frac{\pi}{4}\right) = 2 \sin 3\left(t + \frac{\pi}{12}\right)$$

If you recall our discussion of translations (see page 33), you see why the graph is shifted $\pi/12$ units to the left. Note in particular that $y = 0$ when $t = -\pi/12$ instead of when $t = 0$.

GRAPHING IN THE GENERAL CASE

If

$$y = A \sin(Bt + C) \quad \text{or} \quad y = A \cos(Bt + C)$$

with $B > 0$, all three concepts (period, amplitude, phase shift) make good sense. We have the following formulas.

Period:	$\dfrac{2\pi}{B}$
Amplitude:	$\lvert A \rvert$
Phase shift:	$\dfrac{-C}{B}$

Knowing these three numbers is a great aid in graphing. For example, to graph

$$y = 3 \cos\left(4t - \frac{\pi}{4}\right)$$

we recall the graph of $y = \cos t$ and then modify it using the three numbers.

$$\text{Period:} \qquad \frac{2\pi}{B} = \frac{2\pi}{4} = \frac{\pi}{2}$$

$$\text{Amplitude:} \qquad \lvert A \rvert = \lvert 3 \rvert = 3$$

$$\text{Phase shift:} \qquad -\frac{C}{B} = \frac{\pi/4}{4} = \frac{\pi}{16}$$

The result is shown in Figure 54 on the next page.

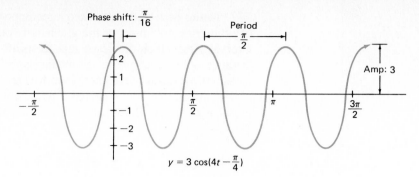

Phase shift: $\frac{\pi}{16}$

Period $\frac{\pi}{2}$

Amp: 3

$y = 3 \cos(4t - \frac{\pi}{4})$

Figure 54

Problem Set 4-5

1. Sketch the graphs of the following equations in the order given. Use the interval $-2\pi \leq t \leq 2\pi$.
   (a) $y = \cos t$ (b) $y = \cos 2t$
   (c) $y = 4 \cos 2t$ (d) $y = 4 \cos(2t + \pi/3)$

2. Sketch the graphs of the following on $-2\pi \leq t \leq 4\pi$.
   (a) $y = \sin t$ (b) $y = \sin \frac{1}{2}t$
   (c) $y = 3 \sin \frac{1}{2}t$ (d) $y = 3 \sin\left(\frac{1}{2}t + \pi/2\right)$

EXAMPLE A (More Graphing) Sketch the graph of $y = 3 \sin(\frac{1}{2}t + \pi/8)$.

Solution. We begin by finding the three key numbers.

$$\text{Period:} \qquad \frac{2\pi}{B} = \frac{2\pi}{\frac{1}{2}} = 4\pi$$

$$\text{Amplitude:} \quad |A| = |3| = 3$$

$$\text{Phase shift:} \quad \frac{-C}{B} = -\frac{\pi/8}{\frac{1}{2}} = -\frac{\pi}{4}$$

Then we draw the graph of Figure 55.

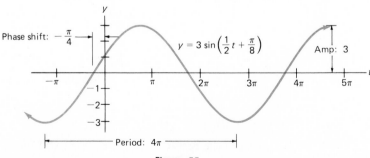

Phase shift: $-\frac{\pi}{4}$

$y = 3 \sin\left(\frac{1}{2}t + \frac{\pi}{8}\right)$

Amp: 3

Period: 4π

Figure 55

In Problems 3–6, find the period, amplitude, and phase shift. Then sketch the graph.

3. (a) $y = 4 \sin 2t$

   (b) $y = 3 \cos\left(t + \dfrac{\pi}{8}\right)$

   (c) $y = \sin\left(4t + \dfrac{\pi}{8}\right)$

   (d) $y = 3 \cos\left(3t - \dfrac{\pi}{2}\right)$

4. (a) $y = \dfrac{1}{2} \cos 3t$

   (b) $y = 3 \sin\left(t - \dfrac{\pi}{6}\right)$

   (c) $y = 2 \sin\left(\dfrac{1}{2}t + \dfrac{\pi}{8}\right)$

   (d) $y = \dfrac{1}{2} \sin(2t - 1)$

5. $y = 3 + 2 \cos\left(\frac{1}{2}t - \pi/16\right)$. *Hint:* The number 3 lifts the graph of $y = 2 \cos(\frac{1}{2}t - \pi/16)$ up 3 units.

6. $y = 4 + 3 \sin(2t + \pi/16)$

EXAMPLE B (Negative A)   Sketch the graph of $y = -3 \cos 2t$.

Solution.   We begin by asking how the graph of $y = -3 \cos 2t$ relates to that of $y = 3 \cos 2t$. Clearly, every y value has the opposite sign, which has the effect of reflecting the graph about the t-axis. Then we calculate the three crucial numbers.

$$\text{Period:} \qquad \frac{2\pi}{B} = \frac{2\pi}{2} = \pi$$

$$\text{Amplitude:} \qquad |A| = |-3| = 3$$

$$\text{Phase shift:} \qquad \frac{-C}{B} = \frac{0}{2} = 0$$

Finally we sketch the graph (Figure 56).

Now sketch the graphs of the equations in Problems 7–12.

7. $y = -2 \sin 3t$

8. $y = -4 \cos \frac{1}{2}t$

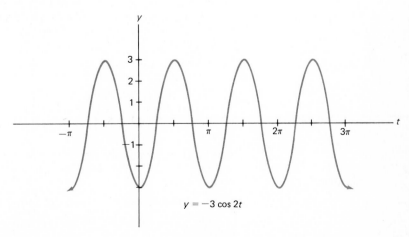

$y = -3 \cos 2t$

Figure 56

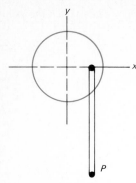

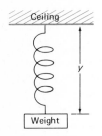

Figure 57

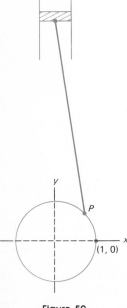

Figure 58

9. $y = \sin(2t - \pi/3)$

10. $y = -\cos(3t + \pi)$

11. $y = -2 \cos(t - \frac{1}{6})$

12. $y = -3 \sin(3t + 3)$

13. A wheel with center at the origin is rotating counterclockwise at 4 radians per second. There is a small hole in the wheel 5 centimeters from the center. If that hole has initial coordinates $(5, 0)$, what will its coordinates be after t seconds?

14. Answer Problem 13 if the hole is initially at $(0, 5)$.

15. A free-hanging shaft, 8 centimeters long, is attached to the wheel of Problem 13 by putting a bolt through the hole (Figure 57). What are the coordinates of P, the bottom point of the shaft, at time t (assuming the shaft continues to hang vertically)?

16. Suppose the wheel of Problem 13 rotates at 3 revolutions per second. What are the coordinates of the hole after t seconds?

MISCELLANEOUS PROBLEMS

17. Find the period, amplitude, and phase shift for each graph.
    (a) $y = \sin 5t$
    (b) $y = \frac{3}{2} \cos(\frac{1}{2}t)$
    (c) $y = 2 \cos(4t - \pi)$
    (d) $y = -4 \sin(3t + 3\pi/4)$

18. Sketch the graphs of the equations in Problem 17 on the interval $-\pi \le t \le 2\pi$.

19. The weight attached to a spring (Figure 58) is bobbing up and down so that

$$y = 8 + 4 \cos\left(\frac{\pi}{2}t + \frac{\pi}{4}\right)$$

where y and t are measured in feet and seconds, respectively. What is the closest the weight gets to the ceiling and when does this first happen for $t > 0$?

20. The equations $x = 2 + 2 \cos 4t$ and $y = 6 + 2 \sin 4t$ give the coordinates of a point moving along the circumference of a circle. Determine the center and radius of the circle. How long does it take for the point to make a complete revolution?

21. Consider the wheel-piston device shown in Figure 59 (which is analogous to the crankshaft and piston in an automobile engine). The wheel has a radius of 1 foot and rotates counterclockwise at 1 radian per second; the connecting rod is 5 feet long. If the point P is initially at $(1, 0)$, find the y-coordinate of Q after t seconds. Assume the x-coordinate is always zero.

22. Redo problem 21, but assume the wheel has radius 2 feet and rotates at 60 revolutions per second and that P is initially at $(2, 0)$. Is Q executing simple harmonic motion in either of these problems?

23. The voltage drop E across the terminals in a certain alternating current circuit is approximately $E = 156 \sin(110\pi t)$, where t is in seconds. What is the maximum voltage drop and what is the **frequency** (number of cycles per second) for this circuit?

24. The carrier wave for the radio wave of a certain FM station has the form $y = A \sin(2\pi \cdot 10^8 t)$, where t is measured in seconds. What is the frequency for this wave?

25. The AM radio wave for a certain station has the form

$$y = 55[1 + .02 \sin(2400\pi t)] \sin(2 \times 10^5 \pi t)$$

Figure 59

(a) Find y when $t = 3$.

(b) Find y when $t = .03216$.

[c] (c) Find y when $t = .0000321$.

26. In predator-prey systems, the number of predators and the number of prey tend to vary periodically. In a certain region with coyotes as predators and rabbits as prey, the rabbit population R varied according to the formula

$$R = 1000 + 150 \sin 2t$$

where t was measured in years after January 1, 1950.

(a) What was the maximum rabbit population?

(b) When was it first reached?

[c] (c) What was the population on January 1, 1953?

27. The number of coyotes C in Problem 26 satisfied

$$C = 200 + 50 \sin(2t - .7)$$

Sketch the graphs of C and R using the same coordinate system and attempt to explain the phase shift in C.

[c] 28. Sketch the graph of $y = 2^{-t} \cos 2t$ for $0 \le t \le 3\pi$. This is an example of damped harmonic motion, which is typical of harmonic motion where there is friction.

29. Use addition laws to write each of the following in the form $A_1 \sin Bt + A_2 \cos Bt$

(a) $4 \sin\left(2t - \dfrac{\pi}{4}\right)$ (b) $3 \cos\left(3t + \dfrac{\pi}{3}\right)$

Note: The same idea would work on any expression of the form $A \sin(Bt + C)$ or $A \cos(Bt + C)$.

30. Determine C so that

$$5 \sin 4t + 12 \cos 4t = 13 \sin(4t + C)$$

31. Suppose that A_1 and A_2 are both positive. Show that

$$A_1 \sin Bt + A_2 \cos Bt = A \sin(Bt + C)$$

where $A = \sqrt{A_1^2 + A_2^2}$ and $C = \tan^{-1}(A_2/A_1)$.

32. Generalize Problem 31 by showing that $A_1 \sin Bt + A_2 \cos Bt$ can always be written in the form $A \sin(Bt + C)$. *Hint:* Choose A as in Problem 31 and let C be the radian measure of an angle that has (A_1, A_2) on its terminal side.

33. Use the result in Problems 31 and 32 to write each of the following in the form $A \sin(Bt + C)$.

(a) $4 \cos 2t + 3 \sin 2t$ (b) $3 \sin 4t - \sqrt{3} \cos 4t$

34. Give an argument to show that

$$A_1 \sin Bt + A_2 \cos Bt, \qquad A_1 A_2 \ne 0, B \ne 0$$

is not a polynomial in t for any choices of A_1, A_2, and B.

35. Find the maximum and minimum values of $\cos t \pm \sin t$.

36. **TEASER**   Prove that $\sin(\cos t) < \cos(\sin t)$ for all t. *Hint:* First show that

$$-\frac{\pi}{2} < \cos t < \frac{\pi}{2} - |\sin t|$$

Jean-Robert Argand (1768-1822)

Though several mathematicians (for example, De Moivre, Euler, Gauss) had thought of complex numbers as points in the plane before Argand, this obscure Swiss bookkeeper gets credit for the idea. In 1806 he wrote a small book on the geometric representation of complex numbers. It was his only contribution to mathematics.

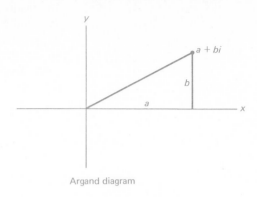

Argand diagram

4-6   Polar Representation of Complex Numbers

Throughout this book we have used the fact that a real number can be thought of as a point on a line. Now we are going to learn that a complex number can be represented as a point in the plane. This simple idea leads rather quickly to the fruitful notion of the polar form for a complex number. This in turn aids in the multiplication and division of complex numbers and greatly facilitates the calculation of powers and roots of complex numbers. Incidently, if you have forgotten the basic facts about complex numbers, you should review the end of Section 1-1 before going on.

COMPLEX NUMBERS AS POINTS IN THE PLANE

Consider a complex number $a + bi$. It is determined by the two real numbers a and b, that is, by the ordered pair (a, b). But (a, b), in turn, determines a point in the plane. That point we now label with the complex number $a + bi$. Thus $2 + 4i$, $2 - 4i$, $-3 + 2i$, and all other complex numbers may be used as labels for points in the plane (Figure 60). The plane with points labeled this way is called the **Argand diagram** or **complex plane.** Note that $3i = 0 + 3i$ labels a point on the y-axis, which we now call the **imaginary axis,** while $4 = 4 + 0i$ corresponds to a point on the x-axis (called the **real axis**).

Recall that the absolute value of a real number a (written $|a|$) is its distance from the origin on the real line. The concept of absolute value is extended to a complex number $a + bi$ by defining

$$|a + bi| = \sqrt{a^2 + b^2}$$

which is also its distance from the origin (Figure 61). Thus while there are only two real numbers with absolute value of 5, namely, -5 and 5, there are infinitely many complex numbers with absolute value 5. They include 5, -5,

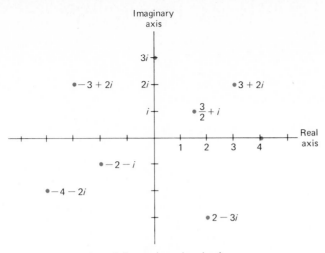

Argand diagram (complex plane)

Figure 60

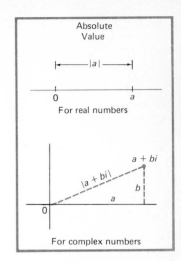

Absolute
Value

For real numbers

For complex numbers

Figure 61

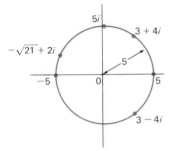

Figure 62

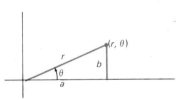

Figure 63

or

$5i, 3 + 4i, 3 - 4i, -\sqrt{21} + 2i$, and, in fact, all complex numbers on a circle of radius 5 centered at the origin (Figure 62).

POLAR FORM

There is another geometric way to describe complex numbers, a way that will prove very useful to us. For the complex number $a + bi$, which we have already identified with the point (a, b) in the plane, let r denote its distance from the origin and let θ be one of the angles that a ray from the origin through the point makes with the positive x-axis. Then from Figure 63 (or one of its analogues in another quadrant), we see that

$$a = r \cos \theta \qquad b = r \sin \theta$$

This means that we can write

$$a + bi = r \cos \theta + (r \sin \theta)i$$

$$\boxed{a + bi = r(\cos \theta + i \sin \theta)}$$

The boxed expression gives the **polar form** of $a + bi$. Notice that r is just the absolute value of $a + bi$; we shall refer to θ as its angle.

To put a number $a + bi$ in polar form, we use the formulas

$$r = \sqrt{a^2 + b^2} \qquad \cos \theta = \frac{a}{r}$$

For example, for $2\sqrt{3} - 2i$,

$$r = \sqrt{(2\sqrt{3})^2 + (-2)^2} = \sqrt{12 + 4} = 4$$

$$\cos\theta = \frac{2\sqrt{3}}{4} = \frac{\sqrt{3}}{2}$$

Since $2\sqrt{3} - 2i$ is in quadrant IV and $\cos\theta = \sqrt{3}/2$, θ can be chosen as an angle of $11\pi/6$ radians or $330°$. Thus

$$2\sqrt{3} - 2i = 4\left(\cos\frac{11\pi}{6} + i\sin\frac{11\pi}{6}\right)$$

$$= 4(\cos 330° + i\sin 330°)$$

For some numbers, finding the polar form is almost trivial. Just picture in your mind (Figure 64) where -6 and $4i$ are located in the complex plane and you will know that

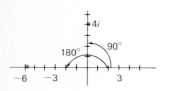

Figure 64

$$-6 = 6(\cos 180° + i\sin 180°)$$

$$4i = 4(\cos 90° + i\sin 90°)$$

Changing from the Cartesian form $a + bi$ to polar form is what we have just illustrated. Going in the opposite direction is much easier. For example, to change the polar form $3(\cos 240° + i\sin 240°)$ to Cartesian form, we simply calculate the sine and cosine of $240°$ and remove the parentheses.

$$3(\cos 240° + i\sin 240°) = 3\left(-\frac{1}{2} + i\frac{-\sqrt{3}}{2}\right)$$

$$= -\frac{3}{2} - \frac{3\sqrt{3}}{2}i$$

MULTIPLICATION AND DIVISION

The polar form is ideally suited for multiplying and dividing complex numbers. Let U and V be complex numbers given in polar form by

$$U = r(\cos\alpha + i\sin\alpha)$$

$$V = s(\cos\beta + i\sin\beta)$$

Then

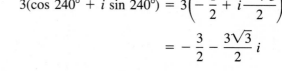

$$U \cdot V = rs[\cos(\alpha + \beta) + i\sin(\alpha + \beta)]$$

$$\frac{U}{V} = \frac{r}{s}[\cos(\alpha - \beta) + i\sin(\alpha - \beta)]$$

In words, to multiply two complex numbers, we multiply their absolute values and add their angles. To divide two complex numbers, we divide their absolute values and subtract their angles (in the correct order). Thus if

$$U = 4(\cos 75° + i\sin 75°)$$

$$V = 3(\cos 60° + i \sin 60°)$$

then

$$U \cdot V = 12(\cos 135° + i \sin 135°)$$

$$\frac{U}{V} = \frac{4}{3}(\cos 15° + i \sin 15°)$$

To establish the multiplication formula we use a bit of trigonometry.

$$U \cdot V = r(\cos \alpha + i \sin \alpha)s(\cos \beta + i \sin \beta)$$

$$= rs(\cos \alpha \cos \beta + i \cos \alpha \sin \beta + i \sin \alpha \cos \beta + i^2 \sin \alpha \sin \beta)$$

$$= rs[(\cos \alpha \cos \beta - \sin \alpha \sin \beta) + i(\sin \alpha \cos \beta + \cos \alpha \sin \beta)]$$

$$= rs[\cos(\alpha + \beta) + i \sin(\alpha + \beta)]$$

The key step was the last one, where we used the addition laws for the cosine and the sine.

You will be asked to establish the division formula in Problem 56.

GEOMETRIC ADDITION AND MULTIPLICATION

Having learned that the complex numbers can be thought of as points in a plane, we should not be surprised that the operations of addition and multiplication have a geometric interpretation. Let U and V be any two complex numbers; that is, let

$$U = a + bi = r(\cos \alpha + i \sin \alpha)$$

$$V = c + di = s(\cos \beta + i \sin \beta)$$

Addition is accomplished algebraically by adding the real parts and imaginary parts separately.

$$U + V = (a + c) + (b + d)i$$

To accomplish the same thing geometrically, we construct the parallelogram that has O, U, and V as three of its vertices (see Figure 65). Then $U + V$ corresponds to the vertex opposite the origin, as you should be able to show by finding the coordinates of this vertex.

To multiply algebraically, we use the polar forms of U and V, adding the angles and multiplying the absolute values.

$$U \cdot V = rs[\cos(\alpha + \beta) + i \sin(\alpha + \beta)]$$

To interpret this geometrically (for the case where α and β are between $0°$ and $180°$), first draw triangle OAU, where A is the point $1 + 0i$. Then construct triangle OVW similar to triangle OAU in the manner indicated in Figure 66. We claim that $W = U \cdot V$. Certainly W has the correct angle, namely, $\alpha + \beta$. Moreover, by similarity of triangles,

$$\frac{\overline{OW}}{\overline{OV}} = \frac{\overline{OU}}{\overline{OA}} = \frac{\overline{OU}}{1}$$

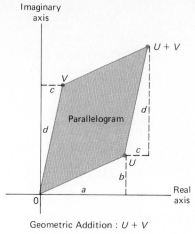

Geometric Addition : $U + V$

Figure 65

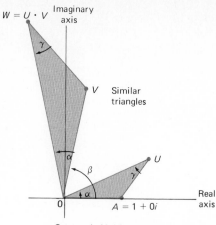

Geometric Multiplication : $U \cdot V$

Figure 66

(Here we are using $\overline{OW}$ for the length of the line segment from O to W.) Thus

$$|W| = \overline{OW} = \overline{OU} \cdot \overline{OV} = |U| \cdot |V|$$

so W also has the correct absolute value.

Problem Set 4-6

In Problems 1–12, plot the given numbers in the complex plane.

1. $2 + 3i$

2. $2 - 3i$

3. $-2 - 3i$

4. $-2 + 3i$

5. 5

6. -6

7. $-4i$

8. $6i$

9. $\frac{3}{5} - \frac{4}{5}i$

10. $-\frac{5}{13} + \frac{12}{13}i$

11. $2\left(\cos \dfrac{\pi}{4} + i \sin \dfrac{\pi}{4}\right)$

12. $3\left(\cos \dfrac{7\pi}{6} + i \sin \dfrac{7\pi}{6}\right)$

13. Find the absolute values of the numbers in Problems 1, 3, 5, 7, 9, and 11.

14. Find the absolute values of the numbers in Problems 2, 4, 6, 8, and 10.

Express each of the following in the form a + bi.

15. $4\left(\cos \dfrac{3\pi}{2} + i \sin \dfrac{3\pi}{2}\right)$

16. $5(\cos \pi + i \sin \pi)$

17. $2(\cos 225° + i \sin 225°)$

18. $\frac{3}{2}(\cos 300° + i \sin 300°)$

Express each of the following in polar form. For example, $1 + i = \sqrt{2} \ (\cos 45° + i \sin 45°)$.

19. -4	20. 9	21. $-5i$	22. $4i$
23. $2 - 2i$	24. $-5 - 5i$	25. $2\sqrt{3} + 2i$	26. $-4\sqrt{3} + 4i$

c 27. $5 + 4i$ c 28. $3 + 2i$

For Problems 29–36 let $u = 2(\cos 140° + i \sin 140°)$, $v = 3(\cos 70° + i \sin 70°)$ and $w = \frac{1}{2}(\cos 55° + i \sin 55°)$. Calculate each product or quotient, leaving your answer in polar form. For example,

$$\frac{u^2}{w} = \frac{u \cdot u}{w} = \frac{4(\cos 280° + i \sin 280°)}{\frac{1}{2}(\cos 55° + i \sin 55°)} = 8(\cos 225° + i \sin 225°)$$

29. uv	30. uw	31. vw	32. uvw
33. u/v	34. uv/w	35. $1/w$	36. $1/v$

EXAMPLE (Finding Products in Two Ways) Find the product

$$(\sqrt{3} + i)(-4 - 4\sqrt{3}i)$$

directly and then by using the polar form.

Solution. *Method 1* We use the definition of multiplication given in Section 1-1 to get

$$(\sqrt{3} + i)(-4 - 4\sqrt{3}i) = (-4\sqrt{3} + 4\sqrt{3}) + (-4 - 12)i$$

$$= -16i$$

Method 2 We change both numbers to polar form, multiply by the method of this section, and finally change to $a + bi$ form.

$$\sqrt{3} + i = 2(\cos 30° + i \sin 30°)$$

$$-4 - 4\sqrt{3}i = 8(\cos 240° + i \sin 240°)$$

$$(\sqrt{3} + i)(-4 - 4\sqrt{3}i) = 16(\cos 270° + i \sin 270°)$$

$$= 16(0 - i)$$

$$= -16i$$

Find each of the following products in two ways, giving your final answer in $a + bi$ form.

37. $(4 - 4i)(2 + 2i)$	38. $(\sqrt{3} + i)(2 - 2\sqrt{3}i)$
39. $(1 + \sqrt{3}i)(1 + \sqrt{3}i)$	40. $(\sqrt{2} + \sqrt{2}i)(\sqrt{2} + \sqrt{2}i)$

Find the following products and quotients, giving your answers in polar form. Start by changing each of the given complex numbers to polar form.

41. $4i(2\sqrt{3} - 2i)$ 42. $(-2i)(5 + 5i)$

43. $\dfrac{4i}{2\sqrt{3} - 2i}$ 44. $\dfrac{-2i}{5 + 5i}$

45. $(2\sqrt{2} - 2\sqrt{2}i)(2\sqrt{2} - 2\sqrt{2}i)$ 46. $(1 - \sqrt{3}i)(1 - \sqrt{3}i)$

MISCELLANEOUS PROBLEMS

47. Plot the given number in the complex plane and find its absolute value.
    (a) $-5 + 12i$ (b) $-4i$ (c) $5(\cos 60° + i \sin 60°)$

48. Express in the form $a + bi$.
    (a) $5[\cos(3\pi/2) + i \sin(3\pi/2)]$ (b) $4(\cos 180° + i \sin 180°)$
    (c) $2(\cos 315° + i \sin 315°)$ (d) $3[\cos(-2\pi/3) + i \sin(-2\pi/3)]$

49. Express in polar form.
    (a) 12 (b) $-\sqrt{2} + \sqrt{2}i$ (c) $-3i$
    (d) $2 - 2\sqrt{3}i$ (e) $4\sqrt{3} + 4i$ (f) $2(\cos 45° - i \sin 45°)$

50. Write in the form $a + bi$.
    (a) $2(\cos 37° + i \sin 37°)8(\cos 113° + i \sin 113°)$
    (b) $6(\cos 123° + i \sin 123°)/[3(\cos 33° + i \sin 33°)]$

51. Perform the indicated operations and write your answer in polar form.
    (a) $1.5(\cos 110° + i \sin 110°)4(\cos 30° + i \sin 30°)2(\cos 20° + i \sin 20°)$

    (b) $\dfrac{12(\cos 115° + i \sin 115°)}{4(\cos 55° + i \sin 55°)(\cos 20° + i \sin 20°)}$

    (c) $\dfrac{(-\sqrt{2} + \sqrt{2}i)(2 - 2\sqrt{3}i)}{4\sqrt{3} + 4i}$ (See Problem 49.)

52. Calculate and write your answer in the form $a + bi$.
    (a) $\dfrac{(-\sqrt{2} + \sqrt{2}i)^2(2 - 2\sqrt{3}i)^3}{4\sqrt{3} + 4i}$ (See Problem 49.)
    (b) $(-1 + \sqrt{3}i)^5(-1 - \sqrt{3}i)^{-4}$

53. In each case, find two values for $z = a + bi$.
    (a) The imaginary part of z is 5 and $|z| = 13$.
    (b) The number z lies on the line $y = x$ and $|z| = 8$.

54. Find four complex numbers $z = a + bi$ that are located on the hyperbola $y^2 - x^2 = 2$ and satisfy $|z| = \sqrt{10}$.

55. Let $u = r(\cos \theta + i \sin \theta)$. Write each of the following in polar form.
    (a) u^3 (b) $\bar{u}$ ($\bar{u}$ is the conjugate of u)
    (c) $u\bar{u}$ (d) $1/u$
    (e) u^{-2} (f) $-u$

56. Prove the division formula: If $U = r(\cos \alpha + i \sin \alpha)$ and $V = s(\cos \beta + i \sin \beta)$, then
$$\frac{U}{V} = \frac{r}{s}[\cos(\alpha - \beta) + i \sin(\alpha - \beta)]$$

57. Let U and V be complex numbers. Give a geometric interpretation for (a) $|U - V|$ and (b) the angle of $U - V$.

58. By expanding $(\cos \theta + i \sin \theta)^3$ in two different ways, derive the formulas.
    (a) $\cos 3\theta = 4 \cos^3 \theta - 3 \cos \theta$
    (b) $\sin 3\theta = -4 \sin^3 \theta + 3 \sin \theta$

59. Let $z_k = 2[\cos(k\pi/4) + i \sin(k\pi/4)]$. Find the exact value of each of the following.
    (a) $z_1 z_2 z_3 \cdots z_8$
    (b) $z_1 + z_2 + z_3 + \cdots + z_8$ (Think geometrically.)

60. **TEASER** Let $z_k = [k/(k + 1)](\cos k° + i \sin k°)$. Find the exact value of the product
$$z_1 z_2 z_3 \cdots z_{179} z_{180}$$

Abraham De Moivre (1667–1754)

Though he was a Frenchman, De Moivre spent most of his life in London. There he became an intimate friend of the great Isaac Newton, inventor of calculus. De Moivre made many contributions to mathematics but his reputation rests most securely on the theorem that bears his name.

$$(\cos \theta + i \sin \theta)^n$$
$$= \cos n\theta + i \sin n\theta$$

De Moivre's Theorem

4-7   Powers and Roots of Complex Numbers

De Moivre's theorem tells us how to raise a complex number of absolute value 1 to an integral power. We can easily extend it to cover the case of any complex number, no matter what its absolute value. Then with a little work, we can use it to find roots of complex numbers. Here we are in for a surprise. Take the number $8i$, for example. After some fumbling around, we find that one of its cube roots is $-2i$, because $(-2i)^3 = -8i^3 = 8i$. We shall find that it has two other cube roots (both nonreal numbers). In fact, we shall see that every nonzero number has exactly three cube roots, four 4th roots, five 5th roots, and so on. To put it in a spectacular way, we claim that any nonzero number, for example $37 + 3.5i$, has 1,000,000 millionth roots.

POWERS OF COMPLEX NUMBERS

To raise the complex number $r(\cos \theta + i \sin \theta)$ to the nth power, n a positive integer, we simply find the product of n factors of $r(\cos \theta + i \sin \theta)$. But from Section 4-6, we know that we multiply complex numbers by multiplying their absolute values and adding their angles. Thus

$$[r(\cos \theta + i \sin \theta)]^n$$
$$= \underbrace{r \cdot r \cdots r}_{n \text{ factors}} [\cos (\underbrace{\theta + \theta + \cdots + \theta}_{n \text{ terms}}) + i \sin (\theta + \theta + \cdots \theta)]$$

In short,

$$[r(\cos \theta + i \sin \theta)]^n = r^n(\cos n\theta + i \sin n\theta)$$

When $r = 1$, this is De Moivre's theorem.

As a first illustration, let us find the 6th power of a complex number that is already in polar form.

$$\left[2\left(\cos\frac{\pi}{6} + i\sin\frac{\pi}{6}\right)\right]^6 = 2^6\left[\cos\left(6\cdot\frac{\pi}{6}\right) + i\sin\left(6\cdot\frac{\pi}{6}\right)\right]$$

$$= 64(\cos\pi + i\sin\pi)$$

$$= 64(-1 + i\cdot 0)$$

$$= -64$$

To find $(1 - \sqrt{3}i)^5$, we could use repeated multiplication of $1 - \sqrt{3}i$ by itself. But how much better to change $1 - \sqrt{3}i$ to polar form and use the boxed formula at the bottom of page 159.

$$1 - \sqrt{3}i = 2(\cos 300° + i\sin 300°)$$

Then

$$(1 - \sqrt{3}i)^5 = 2^5(\cos 1500° + i\sin 1500°)$$

$$= 32(\cos 60° + i\sin 60°)$$

$$= 32\left(\frac{1}{2} + i\frac{\sqrt{3}}{2}\right)$$

$$= 16 + 16\sqrt{3}i$$

THE THREE CUBE ROOTS OF 8i

Because finding roots is tricky, we begin with an example before attempting the general case. We have already noted that $-2i$ is one cube root of $8i$, but now we claim there are two others. How shall we find them? We begin by writing $8i$ in polar form.

$$8i = 8(\cos 90° + i\sin 90°)$$

Finding cube roots is the opposite of cubing. That suggests that we take the real cube root (rather than the cube) of 8 and divide (rather than multiply) the angle 90° by 3. This would give us one cube root

$$2(\cos 30° + i\sin 30°)$$

which reduces to

$$2\left(\frac{\sqrt{3}}{2} + \frac{1}{2}i\right) = \sqrt{3} + i$$

Is this really a cube root of $8i$? For fear that you might be suspicious of the polar form, we will cube it the old-fashioned way and check.

$$(\sqrt{3} + i)^3 = (\sqrt{3} + i)(\sqrt{3} + i)(\sqrt{3} + i)$$

$$= [(3 - 1) + 2\sqrt{3}i](\sqrt{3} + i)$$

$$= 2(1 + \sqrt{3}i)(\sqrt{3} + i)$$

$$= 2(0 + 4i)$$
$$= 8i$$

Of course, the check using polar form is more direct.

$$[2(\cos 30° + i \sin 30°)]^3 = 2^3(\cos 90° + i \sin 90°)$$
$$= 8(0 + i)$$
$$= 8i$$

The process described above yielded one cube root of $8i$ (namely, $\sqrt{3} + i$); there are two others. Let us go back to our representation of $8i$ in polar form. We used the angle 90°; we could as well have used $90° + 360° = 450°$.

$$8i = 8(\cos 450° + i \sin 450°)$$

Now if we take the real cube root of 8 and divide 450° by 3 we get

$$2(\cos 150° + i \sin 150°) = 2\left(-\frac{\sqrt{3}}{2} + \frac{1}{2}i\right) = -\sqrt{3} + i$$

We could again check that this is indeed a cube root of $8i$.

What worked once might work twice. Let us write $8i$ in polar form in a third way, this time adding 2(360°) to its angle of 90°.

$$8i = 8(\cos 810° + i \sin 810°)$$

The corresponding cube root is

$$2(\cos 270° + i \sin 270°) = 2(0 - i) = -2i$$

This does not come as a surprise, since we knew that $-2i$ was one of the cube roots of $8i$.

If we add 3(360°) (that is, 1080°) to 90°, do we get still another cube root of $8i$? No, for if we write

$$8i = 8(\cos 1170° + i \sin 1170°)$$

the corresponding cube root of $8i$ would be

$$2(\cos 390° + i \sin 390°) = 2(\cos 30° + i \sin 30°)$$

But this is the same as the first cube root we found. The truth is that we have found all the cube roots of $8i$, namely, $\sqrt{3} + i$, $-\sqrt{3} + i$, and $-2i$.

Let us summarize. The number $8i$ has three cube roots given by

$$2\left[\cos\left(\frac{90°}{3}\right) + i \sin\left(\frac{90°}{3}\right)\right]$$

$$2\left[\cos\left(\frac{90° + 360°}{3}\right) + i \sin\left(\frac{90° + 360°}{3}\right)\right]$$

$$2\left[\cos\left(\frac{90° + 720°}{3}\right) + i \sin\left(\frac{90° + 720°}{3}\right)\right]$$

We can say the same thing in a shorter way by writing

$$2\left[\cos\left(\frac{90° + k \cdot 360°}{3}\right) + i\sin\left(\frac{90° + k \cdot 360°}{3}\right)\right] \qquad k = 0, 1, 2$$

ROOTS OF COMPLEX NUMBERS

We are ready to generalize. If $u \neq 0$, then

$$u = r(\cos\theta + i\sin\theta)$$

has n distinct nth roots $u_0, u_1, \ldots, u_{n-1}$ given by

$$u_k = \sqrt[n]{r}\left[\cos\left(\frac{\theta + k \cdot 360°}{n}\right) + i\sin\left(\frac{\theta + k \cdot 360°}{n}\right)\right]$$

$$k = 0, 1, 2, \ldots, n - 1$$

Recall that $\sqrt[n]{r}$ denotes the positive real nth root of $r = |u|$. In our example, it was $\sqrt[3]{|8i|} = \sqrt[3]{8} = 2$. To see that each value of u_k is an nth root, simply raise it to the nth power. In each case, you should get u.

The boxed formula assumes that θ is given in degrees. If θ is in radians, the formula takes the following form.

$$u_k = \sqrt[n]{r}\left[\cos\left(\frac{\theta + 2k\pi}{n}\right) + i\sin\left(\frac{\theta + 2k\pi}{n}\right)\right]$$

$$k = 0, 1, 2, \ldots, n - 1$$

A REAL EXAMPLE

Let us use the boxed formula to find the six 6th roots of 64. (Keep in mind that a real number is a special kind of complex number.) Changing to polar form, we write

$$64 = 64(\cos 0° + i\sin 0°)$$

Applying the formula with $r = |64| = 64$, $\theta = 0°$, and $n = 6$ gives

$$u_0 = 2(\cos 0° + i\sin 0°) = 2$$
$$u_1 = 2(\cos 60° + i\sin 60°) = 1 + \sqrt{3}i$$
$$u_2 = 2(\cos 120° + i\sin 120°) = -1 + \sqrt{3}i$$
$$u_3 = 2(\cos 180° + i\sin 180°) = -2$$
$$u_4 = 2(\cos 240° + i\sin 240°) = -1 - \sqrt{3}i$$
$$u_5 = 2(\cos 300° + i\sin 300°) = 1 - \sqrt{3}i$$

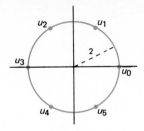

Figure 67

Notice that two of the roots, 2 and −2, are real; the other four are not real.

If you plot these six numbers (Figure 67), you will find that they lie on a circle of radius 2 centered at the origin and that they are equally spaced around the circle. This is typical of what happens in general.

Problem Set 4-7

Find each of the following, leaving your answer in polar form.

1. $\left[2\left(\cos\dfrac{\pi}{4} + i \sin\dfrac{\pi}{4}\right)\right]^3$
2. $\left[3\left(\cos\dfrac{5\pi}{6} + i \sin\dfrac{5\pi}{6}\right)\right]^2$
3. $[\sqrt{5}(\cos 11° + i \sin 11°)]^6$
4. $[\frac{1}{3}(\cos 12.5° + i \sin 12.5°)]^4$
5. $(1 + i)^8$
6. $(1 - i)^4$

Find each of the following powers. Write your answer in a + bi form.

7. $(\cos 36° + i \sin 36°)^{10}$
8. $(\cos 27° + i \sin 27°)^{10}$
9. $(\sqrt{3} + i)^5$
10. $(2 - 2\sqrt{3}i)^4$

Find the nth roots of u for the given u and n, leaving your answers in polar form. Plot these roots in the complex plane.

11. $u = 125(\cos 45° + i \sin 45°)$; $n = 3$
12. $u = 81(\cos 80° + i \sin 80°)$; $n = 4$
13. $u = 64\left(\cos\dfrac{\pi}{2} + i \sin\dfrac{\pi}{2}\right)$; $n = 6$
14. $u = 3^8\left(\cos\dfrac{2\pi}{3} + i \sin\dfrac{2\pi}{3}\right)$; $n = 8$
15. $u = 4(\cos 112° + i \sin 112°)$; $n = 4$
16. $u = 7(\cos 200° + i \sin 200°)$; $n = 5$

Find the nth roots of u for the given u and n. Write your answers in the a + bi form.

17. $u = 16$, $n = 4$
18. $u = -16$, $n = 4$
19. $u = 4i$, $n = 2$
20. $u = -27i$, $n = 3$
21. $u = -4 + 4\sqrt{3}i$, $n = 2$
22. $u = -2 - 2\sqrt{3}i$, $n = 4$

EXAMPLE (Roots of Unity) The *n*th roots of 1, called the **nth roots of unity,** play an important role in advanced algebra. Find the five 5th roots of unity, plot them, and show that four of the roots are powers of the 5th root.

Solution. First we represent 1 in polar form.

$$1 = 1(\cos 0° + i \sin 0°)$$

The five 5th roots are (according to the formula developed in this section)

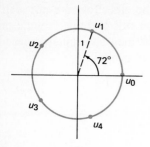

Figure 68

$$u_0 = \cos 0° + i \sin 0° = 1$$

$$u_1 = \cos 72° + i \sin 72°$$

$$u_2 = \cos 144° + i \sin 144°$$

$$u_3 = \cos 216° + i \sin 216°$$

$$u_4 = \cos 288° + i \sin 288°$$

These roots are plotted in Figure 68. They lie on the unit circle and are equally spaced around it. Finally notice that

$$u_1 = u_1$$

$$u_2 = u_1^2$$

$$u_3 = u_1^3$$

$$u_4 = u_1^4$$

$$u_0 = u_1^5$$

Thus all the roots are powers of u_1. These powers of u_1 repeat in cycles of 5. For example, note that

$$u_1^6 = u_1^5 u_1 = u_1$$

$$u_1^7 = u_1^5 u_1^2 = u_1^2$$

In each of the following, find all the nth roots of unity for the given n and plot them in the complex plane.

23. $n = 4$ 24. $n = 6$ 25. $n = 10$ 26. $n = 12$

MISCELLANEOUS PROBLEMS

27. Calculate each of the following, leaving your answer in polar form.
   (a) $[3(\cos 20° + i \sin 20°)]^4$
   (b) $[2.46(\cos 1.54 + i \sin 1.54)]^5$
   (c) $[2(\cos 50° + i \sin 50°)(\cos 30° + i \sin 30°)]^3$
   (d) $\left(\dfrac{8[\cos(2\pi/3) + i \sin(2\pi/3)]}{4[\cos(\pi/4) + i \sin(\pi/4)]} \right)^4$

28. Change to polar form, calculate, and then change back to $a + bi$ form.
   (a) $(1 - \sqrt{3}i)^5$
   (b) $[(\sqrt{3} + i)(2 - 2i)/(-1 + \sqrt{3}i)]^3$

29. Find the five 5th roots of $32(\cos 255° + i \sin 255°)$, giving your answers in polar form.

30. Find the three cube roots of $-4\sqrt{2} - 4\sqrt{2}i$, giving your answers in polar form.

31. Write the eight 8th roots of 1 in $a + bi$ form and calculate their sum and product.

32. Solve the equation $x^3 - 4 - 4\sqrt{3}i = 0$. You may give your answers in polar form.

33. Find the solution to $x^5 + \sqrt{2} - \sqrt{2}i = 0$ with the largest real part. Write your answer in the form $a + bi$.

34. Solve the equation $x^3 + 27 = 0$ in two ways.
   (a) By finding the three cube roots of $-27i$;
   (b) By writing $x^3 + 27 = (x + 3)(x^2 - 3x + 9)$ and using the quadratic formula.

35. Find the six solutions to $x^6 - 1 = 0$ by two different methods.

36. Show that $\cos(\pi/3) + i \sin(\pi/3)$ is a solution to $2x^4 + x^2 + x + 1 = 0$

37. Find all six solutions to $x^6 + x^4 + x^2 + 1 = 0$. *Hint:* The left side can be factored as $(x^2 + 1)(x^4 + 1)$.

38. Show that DeMoivre's theorem is valid when n is a negative integer.

39. If A is a nonreal number, we agree that $\sqrt{A}$ stands for the one of the two square roots with nonnegative real part. For example, the two square roots of $-4 + 4\sqrt{3}i$ are $\sqrt{2} + \sqrt{6}i$ and $-\sqrt{2} - \sqrt{6}i$, but we agree that
$$\sqrt{-4 + 4\sqrt{3}i} = \sqrt{2} + \sqrt{6}i$$
Evaluate.
   (a) $\sqrt{1 + \sqrt{3}i}$ (b) $\sqrt{-1 + \sqrt{3}i}$

40. The quadratic formula is valid even for quadratic equations with nonreal coefficients if we follow the agreement of Problem 39. Solve the following equations.
   (a) $x^2 - 2x + \sqrt{3}i = 0$
   (b) $x^2 - 4ix - 5 + \sqrt{3}i = 0$

41. Let n be an integer that is not divisible by 3. Simplify
$$(-1 + \sqrt{3}i)^n + (-1 - \sqrt{3}i)^n$$
as much as possible.

42. **TEASER**   Let $1, u, u^2, u^3, \ldots, u^{15}$ be the sixteen 16th roots of unity. Calculate each of the following. Look for a simple way in each case.
   (a) $1 + u + u^2 + u^3 + \cdots + u^{15}$
   (b) $1 \cdot u \cdot u^2 \cdot u^3 \cdots u^{15}$
   (c) $(1 - u)(1 - u^2)(1 - u^3) \cdots (1 - u^{15})$
   (d) $(1 + u)(1 + u^2)(1 + u^4)(1 + u^8)(1 + u^{16})$

Chapter Summary

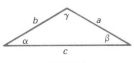

Figure 69

A triangle like the one in Figure 69 that has no right angle is called an oblique triangle. If any three of the six parts α, β, γ, a, b, and c—including at least one side—are given, we can find the remaining parts by using the **law of sines**

$$\frac{a}{\sin \alpha} = \frac{b}{\sin \beta} = \frac{c}{\sin \gamma}$$

and the **law of cosines**

$$a^2 = b^2 + c^2 - 2bc \cos \alpha \qquad \text{(one of three forms)}$$

There is, however, one case (given two sides and an angle opposite one of them) in which there might be no solution or two solutions. We call it the ambiguous case.

Vectors, represented by arrows, play an important role in science because they have both **magnitude** (length) and **direction**. We call two vectors **equivalent** if they have the same magnitude and direction. Vectors can be added (using the parallelogram law) and multiplied by **scalars**, which are real numbers. The special vectors **i** and **j** allow us to write any vector in the form $a\mathbf{i} + b\mathbf{j}$, where a and b are scalars. If $\mathbf{u} = a\mathbf{i} + b\mathbf{j}$ is a vector, its **length** $\|\mathbf{u}\|$ is given by

$$\|\mathbf{u}\| = \sqrt{a^2 + b^2}$$

If $v = c\mathbf{i} + d\mathbf{j}$ is another such vector, the **dot product** $\mathbf{u} \cdot \mathbf{v}$ of **u** and **v** is

$$\mathbf{u} \cdot \mathbf{v} = ac + bd = \|\mathbf{u}\|\,\|\mathbf{v}\| \cos \theta$$

where θ is the smallest positive angle between **u** and **v**. Two vectors are perpendicular if and only if their dot product is zero.

The equations $y = A \sin(Bt + C)$ and $y = A \cos(Bt + C)$ describe a common phenomenon known as **simple harmonic motion.** We can quickly draw the graphs of these equations by making use of three key numbers: the **amplitude,** $|A|$, the **period,** $2\pi/B$, and the **phase shift,** $-C/B$.

A complex number $a + bi$ can be represented geometrically as a point (a, b) in a plane called the **complex plane,** or **Argand diagram.** The horizontal and vertical axes are known as the **real axis** and **imaginary axis,** respectively. The distance from the origin to (a, b) is $\sqrt{a^2 + b^2}$; it is also the absolute value of $a + bi$, denoted as with real numbers by $|a + bi|$. If we let $r = \sqrt{a^2 + b^2}$ and θ be the angle that a ray from the origin through (a, b) makes with the positive x-axis, we obtain the **polar form** of $a + bi$, namely,

$$r(\cos \theta + i \sin \theta)$$

This form facilitates multiplication and division and is especially helpful in finding powers and roots of a number. An important result is the formula

$$[r(\cos \theta + i \sin \theta)]^n = r^n(\cos n\theta + i \sin n\theta)$$

Chapter Review Problem Set

1. Solve each of the following triangles using Table C or a calculator.
   (a) $\alpha = 104.9°$, $\gamma = 36°$, $b = 149$
   (b) $a = 14.6$, $b = 89.2$, $c = 75.8$
   (c) $\gamma = 35°$, $a = 14$, $b = 22$
   (d) $\beta = 48.6°$, $c = 39.2$, $b = 57.6$

2. For the triangle in Figure 70, find x and the area of the triangle.

3. If **u** is a vector 10 units long pointing in the direction N 30° E and $\mathbf{v} = 3\mathbf{i} - 9\mathbf{j}$, calculate $\mathbf{u} + \frac{1}{3}\mathbf{v}$ and write it in the form $a\mathbf{i} + b\mathbf{j}$.

4. If $\mathbf{u} = 3\mathbf{i} - 4\mathbf{j}$ and $\mathbf{v} = 5\mathbf{i} + 12\mathbf{j}$, calculate each of the following.
   (a) $\|\mathbf{u}\|$ (b) $\|\mathbf{v}\|$ (c) $\mathbf{u} \cdot \mathbf{v}$
   (d) The angle θ between **u** and **v**.

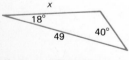

Figure 70

(e) The scalar projection of **u** on **v**.

(f) The vector projection of **u** on **v**.

5. Two men, A and B, are pushing an object along the ground. A pushes with a force **u** of 100 pounds in the direction N(arctan $\frac{4}{3}$)E; B pushes with a force **v** of 40 pounds straight east.

   (a) Find the resultant force $\mathbf{w} = \mathbf{u} + \mathbf{v}$ and write it in the form $a\mathbf{i} + b\mathbf{j}$.

   (b) Calculate the work done by A when the object is moved a distance of 2 feet in the direction of **w**.

6. Find the period, amplitude, and phase shift for the graph of each equation.

   (a) $y = \cos 2t$ (b) $y = 3 \cos 4t$

   (c) $y = 2 \sin(3t - \pi/2)$ (d) $y = -2 \sin(\frac{1}{2}t + \pi)$

7. Sketch the graphs of the equations in Problem 6 on $-\pi \le t \le \pi$.

8. A wheel of radius 4 feet with center at the origin is rotating counterclockwise at $3\pi/4$ radians per second (Figure 71). If a paint speck P has coordinates $(-4, 0)$ initially, what will its coordinates be after t seconds?

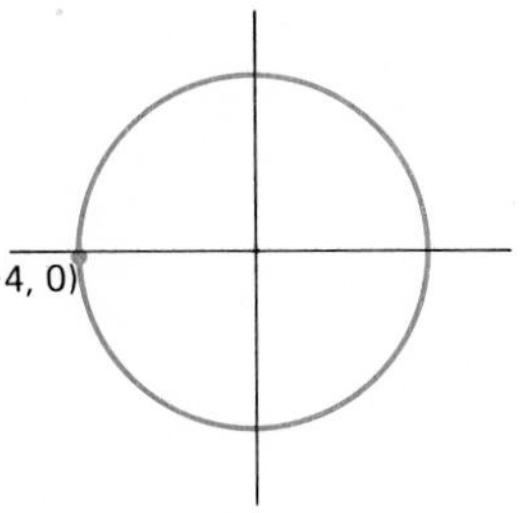

9. A point is moving in a straight line according to the equation $x = 3 \cos(5t + 3\pi)$, where x is in feet and t in seconds.

   (a) What is the period of the motion?

   (b) Find the initial position of the point relative to the point $x = 0$.

   (c) When is $x = 3$ for the first time?

10. Plot the following numbers in the complex plane.

    (a) $3 - 4i$ (b) -6 (c) $5i$

    (d) $3\left(\cos\dfrac{3\pi}{4} + i \sin\dfrac{3\pi}{4}\right)$ (e) $4(\cos 300° + i \sin 300°)$

11. Find the absolute value of each number in Problem 10.

12. Express $4(\cos 150° + i \sin 150°)$ in the form $a + bi$.

13. Express in polar form.

    (a) $3i$ (b) -6 (c) $-1 - i$ (d) $2\sqrt{3} - 2i$

14. Let

$$u = 8(\cos 105° + i \sin 105°)$$

$$v = 4(\cos 40° + i \sin 40°)$$

Calculate each of the following, leaving your answer in polar form.

    (a) uv (b) u/v (c) u^3 (d) u^2v^3

15. Find all the 6th roots of

$$2^6(\cos 120° + i \sin 120°)$$

leaving your answers in polar form.

16. Find the five solutions to $x^5 - 1 = 0$.

$P(-4, 0)$

Figure 71

The method of logarithms, by reducing to a few days the labor of many months, doubles as it were, the life of the astronomer, besides freeing him from the errors and disgust inseparable—from long calculation.

P. S. Laplace

CHAPTER 5

Exponential and Logarithmic Functions

Some Sense—Some Nonsense

One of the authors once asked a student to write the definition of $2^{4.6}$ on the blackboard. After thinking deeply for a minute, he wrote:

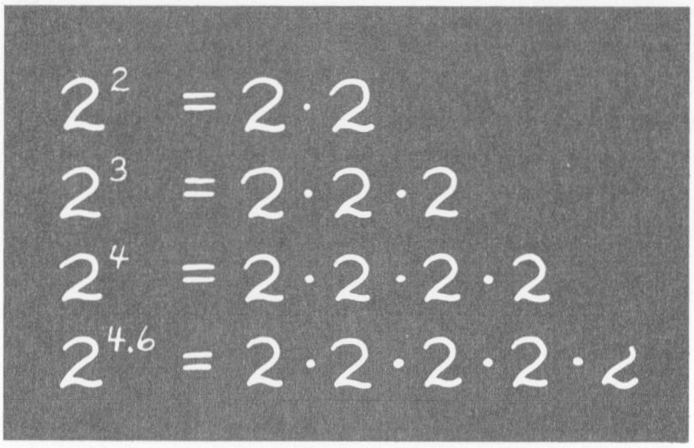

2² = 2·2
2³ = 2·2·2
2⁴ = 2·2·2·2
2^4.6 = 2·2·2·2·2

5-1 Exponents and Exponential Functions

After you have criticized the student mentioned above, ask yourself how you would define $2^{4.6}$. Of course, integral powers of 2 make perfectly good sense. Undoubtedly, you know that $2^3 = 2 \cdot 2 \cdot 2$, $2^0 = 1$, $2^{-3} = 1/2^3$, and more generally for n a positive integer,

$$a^n = a \cdot a \cdot a \cdots a \qquad (n \text{ factors})$$

$$a^0 = 1$$

$$a^{-n} = \frac{1}{a^n}$$

These definitions lead to the familiar rules for exponents listed in Figure 1. Now we want to attach meaning to powers like $2^{1/2}$, $2^{4.6}$, and even 2^{π}. Moreover, we want to do this in such a way that the rules for exponents continue to hold.

Rules for Exponents

1. $a^m a^n = a^{m+n}$
2. $\dfrac{a^m}{a^n} = a^{m-n}$
3. $(a^m)^n = a^{mn}$

Figure 1

RATIONAL EXPONENTS

We assume throughout this section that $a > 0$. If n is any positive integer, we want

$$(a^{1/n})^n = a^{(1/n) \cdot n} = a^1 = a$$

But we know that $(\sqrt[n]{a})^n = a$. Thus we define

$$a^{1/n} = \sqrt[n]{a}$$

Recall that if n is even and $a \geq 0$, $\sqrt[n]{a}$ denotes the nonnegative nth root of a. Thus $\sqrt[4]{16} = 2$ even though both -2 and 2 are fourth roots of 16. We conclude that $(16)^{1/4} = \sqrt[4]{16} = 2$. Also, $(27)^{1/3} = 3$ and $(.09)^{1/2} = .3$.

Next, if m and n are positive integers, we want

$$(a^{1/n})^m = a^{m/n} \quad \text{and} \quad (a^m)^{1/n} = a^{m/n}$$

This forces us to define

$$a^{m/n} = (\sqrt[n]{a})^m = \sqrt[n]{a^m}$$

Accordingly,

$$2^{3/2} = (\sqrt{2})^3 = \sqrt{2}\sqrt{2}\sqrt{2} = 2\sqrt{2}$$

and

$$27^{2/3} = (\sqrt[3]{27})^2 = 3^2 = 9$$

Lastly, we define

$$a^{-m/n} = \frac{1}{a^{m/n}}$$

so that

$$2^{-1/2} = \frac{1}{2^{1/2}} = \frac{1}{\sqrt{2}}$$

and

$$4^{-3/2} = \frac{1}{4^{3/2}} = \frac{1}{(\sqrt{4})^3} = \frac{1}{8}$$

We have just succeeded in defining a^x for all rational numbers x (recall that a rational number is a ratio of two integers). What is more important is that we have done it in such a way that the rules of exponents still hold. Incidentally, we can now answer the question in our opening display.

$$2^{4.6} = 2^4 2^{.6} = 2^4 2^{6/10} = 16(\sqrt[10]{2})^6$$

For simplicity, we have assumed that a is positive in our discussion of $a^{m/n}$. But we should point out that the definition of $a^{m/n}$ given above is also appropriate for the case in which a is negative and n is odd. For example,

$$(-27)^{2/3} = (\sqrt[3]{-27})^2 = (-3)^2 = 9$$

REAL EXPONENTS

Irrational powers such as 2^π and $3^{\sqrt{2}}$ are intrinsically more difficult to define than are rational powers. Rather than attempt a technical definition, we ask you to consider what 2^π might mean. The decimal expansion of π is $3.14159\ldots$. Thus we could look at the sequence of rational powers

$$2^3, \ 2^{3.1}, \ 2^{3.14}, \ 2^{3.141}, \ 2^{3.1415}, \ 2^{3.14159}, \ \ldots$$

As you should suspect, when the exponents get closer and closer to π, the corresponding powers of 2 get closer and closer to a definite number. We shall call the number 2^π.

The process of starting with integral exponents and then extending to rational exponents and finally to real exponents can be clarified by means of three graphs. Note the table of values (Figure 2) at the left.

The first graph suggests a curve rising from left to right. The second graph makes the suggestion stronger. The third graph leaves nothing to the imagination; it is a continuous curve and it shows 2^x for all values of x, rational and irrational. As x increases in the positive direction, the values of 2^x increase without bound; in the negative direction, the values of 2^x approach 0. Notice that 2^π is a little less than 9; its value correct to seven decimal places is

$$2^\pi = 8.8249778$$

See if your calculator gives this value.

x	2^x
-3	$\dfrac{1}{8}$
-2	$\dfrac{1}{4}$
-1	$\dfrac{1}{2}$
0	1
$\dfrac{1}{2}$	$\sqrt{2} \approx 1.4$
1	2
$\dfrac{3}{2}$	$2\sqrt{2} \approx 2.8$
2	4
3	8

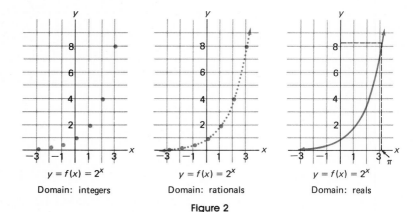

$y = f(x) = 2^x$
Domain: integers

$y = f(x) = 2^x$
Domain: rationals

$y = f(x) = 2^x$
Domain: reals

Figure 2

EXPONENTIAL FUNCTIONS

The function $f(x) = 2^x$, graphed above, is one example of an exponential function. But what has been done with 2 can be done with any positive real number a. In general, the formula

$$f(x) = a^x$$

determines a function called an **exponential function with base a.** Its domain is the set of all real numbers and its range is the set of positive numbers.

Let us see what effect the size of a has on the graph of $f(x) = a^x$. We choose $a = 2$, $a = 3$, $a = 5$, and $a = \frac{1}{3}$, showing all four graphs in Figure 3. The graph of $f(x) = 3^x$ looks much like the graph of $f(x) = 2^x$, although it rises more rapidly. The graph of $f(x) = 5^x$ is even steeper. All three of these functions are *increasing functions*, meaning that the values of $f(x)$ increase as x increases; more formally, $x_2 > x_1$ implies $f(x_2) > f(x_1)$. The function $f(x) = (\frac{1}{3})^x$, on the other hand, is a *decreasing function*. In fact, you can get the graph of $f(x) = (\frac{1}{3})^x$ by reflecting the graph of $f(x) = 3^x$ about the y-axis. This is because $(\frac{1}{3})^x = 3^{-x}$.

We can summarize what is suggested by our discussion as follows.

If $a > 1$, $f(x) = a^x$ is an increasing function.
If $0 < a < 1$, $f(x) = a^x$ is a decreasing function.

In both of these cases, the graph of f has the x-axis as an asymptote. The case $a = 1$ is not very interesting since it yields the constant function $f(x) = 1$.

x	3^x	$(\frac{1}{3})^x$
-2	1/9	9
-1	1/3	3
0	1	1
1	3	1/3
2	9	1/9

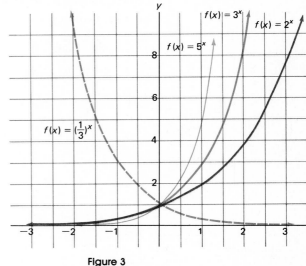

Figure 3

PROPERTIES OF EXPONENTIAL FUNCTIONS

It is easy to describe the main properties of exponential functions, since they include the three rules already mentioned. Perhaps it is worth restating them, since we do want to emphasize that they now hold for all *real* exponents x and y (at least for the case where a and b are both positive).

CAUTION

$(-8)^{-1/3} = 8^{1/3} = 2$

$(-8)^{-1/3} = \dfrac{1}{(-8)^{1/3}} = \dfrac{1}{-2}$

1. $a^x a^y = a^{x+y}$

2. $\dfrac{a^x}{a^y} = a^{x-y}$

3. $(a^x)^y = a^{xy}$

4. $(ab)^x = a^x b^x$

5. $\left(\dfrac{a}{b}\right)^x = \dfrac{a^x}{b^x}$

Here are a number of examples that are worth studying.

$$3^{1/2}3^{3/4} = 3^{1/2+3/4} = 3^{5/4}$$

$$\frac{\pi^4}{\pi^{5/2}} = \pi^{4-5/2} = \pi^{3/2}$$

$$(2^{\sqrt{3}})^4 = 2^{4\sqrt{3}}$$

$$(5^{\sqrt{2}}5^{1-\sqrt{2}})^{-3} = (5^1)^{-3} = 5^{-3}$$

Problem Set 5-1

Write each of the following as a power of 7.

1. $\sqrt[3]{7}$ 2. $\sqrt[5]{7}$ 3. $\sqrt[3]{7^2}$ 4. $\sqrt[5]{7^3}$ 5. $\dfrac{1}{\sqrt[3]{7}}$

6. $\dfrac{1}{\sqrt[5]{7}}$ 7. $\dfrac{1}{\sqrt[3]{7^2}}$ 8. $\dfrac{1}{\sqrt[5]{7^3}}$ 9. $7\sqrt[3]{7}$ 10. $7\sqrt[5]{7}$

Rewrite each of the following using exponents instead of radicals. For example,
$\sqrt[5]{x^3} = x^{3/5}$.

11. $\sqrt[3]{x^2}$ 12. $\sqrt[4]{x^3}$ 13. $x^2\sqrt{x}$ 14. $x\sqrt[3]{x}$
15. $\sqrt{(x + y)^3}$ 16. $\sqrt[3]{(x + y)^2}$ 17. $\sqrt{x^2 + y^2}$ 18. $\sqrt[3]{x^3 + 8}$

Rewrite each of the following using radicals instead of fractional exponents. For
example, $(xy^2)^{3/7} = \sqrt[7]{x^3y^6}$

19. $4^{2/3}$ 20. $10^{3/4}$ 21. $8^{-3/2}$
22. $12^{-5/6}$ 23. $(x^4 + y^4)^{1/4}$ 24. $(x^2 + xy)^{1/2}$
25. $(x^2y^3)^{2/5}$ 26. $(3ab^2)^{2/3}$ 27. $(x^{1/2} + y^{1/2})^{1/2}$
28. $(x^{1/3} + y^{2/3})^{1/3}$

Simplify each of the following. Give your answer without any exponents.

29. $25^{1/2}$ 30. $27^{1/3}$
31. $8^{2/3}$ 32. $16^{3/2}$
33. $9^{-3/2}$ 34. $64^{-2/3}$
35. $(-.008)^{2/3}$ 36. $(-.027)^{5/3}$
37. $(.0025)^{3/2}$ 38. $(1.44)^{3/2}$
39. $5^{2/3}5^{-5/3}$ 40. $4^{3/4}4^{-1/4}$
41. $16^{7/6}16^{-5/6}16^{-4/3}$ 42. $9^29^{2/3}9^{-7/6}$
43. $(8^2)^{-2/3}$ 44. $(4^{-3})^{3/2}$

EXAMPLE A (Simplifying Expressions Involving Exponents) Simplify and
write the answer without negative exponents.

(a) $\dfrac{x^{1/3}(8x)^{-2/3}}{x^{-3/4}}$ (b) $\left(\dfrac{2x^{-1/2}}{y}\right)^{4}\left(\dfrac{x}{y}\right)^{-1}(3x^{10/3})$

Solution.

(a) $\dfrac{x^{1/3}(8x)^{-2/3}}{x^{-3/4}} = x^{1/3}8^{-2/3}x^{-2/3}x^{3/4} = \dfrac{x^{1/3-2/3+3/4}}{8^{2/3}} = \dfrac{x^{5/12}}{4}$

(b) $\left(\dfrac{2x^{-1/2}}{y}\right)^{4}\left(\dfrac{x}{y}\right)^{-1}(3x^{10/3}) = \left(\dfrac{16x^{-2}}{y^{4}}\right)\left(\dfrac{y}{x}\right)(3x^{10/3})$

$$= \dfrac{48x^{-2-1+10/3}}{y^{4-1}} = \dfrac{48x^{1/3}}{y^{3}}$$

Simplify, writing your answer without negative exponents.

45. $(3a^{1/2})(-2a^{3/2})$ 46. $(2x^{3/4})(5x^{-3/4})$

47. $(2^{1/2}x^{-2/3})^{6}$ 48. $(\sqrt{3}x^{-1/4}y^{3/4})^{4}$

49. $(xy^{-2/3})^{3}(x^{1/2}y)^{2}$ 50. $(a^{2}b^{-1/4})^{2}(a^{-1/3}b^{1/2})^{3}$

51. $\dfrac{(2x^{-1}y^{2/3})^{2}}{x^{2}y^{-2/3}}$ 52. $\left(\dfrac{a^{1/2}b^{1/3}}{c^{5/6}}\right)^{12}$

53. $\left(\dfrac{x^{-2}y^{3/4}}{x^{1/2}}\right)^{12}$ 54. $\dfrac{x^{1/3}y^{-3/4}}{x^{-2/3}y^{1/2}}$

55. $y^{2/3}(2y^{4/3} - y^{-5/3})$ 56. $x^{-3/4}\left(-x^{7/4} + \dfrac{2}{\sqrt[4]{x}}\right)$

57. $(x^{1/2} + y^{1/2})^{2}$ 58. $(a^{3/2} + \pi)^{2}$

EXAMPLE B (Combining Fractions)   Perform the following addition.

$$\dfrac{(x+1)^{2/3}}{x} + \dfrac{1}{(x+1)^{1/3}}$$

Solution.

$$\dfrac{(x+1)^{2/3}}{x} + \dfrac{1}{(x+1)^{1/3}} = \dfrac{(x+1)^{2/3}(x+1)^{1/3}}{x(x+1)^{1/3}} + \dfrac{x}{x(x+1)^{1/3}}$$

$$= \dfrac{x+1+x}{x(x+1)^{1/3}} = \dfrac{2x+1}{x(x+1)^{1/3}}$$

Combine the fractions in each of the following.

59. $\dfrac{(x+2)^{4/5}}{3} + \dfrac{2x}{(x+2)^{1/5}}$ 60. $\dfrac{(x-3)^{1/3}}{4} - \dfrac{1}{(x-3)^{2/3}}$

61. $(x^{2}+1)^{1/3} - \dfrac{2x^{2}}{(x^{2}+1)^{2/3}}$ 62. $(x^{2}+2)^{1/4} + \dfrac{x^{2}}{(x^{2}+2)^{3/4}}$

EXAMPLE C (Mixing Radicals of Different Orders)   Express $\sqrt{2}\,\sqrt[3]{5}$ using just one radical.

Solution.   Square roots and cube roots mix about as well as oil and water, but exponents can serve as a blender. They allow us to write both $\sqrt{2}$ and $\sqrt[3]{5}$ as sixth roots.

$$\sqrt{2}\,\sqrt[3]{5} = 2^{1/2} \cdot 5^{1/3}$$
$$= 2^{3/6} \cdot 5^{2/6}$$
$$= (2^3 \cdot 5^2)^{1/6}$$
$$= \sqrt[6]{200}$$

Express each of the following in terms of at most one radical in simplest form.

63. $\sqrt{2}\,\sqrt[3]{2}$ 64. $\sqrt[3]{2}\,\sqrt[3]{2}$ 65. $\sqrt[4]{2}\,\sqrt[6]{x}$

66. $\sqrt[3]{5}\,\sqrt{x}$ 67. $\sqrt[3]{x}\,\sqrt{x}$ 68. $\sqrt{x\sqrt[3]{x}}$

c *Use a calculator to find an approximate value of each of the following.*

69. $2^{1.34}$ 70. $2^{-.79}$ 71. $\pi^{1.34}$ 72. π^{π}

73. $(1.46)^{\sqrt{2}}$ 74. $\pi^{\sqrt{2}}$ 75. $(.9)^{50.2}$ 76. $(1.01)^{50.2}$

Sketch the graph of each of the following functions.

77. $f(x) = 4^x$ 78. $f(x) = 4^{-x}$ 79. $f(x) = (\tfrac{2}{3})^x$

80. $f(x) = (\tfrac{2}{3})^{-x}$ 81. $f(x) = \pi^x$ 82. $f(x) = (\sqrt{2})^x$

MISCELLANEOUS PROBLEMS

83. Rewrite using exponents in place of radicals and simplify.
   (a) $\sqrt[5]{b^3}$ (b) $\sqrt[8]{x^4}$ (c) $\sqrt[3]{a^2 + 2ab + b^2}$

84. Simplify.
   (a) $(32)^{-6/5}$ (b) $(-.008)^{2/3}$ (c) $(5^{-1/2}\,5^{3/4}\,5^{1/8})^{16}$

85. Simplify, writing your answer without either radicals or negative exponents.
   (a) $(27)^{2/3}(.0625)^{-3/4}$ (b) $\sqrt[3]{4}\,\sqrt{2} + \sqrt[6]{2}$
   (c) $\sqrt[3]{a^2}\,\sqrt[4]{a^3}$ (d) $\sqrt{a\sqrt[3]{a^2}}$
   (e) $[a^{3/2} + a^{-3/2}]^2$ (f) $[a^{1/4}(a^{-5/4} + a^{3/4})]^{-1}$
   (g) $\left(\dfrac{\sqrt[3]{a^3 b^2}}{\sqrt[4]{a^6 b^3}}\right)^{-1}$ (h) $\left(\dfrac{a^{-2} b^{2/3}}{b^{-1/2}}\right)^{-4}$
   (i) $\left[\dfrac{(27)^{4/3} - (27)^0}{(3^2 + 4^2)^{1/2}}\right]^{3/4}$ (j) $(16a^2 b^3)^{3/4} - 4ab^2(a^2 b)^{1/4}$
   (k) $(\sqrt{3})^{3\sqrt{3}} - (3\sqrt{3})^{\sqrt{3}} + (\sqrt{3}^{\sqrt{3}})^{\sqrt{3}}$
   (l) $(a^{1/3} - b^{1/3})(a^{2/3} + a^{1/3}b^{1/3} + b^{2/3})$

86. Combine and simplify, writing your answer without negative exponents.
   (a) $4x^2(x^2 + 2)^{-2/3} - 3(x^2 + 2)^{1/3}$
   (b) $x^3(x^3 - 1)^{-3/4} - (x^3 - 1)^{1/4}$

87. Solve for x.
   (a) $4^{x+1} = (1/2)^{2x}$ (b) $5^{x^2-x} = 25$
   (c) $2^{4x}4^{x-3} = (64)^{x-1}$ (d) $(x^2 + x + 4)^{3/4} = 8$

(e) $x^{2/3} - 3x^{1/3} = -2$ (f) $2^{2x} - 2^{x+1} - 8 = 0$

88. Using the same axes, sketch the graph of each of the following.
   (a) $f(x) = 2^x$ (b) $g(x) = -2^x$ (c) $h(x) = 2^{-x}$
   (d) $k(x) = 2^x + 2^{-x}$ (e) $m(x) = 2^{x-4}$

89. Sketch the graph of $f(x) = 2^{-|x|}$.

90. Using the same axes, sketch the graphs of $f(x) = x^{\pi}$ and $g(x) = \pi^x$ on the interval $2 \le x \le 3.5$. One solution of $x^{\pi} = \pi^x$ is π. Use your graphs to help you find another one (approximately).

91. Give a simple argument to show that an exponential function $f(x) = a^x (a > 0, a \ne 1)$ is not equivalent to any polynomial function.

92. **TEASER** If a and b are irrational, does it follow that a^b is irrational? *Hint:* Consider $\sqrt{2}^{\sqrt{2}}$ and $(\sqrt{2}^{\sqrt{2}})^{\sqrt{2}}$.

A Packing Problem

World population is growing at about 2 percent per year. If this continues indefinitely, how long will it be until we are all packed together like sardines? In answering, assume that "sardine packing" for humans is one person per square foot of land area.

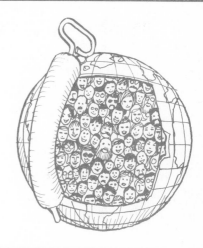

5-2   Exponential Growth and Decay

The phrase *exponential growth* is used repeatedly by professors, politicians, and pessimists. Population, energy use, mining of ores, pollution, and the number of books about these things are all said to be growing exponentially. Most people probably do not know what exponential growth means, except that they have heard it guarantees alarming consequences. For students of this book, it is easy to explain its meaning. For y to grow exponentially with time t means that it satisfies the relationship

$$y = Ca^t$$

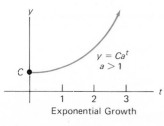

Exponential Growth

Figure 4

for constants C and a, with $C > 0$ and $a > 1$ (see Figure 4). Why should so many ingredients of modern society behave this way? The basic cause is population growth.

POPULATION GROWTH

Figure 5

Simple organisms reproduce by cell division. If, for example, there is one cell today, that cell may split so that there are two cells tomorrow. Then each of those cells may divide giving four cells the following day (Figure 5). As this process continues, the numbers of cells on successive days form the sequence

$$1, 2, 4, 8, 16, 32, \ldots$$

If we start with 100 cells and let $f(t)$ denote the number present t days from now, we have the results indicated in the table of Figure 6.

t	0	1	2	3	4	5
$f(t)$	100	200	400	800	1600	3200

Figure 6

It seems that

$$f(t) = (100)2^t$$

A perceptive reader will ask if this formula is really valid. Does it give the right answer when $t = 5.7$? Is not population growth a discrete process, occurring in unit amounts at distinct times, rather than a continuous process as the formula implies? The answer is that the exponential growth model provides a very good approximation to the growth of simple organisms, provided the initial population is large.

The mechanism of reproduction is different (and more interesting) for people, but the pattern of population growth is similar. World population is presently growing at about 2 percent per year. In 1975, there were about 4 billion people. Accordingly, the population in 1976 in billions was $4 + 4(.02) = 4(1.02)$, in 1977 it was $4(1.02)^2$, in 1978 it was $4(1.02)^3$, and so on. If this trend continues, there will be $4(1.02)^{20}$ billion people in the world in 1995, that is, 20 years after 1975. It appears that world population obeys the formula.

$$p(t) = 4(1.02)^t$$

where $p(t)$ represents the number of people (in billions) t years after 1975.

In general, if $A(t)$ is the amount at time t of a quantity growing exponentially at the rate of r (written as a decimal), then

$$\boxed{A(t) = A(0)(1 + r)^t}$$

CAUTION

The rate is $m\%$
$A(t) = A(0)(1 + m)^t$

The rate is $m\%$
$A(t) = A(0)\left(1 + \dfrac{m}{100}\right)^t$

Here $A(0)$ is the initial amount—that is, the amount present at $t = 0$.

t	$(1.02)^t$
5	1.104
10	1.219
15	1.346
20	1.486
25	1.641
30	1.811
35	2.000
40	2.208
45	2.438
50	2.692
55	2.972
60	3.281
65	3.623
70	4.000
75	4.416
80	4.875
85	5.383
90	5.943

Figure 7

DOUBLING TIMES

One way to get a feeling for the spectacular nature of exponential growth is via the concept of **doubling time;** this is the length of time required for an exponentially growing quantity to double in size. It is easy to show that if a quantity doubles in an initial time interval of length T, it will double in size in *any* time interval of length T. Consider the world population problem as an example. By the table of Figure 7, $(1.02)^{35} \approx 2$, so world population doubles in 35 years. Since it was 4 billion in 1975, it should be 8 billion in 2010, 16 billion in 2045, and so on. This alarming information is displayed graphically in Figure 8.

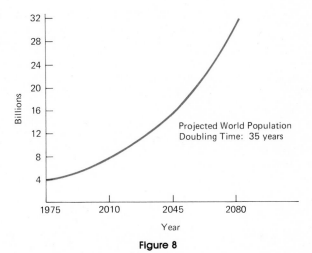

Figure 8

Now we can answer the question about sardine packing in our opening display. There are slightly more than 1,000,000 billion square feet of land area on the surface of the earth. Sardine packing for humans is about 1 square foot per person. Thus we are asking when $4(1.02)^t$ billion will equal 1,000,000 billion. This leads to the equation

$$(1.02)^t = 250,000$$

To solve this exponential equation, we use the following approximations

$$(1.02)^{35} \approx 2 \qquad 250,000 \approx 2^{18}$$

Our equation can then be rewritten as

$$[(1.02)^{35}]^{t/35} = 2^{18}$$

or

$$2^{t/35} = 2^{18}$$

We conclude that

$$\frac{t}{35} = 18$$

$$t = (18)(35) = 630$$

Thus, after about 630 years, we will be packed together like sardines. If it is any comfort, war, famine, or birth control will change population growth patterns before then.

COMPOUND INTEREST

One of the best practical illustrations of exponential growth is money earning compound interest. Suppose that Amy puts $1000 in a bank today at 8 percent interest compounded annually. Then at the end of one year the bank adds the interest of $(.08)(1000) = \$80$ to her $1000, giving her a total of $1080. But note that $1080 = 1000(1.08)$. During the second year, $1080 draws interest. At the end of that year, the bank adds $(.08)(1080)$ to the account, bringing the total to

$$1080 + (.08)(1080) = (1080)(1.08)$$
$$= 1000(1.08)(1.08)$$
$$= 1000(1.08)^2$$

Continuing in this way, we see that Amy's account will have grown to $1000(1.08)^3$ by the end of 3 years, $1000(1.08)^4$ by the end of 4 years, and so on. By the end of 15 years, it will have grown to

$$1000(1.08)^{15} \approx 1000(3.172169)$$
$$= \$3172.17$$

To calculate $(1.08)^{15}$, we used the table at the end of the problem set. We could have used a calculator.

How long would it take for Amy's money to double—that is, when will

$$1000(1.08)^t = 2000$$

This will occur when $(1.08)^t = 2$. According to the table just mentioned, this happens at $t \approx 9$, or in about 9 years.

EXPONENTIAL DECAY

Fortunately, not all things grow; some decline or decay. In fact, some things—notably the radioactive elements—decay exponentially. This means that the amount y present at time t satisfies

$$y = Ca^t$$

for some constants C and a with $C > 0$ and $0 < a < 1$ (see Figure 9).

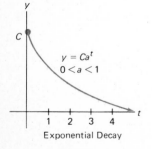

$y = Ca^t$
$0 < a < 1$

Exponential Decay

Figure 9

Here an important idea is that of **half-life,** the time required for half of a substance to disappear. For example, radium decays with a half-life of 1620 years. Thus if 1000 grams of radium are present now, 1620 years from now 500 grams will be present, $2(1620) = 3240$ years from now only 250 grams will be present, and so on.

The precise nature of radioactive decay is used to date old objects. If an object contains radium and lead (the product to which radium decays) in the ratio 1 to 3, then it is believed that an original amount of pure radium has decayed to $\frac{1}{4}$ its original size. The object must be two half-lives, or 3240 years, old. Two important assumptions have been made: (1) decay of radium is exactly exponential over long periods of time; and (2) no lead was originally present. Recent research raises some question about the correctness of such assumptions.

Problem Set 5-2

1. In each of the following, indicate whether y grows exponentially or decays exponentially with t.

   (a) $y = 128\left(\frac{1}{2}\right)^t$

   (b) $y = 5\left(\frac{5}{3}\right)^t$

   (c) $y = 4(10)^9(1.03)^t$

   (d) $y = 1000(.99)^t$

2. Find the values of y corresponding to $t = 0$, $t = 1$, and $t = 2$ for each case in Problem 1.

3. Use the table at the end of the problem set or a calculator with a $\boxed{y^x}$ key to find each value.

   (a) $(1.08)^{20}$ (b) $(1.12)^{25}$ (c) $1000(1.04)^{40}$ (d) $2000(1.02)^{80}$

4. Evaluate

   (a) $(1.01)^{100}$ (b) $(1.02)^{40}$ (c) $100(1.12)^{50}$ (d) $500(1.04)^{30}$

5. Silver City's present population of 1000 is expected to grow exponentially over the next 10 years at 4 percent per year. How many people will it have at the end of that time? *Hint:* $A(t) = A(0)[1 + r]^t$.

6. The value of houses in Longview is said to be growing exponentially at 12 percent per year. What will a house valued at $100,000 today be worth after 8 years?

7. Under the assumptions concerning world population used in this section, what will be the approximate number of people on earth in each year?

   (a) 1990 (that is, 15 years after 1975)

   (b) 2000

   (c) 2065

8. A certain radioactive substance has a half-life of 40 minutes. What fraction of an initial amount of this substance will remain after 1 hour, 20 minutes (that is, after 2 half-lives)? After 2 hours, 40 minutes?

EXAMPLE A (Compound Interest)   Roger put $1000 in a money market fund at 15 percent interest compounded annually. How much was it worth after 4 years?

Solution.   We will have to use a calculator since $(1.15)^n$ is not in our table. The answer is

$$1000(1.15)^4 = \$1749.01$$

9. If you put $100 in the bank for 8 years, how much will it be worth at the end of that time at
   (a) 8 percent compounded annually;
   (b) 12 percent compounded annually?

10. If you invest $500 in the bank today, how much will it be worth after 25 years at
    (a) 8 percent compounded annually;
    (b) 4 percent compounded annually;
    (c) 12 percent compounded annually?

11. If you put $3500 in the bank today, how much will it be worth after 40 years at
    (a) 8 percent compounded annually;
    (b) 12 percent compounded annually?

12. Approximately how long will it take for money to accumulate to twice its value if
    (a) it is invested at 8 percent compounded annually;
    (b) it is invested at 12 percent compounded annually?

13. Suppose that you invest P dollars at r percent compounded annually. Write an expression for the amount accumulated after n years.

EXAMPLE B (More on Compound Interest)   If $1000 is invested at 8 percent compounded quarterly, find the accumulated amount after 15 years.

Solution.   Interest calculated at 2 percent ($\frac{1}{4}$ of 8 percent) is converted to principal every 3 months. By the end of the first 3-month period, the account has grown to $1000(1.02) = \$1020$; by the end of the second 3-month period, it has grown to $1000(1.02)^2$; and so on. The accumulated amount after 15 years, or 60 conversion periods, is

$$1000(1.02)^{60} \approx 1000(3.28103)$$

$$= \$3281.03$$

Suppose more generally that P dollars is invested at a rate r (written as a decimal), which is compounded m times per year. Then the accumulated amount A after t years is given by

$$A = P\left(1 + \frac{r}{m}\right)^{tm}$$

In our example

$$A = 1000\left(1 + \frac{.08}{4}\right)^{15\cdot4} = 1000(1.02)^{60}$$

Find the accumulated amount for the indicated initial principal, compound interest rate, and total time period.

14. $2000; 8 percent compounded annually; 15 years

15. $5000; 8 percent compounded semiannually; 5 years

16. $5000; 12 percent compounded monthly; 5 years

[c] 17. $3000; 9 percent compounded annually; 10 years

[c] 18. $3000; 9 percent compounded semiannually; 10 years

[c] 19. $3000; 9 percent compounded quarterly; 10 years

[c] 20. $3000; 9 percent compounded monthly; 10 years

[c] 21. $1000; 8 percent compounded monthly; 10 years

[c] 22. $1000; 8 percent compounded daily; 10 years   *Hint:* Assume there are 365 days in a year, so that the interest rate per day is .08/365.

MISCELLANEOUS PROBLEMS

23. Let $y = 5400(2/3)^t$. Evaluate y for $t = -1, 0, 1, 2,$ and 3.

24. For what value of t in Problem 23 is $y = 3200/3$?

25. If $(1.023)^T = 2$, find the value of $100(1.023)^{3T}$.

26. If $(.67)^H = \frac{1}{2}$, find the value of $32(.67)^{4H}$.

[c] 27. **Suppose the population of a certain city follows the formula**

$$p(t) = 4600(1.016)^t$$

where $p(t)$ is the population t years after 1980.
(a) What will the population be in 2020? In 2080?
(b) Experiment with your calculator to find the doubling time for this population.

28. The number of bacteria in a certain culture is known to triple every hour. Suppose the count at 12:00 noon is 162,000. What was the count at 11:00 A.M.? At 8:00 A.M.?

29. If $100 is invested today, how much will it be worth after 5 years at 8 percent interest if interest is:
(a) compounded annually;
(b) compounded quarterly;
[c] (c) compounded monthly;
[c] (d) compounded daily? (There are 365 days in a year.)

30. How long does it take money to double if invested at 12 percent compounded monthly? (Use the compound interest table.)

31. About how long does it take an exponentially growing population to double if its rate of growth is:
(a) 8 percent per year (use the compound interest table);
[c] (b) 6.5 percent per year? (Experiment with your calculator.)

[c] 32. A manufacturer of radial tires found that the percentage P of tires still usable after being driven m miles was given by

$$P = 100(2.71)^{-.000025m}$$

What percentage of tires are still usable at 80,000 miles?

33. A certain radioactive element has a half-life of 1690 years. Starting with 30 milligrams, there will be $q(t)$ milligrams left after t years, where $q(t) = 30(1/2)^{kt}$.

(a) Determine the constant k.

c (b) How much will be left after 2500 years?

34. One method of depreciation allowed by IRS is the double-declining-balance method. In this method, the original value C of an item is depreciated each year by $100(2/N)$ percent of its value at the beginning of that year, N being the useful life of the item.

(a) Write a formula for the value V of the item after n years.

c (b) If an item cost $10,000 and has a useful life of 15 years, calculate its value after 10 years; after 15 years.

(c) Does the value of an item ever become zero by this depreciation method?

c 35. (Carbon dating)   All living things contain carbon 12, which is a stable element, and carbon 14, which is radioactive. While a plant or animal is alive, the ratio of these two isotopes of carbon remains unchanged, since carbon 14 is constantly renewed; but after death, no more carbon 14 is absorbed. The half-life of carbon 14 is 5730 years. Bones from a human body were found to contain only 76 percent of the carbon 14 in living bones. How long before did the person die?

c 36. Manhattan Island is said to have been bought from the Indians by Peter Minuit in 1626 for $24. If, instead of making this purchase, Minuit had put the money in a savings account drawing interest at 6 percent compounded annually, what would that account be worth in the year 2000?

c 37. Hamline University was founded in 1854 with a gift of $25,000 from Bishop Hamline of the Methodist Church. Suppose that Hamline University had wisely put $10,000 of this gift in an endowment drawing 10 percent interest compounded annually, promising not to touch it until 1988 (exactly 134 years later). How much could it then withdraw each year and still maintain this endowment at the 1988 level?

38. **TEASER** Suppose one water lily growing exponentially at the rate of 8 percent per day is able to cover a certain pond in 50 days. How long would it take 10 of these lilies to cover the pond?

TABLE 1 Compound Interest Table

n	$(1.01)^n$	$(1.02)^n$	$(1.04)^n$	$(1.08)^n$	$(1.12)^n$
1	1.01000000	1.02000000	1.04000000	1.08000000	1.12000000
2	1.02010000	1.04040000	1.08160000	1.16640000	1.25440000
3	1.03030100	1.06120800	1.12486400	1.25971200	1.40492800
4	1.04060401	1.08243216	1.16985856	1.36048896	1.57351936
5	1.05101005	1.10408080	1.21665290	1.46932808	1.76234168
6	1.06152015	1.12616242	1.26531902	1.58687432	1.97382269
7	1.07213535	1.14868567	1.31593178	1.71382427	2.21068141
8	1.08285671	1.17165938	1.36856905	1.85093021	2.47596318
9	1.09368527	1.19509257	1.42331181	1.99900463	2.77307876
10	1.10462213	1.21899442	1.48024428	2.15892500	3.10584821
11	1.11566835	1.24337431	1.53945406	2.33163900	3.47854999
12	1.12682503	1.26824179	1.60103222	2.51817012	3.89597599
15	1.16096896	1.34586834	1.80094351	3.17216911	5.47356576
20	1.22019004	1.48594740	2.19112314	4.66095714	9.64629309
25	1.28243200	1.64060599	2.66583633	6.84847520	17.00006441
30	1.34784892	1.81136158	3.24339751	10.06265689	29.95992212
35	1.41660276	1.99988955	3.94608899	14.78534429	52.79961958
40	1.48886373	2.20803966	4.80102063	21.72452150	93.05097044
45	1.56481075	2.43785421	5.84117568	31.92044939	163.98760387
50	1.64463182	2.69158803	7.10668335	46.90161251	289.00218983
55	1.72852457	2.97173067	8.64636692	68.91385611	509.32060567
60	1.81669670	3.28103079	10.51962741	101.25706367	897.59693349
65	1.90936649	3.62252311	12.79873522	148.77984662	1581.87249060
70	2.00676337	3.99955822	15.57161835	218.60640590	2787.79982770
75	2.10912847	4.41583546	18.94525466	321.20452996	4913.05584077
80	2.21671522	4.87543916	23.04979907	471.95483426	8658.48310008
85	2.32978997	5.38287878	28.04360494	693.45648897	15259.20568055
90	2.44863267	5.94313313	34.11933334	1018.91508928	26891.93422336
95	2.57353755	6.56169920	41.51138594	1497.12054855	47392.77662369
100	2.70481383	7.24464612	50.50494818	2199.76125634	83522.26572652

An active participant in the political and religious battles of his day, the Scot. John Napier amused himself by studying mathematics and science. He was interested in reducing the work involved in the calculations of spherical trigonometry, especially as they applied to astronomy. In 1614 he published a book containing the idea that made him famous. He gave it the name *logarithm*.

John Napier
(1550–1617)

5-3   Logarithms and Logarithmic Functions

Napier's approach to logarithms is out of style, but the goal he had in mind is still worth considering. He hoped to replace multiplications by additions. He thought additions were easier to do, and he was right.

Consider the exponential function $f(x) = 2^x$ and recall that

$$2^x \cdot 2^y = 2^{x+y}$$

On the left, we have a multiplication and on the right, an addition. If we are to fulfill Napier's objective, we want logarithms to behave like exponents. That suggests a definition. The logarithm of N to the base 2 is the exponent to which 2 must be raised to yield N. That is,

$$\log_2 N = x \quad \text{if and only if} \quad 2^x = N$$

Thus

$$\log_2 4 = 2 \quad \text{since} \quad 2^2 = 4$$
$$\log_2 8 = 3 \quad \text{since} \quad 2^3 = 8$$
$$\log_2 \sqrt{2} = \tfrac{1}{2} \quad \text{since} \quad 2^{1/2} = \sqrt{2}$$

and, in general,

$$\log_2(2^x) = x \quad \text{since} \quad 2^x = 2^x$$

Has Napier's goal been achieved? Does the logarithm turn a product into a sum? Yes, for note that

$$\log_2(2^x \cdot 2^y) = \log_2(2^{x+y}) \qquad \text{(property of exponents)}$$
$$= x + y \qquad \text{(definition of } \log_2 \text{)}$$
$$= \log_2(2^x) + \log_2(2^y)$$

Thus

$$\log_2(2^x \cdot 2^y) = \log_2(2^x) + \log_2(2^y)$$

which has the form

$$\log_2(M \cdot N) = \log_2 M + \log_2 N$$

THE GENERAL DEFINITION

What has been done for 2 can be done for any base $a > 1$. The **logarithm of N to the base a** is the exponent x to which a must be raised to yield N. Thus

$$\log_a N = x \quad \text{if and only if} \quad a^x = N$$

Now we can calculate many kinds of logarithms.

$$\log_4 16 = 2 \quad \text{since} \quad 4^2 = 16$$
$$\log_{10} 1000 = 3 \quad \text{since} \quad 10^3 = 1000$$
$$\log_{10}(.001) = -3 \quad \text{since} \quad 10^{-3} = \frac{1}{1000} = .001$$

What is $\log_{10} 7$? We are not ready to answer that yet, except to say it is a number x satisfying $10^x = 7$ (see Section 5-6).

We point out that negative numbers and zero do not have logarithms. Suppose -4 and 0 did have logarithms, that is, suppose

$$\log_a(-4) = m \quad \text{and} \quad \log_a 0 = n$$

Then

$$a^m = -4 \quad \text{and} \quad a^n = 0$$

But that is impossible; we learned earlier that a^x is always positive.

PROPERTIES OF LOGARITHMS

There are three main properties of logarithms.

PROPERTIES OF LOGARITHMS

1. $\log_a(M \cdot N) = \log_a M + \log_a N$
2. $\log_a(M/N) = \log_a M - \log_a N$
3. $\log_a(M^p) = p \log_a M$

To establish Property 1, let

$$x = \log_a M \quad \text{and} \quad y = \log_a N$$

Then, by definition,

$$M = a^x \quad \text{and} \quad N = a^y$$

so that

$$M \cdot N = a^x \cdot a^y = a^{x+y}$$

Thus $x + y$ is the exponent to which a must be raised to yield $M \cdot N$, that is,

$$\log_a(M \cdot N) = x + y = \log_a M + \log_a N$$

Properties 2 and 3 are demonstrated in a similar fashion.

THE LOGARITHMIC FUNCTION

The function determined by

$$g(x) = \log_a x$$

is called the **logarithmic function with base a.** We can get a feeling for the behavior of this function by drawing its graph for $a = 2$ (Figure 10).

Several properties of $y = \log_2 x$ are apparent from this graph. The domain consists of all positive real numbers. If $0 < x < 1$, $\log_2 x$ is negative; if $x > 1$, $\log_2 x$ is positive. The y-axis is a vertical asymptote of the graph since very small positive x values yield large negative y values. Although $\log_2 x$ continues to increase as x increases, even this small part of the complete graph indicates how slowly it grows for large x. In fact, by the time x reaches 1,000,000, $\log_2 x$ is still loafing along at about 20. In this sense, it behaves in a manner opposite to the exponential function 2^x, which grows more and more rapidly as x increases. There is a good reason for this opposite behavior; the two functions are inverses of each other.

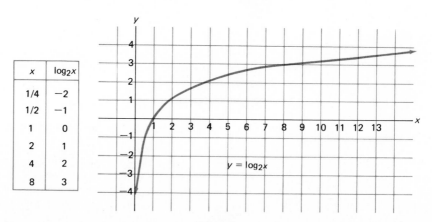

x	$\log_2 x$
1/4	−2
1/2	−1
1	0
2	1
4	2
8	3

Figure 10

INVERSE FUNCTIONS

We begin by emphasizing two facts that you must not forget.

$$a^{\log_a x} = x$$
$$\log_a(a^x) = x$$

For example, $2^{\log_2 7} = 7$ and $\log_2(2^{-19}) = -19$. Both of these facts are direct consequences of the definition of logarithms; the second is also a special case of Property 3, stated earlier. What these facts tell us is that the logarithmic and exponential functions undo each other.

Let us put it in the language of Section 1-5. If $f(x) = a^x$ and $g(x) = \log_a x$, then

$$f(g(x)) = f(\log_a x) = a^{\log_a x} = x$$

and

$$g(f(x)) = g(a^x) = \log_a(a^x) = x$$

Thus g is really f^{-1}. This fact also tells us something about the graphs of g and f: They are simply reflections of each other about the line $y = x$ (Figure 11).

Note finally that $f(x) = a^x$ has the set of all real numbers as its domain and the positive real numbers as its range. Thus its inverse $f^{-1}(x) = \log_a x$ has domain consisting of the positive real numbers and range consisting of all real numbers. We emphasize again a fact that is important to remember. *Negative numbers and zero do not have logarithms.*

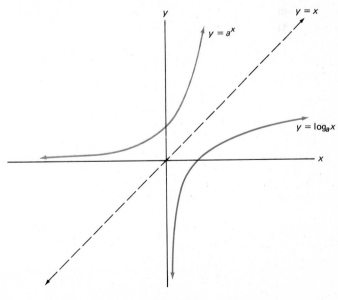

Figure 11

Problem Set 5-3

Write each of the following in logarithmic form. For example, $3^4 = 81$ can be written as $\log_3 81 = 4$.

1. $4^3 = 64$
2. $7^3 = 343$
3. $27^{1/3} = 3$
4. $16^{1/4} = 2$
5. $4^0 = 1$
6. $81^{-1/2} = \frac{1}{9}$
7. $125^{-2/3} = \frac{1}{25}$
8. $2^{9/2} = 16\sqrt{2}$
9. $10^{\sqrt{3}} = a$
10. $5^{\sqrt{2}} = b$
11. $10^a = \sqrt{3}$
12. $b^x = y$

Write each of the following in exponential form. For example, $\log_5 125 = 3$ can be written as $5^3 = 125$.

13. $\log_5 625 = 4$
14. $\log_6 216 = 3$
15. $\log_4 8 = \frac{3}{2}$
16. $\log_{27} 9 = \frac{2}{3}$
17. $\log_{10}(.01) = -2$
18. $\log_3(\frac{1}{27}) = -3$
19. $\log_c c = 1$
20. $\log_b N = x$
21. $\log_c Q = y$

Determine the value of each of the following logarithms.

22. $\log_4 16$
23. $\log_5 25$
24. $\log_7 \frac{1}{7}$
25. $\log_3 \frac{1}{3}$
26. $\log_4 2$
27. $\log_{27} 3$
28. $\log_{10}(10^{-6})$
29. $\log_{10}(.0001)$
30. $\log_8 1$
31. $\log_3 1$
32. $\log_{100} 1000$
33. $\log_8 16$

Find the value of c in each of the following.

34. $\log_c 25 = 2$
35. $\log_c 8 = 3$
36. $\log_4 c = -\frac{1}{2}$
37. $\log_9 c = -\frac{3}{2}$
38. $\log_2(2^{5.6}) = c$
39. $\log_3(3^{-2.9}) = c$
40. $8^{\log_8 11} = c$
41. $5^{2 \log_5 7} = c$
42. $3^{4 \log_3 2} = c$

Given $\log_{10} 2 = .301$ and $\log_{10} 3 = .477$, calculate each of the following without the use of tables. For example, in Problem 43, $\log_{10} 6 = \log_{10} 2 \cdot 3 = \log_{10} 2 + \log_{10} 3$, and in Problem 50, $\log_{10} 54 = \log_{10} 2 \cdot 3^3 = \log_{10} 2 + \log_{10} 3^3 = \log_{10} 2 + 3 \log_{10} 3$.

43. $\log_{10} 6$
44. $\log_{10} \frac{3}{2}$
45. $\log_{10} 16$
46. $\log_{10} 27$
47. $\log_{10} \frac{1}{4}$
48. $\log_{10} \frac{1}{27}$
49. $\log_{10} 24$
50. $\log_{10} 54$
51. $\log_{10} \frac{8}{9}$
52. $\log_{10} \frac{3}{8}$
53. $\log_{10} 5$
54. $\log_{10} \sqrt[3]{3}$

© *Your scientific calculator has a $\log_{10}$ key (which may be abbreviated log). Use it to find each of the following.*

55. $\log_{10} 34$
56. $\log_{10} 1417$
57. $\log_{10}(.0123)$
58. $\log_{10}(.3215)$
59. $\log_{10} 9723$
60. $\log_{10}(\frac{21}{312})$

CAUTION

$\log_b 6 + \log_b 4 = \log_b 10$
$\log_b 30 - \log_b 5 = \log_b 25$

$\log_b 6 + \log_b 4 = \log_b 24$
$\log_b 30 - \log_b 5 = \log_b 6$

EXAMPLE A (Combining Logarithms)   Write the following expression as a single logarithm.

$$2 \log_{10} x + 3 \log_{10}(x + 2) - \log_{10}(x^2 + 5)$$

Solution. We use the properties of logarithms to rewrite this as

$$\log_{10} x^2 + \log_{10}(x + 2)^3 - \log_{10}(x^2 + 5) \qquad \text{(Property 3)}$$

$$= \log_{10} x^2(x + 2)^3 - \log_{10}(x^2 + 5) \qquad \text{(Property 1)}$$

$$= \log_{10}\left[\frac{x^2(x + 2)^3}{x^2 + 5}\right] \qquad \text{(Property 2)}$$

Write each of the following as a single logarithm.

61. $3 \log_{10}(x + 1) + \log_{10}(4x + 7)$
62. $\log_{10}(x^2 + 1) + 5 \log_{10} x$
63. $3 \log_2(x + 2) + \log_2 8x - 2 \log_2(x + 8)$
64. $2 \log_5 x - 3 \log_5(2x + 1) + \log_5(x - 4)$
65. $\frac{1}{2} \log_6 x + \frac{1}{3} \log_6(x^3 + 3)$
66. $-\frac{2}{3} \log_3 x + \frac{5}{2} \log_3(2x^2 + 3)$

EXAMPLE B (Solving Logarithmic Equations) Solve the equation

$$\log_2 x + \log_2(x + 2) = 3$$

Solution. First we note that we must have $x > 0$ so that both logarithms exist. Next we rewrite the equation using the first property of logarithms and then the definition of a logarithm.

$$\log_2 x(x + 2) = 3$$

$$x(x + 2) = 2^3$$

$$x^2 + 2x - 8 = 0$$

$$(x + 4)(x - 2) = 0$$

We reject $x = -4$ (because $-4 < 0$) and keep $x = 2$. To make sure that 2 is a solution, we substitute 2 for x in the original equation.

$$\log_2 2 + \log_2(2 + 2) \stackrel{?}{=} 3$$

$$1 + 2 = 3$$

Solve each of the following equations.

67. $\log_7(x + 2) = 2$
68. $\log_5(3x + 2) = 1$
69. $\log_2(x + 3) = -2$
70. $\log_4(\frac{1}{64}x + 1) = -3$
71. $\log_2 x - \log_2(x - 2) = 3$
72. $\log_3 x - \log_3(2x + 3) = -2$
73. $\log_2(x - 4) + \log_2(x - 3) = 1$
74. $\log_{10} x + \log_{10}(x - 3) = 1$

EXAMPLE C (Change of Base) In Problems 89 and 90, you will be asked to establish the **change-of-base formula**

$$\log_b x = \frac{\log_a x}{\log_a b}$$

Use this formula and the $\boxed{\log_{10}}$ key on a calculator to find $\log_2 13$.

Solution.

$$\log_2 13 = \frac{\log_{10} 13}{\log_{10} 2} \approx \frac{1.1139433}{.30103} \approx 3.7004$$

□c *Use the method of Example C to find the following.*

75. $\log_2 128$ 76. $\log_3 128$ 77. $\log_3 82$

78. $\log_5 110$ 79. $\log_6 39$ 80. $\log_2(.26)$

MISCELLANEOUS PROBLEMS

81. Find the value of x in each of the following.
    (a) $x = \log_6 36$ (b) $x = \log_4 2$ (c) $\log_{25} x = \frac{3}{2}$
    (d) $\log_4 x = \frac{5}{2}$ (e) $\log_x 10\sqrt{10} = \frac{3}{2}$ (f) $\log_x \frac{1}{8} = -\frac{3}{2}$

82. Write each of the following as a single logarithm.
    (a) $3 \log_2 5 - 2 \log_2 7$
    (b) $\frac{1}{2} \log_5 64 + \frac{1}{3} \log_5 27 - \log_5(x^2 + 4)$
    (c) $\frac{2}{3} \log_{10}(x + 5) + 4 \log_{10} x - 2 \log_{10}(x - 3)$

83. Evaluate $\dfrac{(\log_{27} 3)(\log_{27} 9)(3^{2\log_3 2})}{\log_3 27 - \log_3 9 + \log_3 1}$

84. Solve for x.
    (a) $2(\log_4 x)^2 + 3\log_4 x - 2 = 0$
    (b) $(\log_x 8)^2 - \log_x 8 - 6 = 0$
    (c) $\log_x \sqrt{3} + \log_x 3^5 + \log_x(\frac{1}{27}) = \frac{5}{4}$

85. Solve for x.
    (a) $\log_5(2x - 1) = 2$
    (b) $\log_4\left(\dfrac{x - 2}{2x + 3}\right) = 0$
    (c) $\log_4(x - 2) - \log_4(2x + 3) = 0$
    (d) $\log_{10} x + \log_{10}(x - 15) = 2$
    (e) $\dfrac{\log_2(x + 1)}{\log_2(x - 1)} = 2$
    (f) $\log_8[\log_4(\log_2 x)] = 0$

86. Solve for x.
    (a) $2^{\log_2 x} = 16$ (b) $2^{\log_x 2} = 16$ (c) $x^{\log_2 x} = 16$
    (d) $\log_2 x^2 = 2$ (e) $(\log_2 x)^2 = 1$ (f) $x = (\log_2 x)^{\log_2 x}$

87. Solve for y in terms of x.
    (a) $\log_a(x + y) = \log_a x + \log_a y$
    (b) $x = \log_a(y + \sqrt{y^2 - 1})$

88. Show that $f(x) = \log_a(x + \sqrt{1 + x^2})$ is an odd function.

89. Show that $\log_2 x = \log_{10} x \log_2 10$, where $x > 0$. *Hint:* Let $\log_{10} x = c$. Then $x = 10^c$. Next take $\log_2$ of both sides.

90. Use the technique outlined in Problem 89 to show that for $a > 0$, $b > 0$, and $x > 0$,

$$\log_a x = \log_b x \log_a b$$

This is equivalent to the change-of-base formula of Example C.

91. Show that $\log_a b = 1/\log_b a$, where a and b are positive.
92. If $\log_b N = 2$, find $\log_{1/b} N$.
93. Graph the equations $y = 3^x$ and $y = \log_3 x$ using the same coordinate axes.
94. Find the solution set for each of the following inequalities.
    (a) $\log_2 x < 0$ (b) $\log_{10} x \geq -1$
    (c) $2 < \log_3 x < 3$ (d) $-2 \leq \log_{10} x \leq -1$
    (e) $2^x > 10$ (f) $2^x < 3^x$
95. Sketch the graph of each of the following functions using the same coordinate axes.
    (a) $f(x) = \log_2 x$ (b) $g(x) = \log_2(x + 1)$ (c) $h(x) = 3 + \log_2 x$
96. **TEASER** Let log represent $\log_{10}$. Evaluate.
    (a) $\log \dfrac{1}{2} + \log \dfrac{2}{3} + \log \dfrac{3}{4} + \cdots + \log \dfrac{98}{99} + \log \dfrac{99}{100}$
    (b) $\log \dfrac{3}{4} + \log \dfrac{8}{9} + \log \dfrac{15}{16} + \cdots + \log \dfrac{99^2 - 1}{99^2} + \log \dfrac{100^2 - 1}{100^2}$
    (c) $\log_2 3 \cdot \log_3 4 \cdot \log_4 5 \cdot \log_5 6 \cdots \log_{63} 64$

The collected works of this brilliant Swiss mathematician will fill 74 volumes when completed. No other person has written so profusely on mathematical topics. Remarkably, 400 of his research papers were written after he was totally blind. One of his contributions was the introduction of the number $e = 2.71828 \ldots$ as the base for natural logarithms.

Leonhard Euler
1707-1783

5-4 Natural Logarithms and Applications

Napier invented logarithms to simplify arithmetic calculations. Computers and calculators have reduced that application to minor significance, though we shall discuss such a use of logarithms later in this chapter. Here we have in mind deeper applications such as solving exponential equations, defining power functions, and modeling physical phenomena.

In order to make any progress, we shall need an easy way to calculate logarithms. Fortunately, this has been done for us as tables of logarithms to several bases are available. For our purposes in this section, one base is as good as another. Base 10 would be an appropriate choice, but we would rather defer discussion of logarithms to base 10 (common logarithms) until Section 5-5. We have chosen rather to introduce you to the number e (after Euler), which is used as a base of logarithms in all advanced mathematics courses. You will see the importance of logarithms to this base (**natural logarithms**) when you study calculus (see also Figure 12). An approximate value of e is

$$e \approx 2.71828$$

and, like π, e is an irrational number. Page 195 shows a table of values of natural logarithms. In this table, $\log_e$ is denoted by ln, a practice we shall follow from now on.

Since ln denotes a genuine logarithm function, we have as in Section 5-3

$$\ln N = x \quad \text{if and only if} \quad e^x = N$$

and consequently

$$\ln e^x = x \quad \text{and} \quad e^{\ln N} = N$$

Moreover the three properties of logarithms hold.

1. $\ln (MN) = \ln M + \ln N$
2. $\ln(M/N) = \ln M - \ln N$
3. $\ln (N^p) = p \ln N$

SOLVING EXPONENTIAL EQUATIONS

Consider first the simple equation

$$5^x = 1.7$$

We call it an *exponential equation* because the unknown is in the exponent. To solve it, we take natural logarithms of both sides.

$$5^x = 1.7$$

$$\ln(5^x) = \ln 1.7$$

$$x \ln 5 = \ln 1.7 \qquad \text{(Property 3)}$$

$$x = \frac{\ln 1.7}{\ln 5}$$

$$x \approx \frac{.531}{1.609} \approx .330 \quad \text{(table of ln valves)}$$

We point out that the last step can also be done on a scientific calculator, which has a key for calculating natural logarithms.

TABLE 2   Table of Natural Logarithms

x	$\ln x$	x	$\ln x$	x	$\ln x$
		4.0	1.386	8.0	2.079
0.1	−2.303	4.1	1.411	8.1	2.092
0.2	−1.609	4.2	1.435	8.2	2.104
0.3	−1.204	4.3	1.459	8.3	2.116
0.4	−0.916	4.4	1.482	8.4	2.128
0.5	−0.693	4.5	1.504	8.5	2.140
0.6	−0.511	4.6	1.526	8.6	2.152
0.7	−0.357	4.7	1.548	8.7	2.163
0.8	−0.223	4.8	1.569	8.8	2.175
0.9	−0.105	4.9	1.589	8.9	2.186
1.0	0.000	5.0	1.609	9.0	2.197
1.1	0.095	5.1	1.629	9.1	2.208
1.2	0.182	5.2	1.649	9.2	2.219
1.3	0.262	5.3	1.668	9.3	2.230
1.4	0.336	5.4	1.686	9.4	2.241
1.5	0.405	5.5	1.705	9.5	2.251
1.6	0.470	5.6	1.723	9.6	2.262
1.7	0.531	5.7	1.740	9.7	2.272
1.8	0.588	5.8	1.758	9.8	2.282
1.9	0.642	5.9	1.775	9.9	2.293
2.0	0.693	6.0	1.792	10	2.303
2.1	0.742	6.1	1.808	20	2.996
2.2	0.788	6.2	1.825	30	3.401
2.3	0.833	6.3	1.841	40	3.689
2.4	0.875	6.4	1.856	50	3.912
2.5	0.916	6.5	1.872	60	4.094
2.6	0.956	6.6	1.887	70	4.248
2.7	0.993	6.7	1.902	80	4.382
2.8	1.030	6.8	1.917	90	4.500
2.9	1.065	6.9	1.932	100	4.605
3.0	1.099	7.0	1.946		
3.1	1.131	7.1	1.960		
3.2	1.163	7.2	1.974	e	1.000
3.3	1.194	7.3	1.988		
3.4	1.224	7.4	2.001	π	1.145
3.5	1.253	7.5	2.015		
3.6	1.281	7.6	2.028		
3.7	1.308	7.7	2.041		
3.8	1.335	7.8	2.054		
3.9	1.361	7.9	2.067		

To find the natural logarithm of a number N which is either smaller than 0.1 or larger than 10, write N in scientific notation, that is, write $N = c \times 10^k$. Then $\ln N = \ln c + k \ln 10 = \ln c + k(2.303)$. A more complete table of natural logarithms appears as Table C of the Appendix.

Here is a more complicated example.

$$5^{2x-1} = 7^{x+2}$$

Begin by taking natural logarithms of both sides and then solve for x.

$$\ln(5^{2x-1}) = \ln(7^{x+2})$$

$$(2x - 1) \ln 5 = (x + 2) \ln 7 \qquad \text{(Property 3)}$$

$$2x \ln 5 - \ln 5 = x \ln 7 + 2 \ln 7$$

$$2x \ln 5 - x \ln 7 = \ln 5 + 2 \ln 7$$

$$x(2 \ln 5 - \ln 7) = \ln 5 + 2 \ln 7$$

$$x = \frac{\ln 5 + 2 \ln 7}{2 \ln 5 - \ln 7}$$

$$\approx \frac{1.609 + 2(1.946)}{2(1.609) - 1.946} \quad \text{(table of ln values or a calculator)}$$

$$\approx 4.325$$

You get a more accurate answer, 4.321, if you use a calculator all the way.

THE GRAPHS OF ln x AND e^x

We have already pointed out that $\ln e^x = x$ and $e^{\ln x} = x$. Thus $f(x) = \ln x$ and $g(x) = e^x$ are inverse functions, which means that their graphs are reflections of each other across the line $y = x$. They are shown in Figure 13.

x	ln x
.1	−2.3
.5	− .7
1	0
2	.7
3	1.1
4	1.4
5	1.6

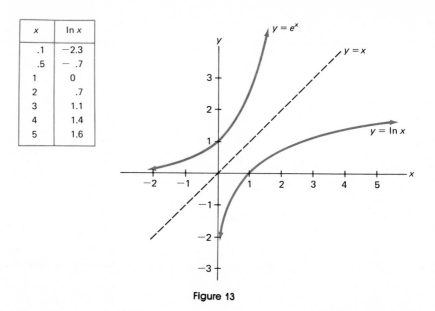

Figure 13

Since $\ln x$ is not defined for $x \leq 0$, it is of some interest to consider $\ln|x|$, which is defined for all x except 0. Its graph is shown in Figure 14. Note the symmetry with respect to the y-axis.

EXPONENTIAL FUNCTIONS VERSUS POWER FUNCTIONS

Look closely at the following formulas:

$$f(x) = 2^x \qquad f(x) = x^2$$

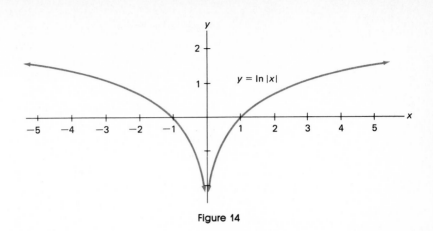

Figure 14

They are very different, yet easily confused. The first is an exponential function, while the second is called a power function. Both grow rapidly for large x, but the exponential function ultimately gets far ahead (see the graphs in Figure 15).

The situation described above is a special instance of two very general classes of functions.

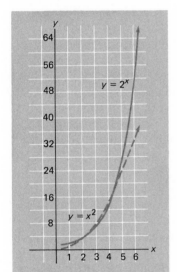

Figure 15

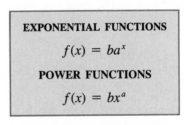

EXPONENTIAL FUNCTIONS

$$f(x) = ba^x$$

POWER FUNCTIONS

$$f(x) = bx^a$$

CURVE FITTING

A recurring theme in science is to fit a mathematical curve to a set of experimental data. Suppose that a scientist, studying the relationship between two variables x and y, obtained the data plotted in Figure 16. In searching for curves to fit these data, the scientist naturally thought of exponential curves and power curves. How did he or she decide if either was appropriate? The scientist took logarithms. Let us see why.

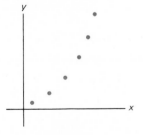

Figure 16

MODEL 1	*MODEL 2*
$y = ba^x$	$y = bx^a$
$\ln y = \ln b + x \ln a$	$\ln y = \ln b + a \ln x$
$Y = B + Ax$	$Y = B + aX$

Here the scientist made the substitutions $Y = \ln y$, $B = \ln b$, $A = \ln a$, and $X = \ln x$.

In both cases, the final result is a linear equation. But note the difference. In the first case, $\ln y$ is a linear function of x, whereas in the second case, ln y is a linear function of ln x. These considerations suggest the following procedures. Make two additional plots of the data. In the first, plot ln y against x, and in the second, plot ln y against ln x. If the first plotting gives data nearly along a straight line, Model 1 is appropriate; if the second does, then Model 2 is appropriate. If neither plot approximates a straight line, our scientist should look for a different and perhaps more complicated model.

We have used natural logarithms in the discussion above; we could also have used common logarithms (logarithms to the base 10). In the latter case, special kinds of graph paper are available to simplify the curve fitting process. On semilog paper, the vertical axis has a logarithmic scale; on log-log paper, both axes have logarithmic scales. The xy-data can be plotted *directly* on this paper. If semilog paper gives an (approximately) straight line, Model 1 is indicated; if log-log paper does so, then Model 2 is appropriate. You will have ample opportunity to use these special kinds of graph paper in your science courses.

LOGARITHMS AND PHYSIOLOGY

The human body appears to have a built-in logarithmic calculator. What do we mean by this statement?

In 1834, the German physiologist E. Weber noticed an interesting fact. Two heavy objects must differ in weight by considerably more than two light objects if a person is to perceive a difference between them. Other scientists noted the same phenomenon when human subjects tried to differentiate loudness of sounds, pitches of musical tones, brightness of light, and so on. Experiments suggested that people react to stimuli on a logarithmic scale, a result formulated as the Weber-Fechner law (see Figure 17).

$$S = C \ln\left(\frac{R}{r}\right)$$

Here R is the actual intensity of the stimulus, r is the threshold value (smallest value at which the stimulus is observed), C is a constant depending on the type of stimulus, and S is the perceived intensity of the stimulus. Note that a change in R is not as perceptible for large R as for small R because as R increases, the graph of the logarithmic functions gets steadily flatter.

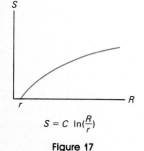

$S = C \ln(\frac{R}{r})$

Figure 17

Problem Set 5-4

In Problems 1–8, find the value of each natural logarithm.

1. ln e
2. $\ln(e^2)$
3. ln 1
4. $\ln\left(\frac{1}{e}\right)$

5. $\ln\sqrt{e}$
6. $\ln(e^{1.1})$
7. $\ln\left(\frac{1}{e^3}\right)$
8. $\ln(e^n)$

CAUTION

$$\ln\frac{4}{3} = \frac{\ln 4}{\ln 3}$$
$$= \frac{1.386}{1.099} = 1.261$$

$$\ln\frac{4}{3} = \ln 4 - \ln 3$$
$$= 1.386 - 1.099 = .287$$

For Problems 9–14, find each value. Assume that ln a = 2.5 and ln b = −.4.

9. $\ln(ae)$

10. $\ln\left(\dfrac{e}{b}\right)$

11. $\ln\sqrt{b}$

12. $\ln(a^2 b^{10})$

13. $\ln\left(\dfrac{1}{a^3}\right)$

14. $\ln(a^{4/5})$

Use the table of natural logarithms to calculate the values in Problems 15–20.

15. $\ln 120 = \ln 60 + \ln 2$

16. $\ln 150$

17. $\ln 690$

18. $\ln 84$

19. $\ln \frac{6}{5}$

20. $\ln 20{,}000$

In Problems 21–26, use the natural logarithm table to find N.

21. $\ln N = 2.208$

22. $\ln N = 1.808$

23. $\ln N = -.105$

24. $\ln N = -.916$

25. $\ln N = 4.500$

26. $\ln N = 9.000$

© Use your calculator ($\boxed{\ln}$ or $\boxed{\log_e}$ key) to find each of the following.

27. $\ln 4.31$

28. $\ln 517$

29. $\ln(.127)$

30. $\ln(.00424)$

31. $\ln\left(\dfrac{6.71}{42.3}\right)$

32. $\ln\sqrt{457}$

33. $\dfrac{\ln 6.71}{\ln 42.3}$

34. $\sqrt{\ln 457}$

35. $\ln(51.4)^3$

36. $\ln(31.2 + 43.1)$

37. $(\ln 51.4)^3$

38. $3 \ln 51.4$

© Use your calculator to find N in each of the following. Hint: ln N = 5.1 if and only if $N = e^{5.1}$. On some calculators, there is an $\boxed{e^x}$ key; on others, you use the two keys $\boxed{\text{INV}}$ $\boxed{\ln}$.

39. $\ln N = 2.12$

40. $\ln N = 5.63$

41. $\ln N = -.125$

42. $\ln N = .00257$

43. $\ln\sqrt{N} = 3.41$

44. $\ln N^3 = .415$

EXAMPLE A (Exponential Equations)   Solve $4^{3x-2} = 15$ for x.

Solution.   Take natural logarithms of both sides and then solve for x. Complete the solution using the ln table or a calculator.

$$\ln 4^{3x-2} = \ln 15$$

$$(3x - 2)\ln 4 = \ln 15$$

$$3x \ln 4 - 2 \ln 4 = \ln 15$$

$$3x \ln 4 = 2 \ln 4 + \ln 15$$

$$x = \frac{2 \ln 4 + \ln 15}{3 \ln 4}$$

$$x \approx 1.32$$

© Solve for x using the method above.

45. $3^x = 20$

46. $5^x = 40$

47. $2^{x-1} = .3$

48. $4^x = 3^{2x-1}$

49. $(1.4)^{x+2} = 19.6$

50. $5^x = \frac{1}{2}(4^x)$

SECTION 5-4   Natural Logarithms and Applications   199

EXAMPLE B (Doubling Time)   How long will it take money to double in value if it is invested at 9.5 percent interest compounded annually?

Solution.   For convenience, consider investing $1. From Section 5-2, we know that this dollar will grow to $(1.095)^t$ dollars after t years. Thus we must solve the exponential equation $(1.095)^t = 2$. This we do by the method of Example A, using a calculator. The result is

$$t = \frac{\ln 2}{\ln 1.095} \approx 7.64$$

C **51.** How long would it take money to double at 12 percent compounded annually?

C **52.** How long would it take money to double at 12 percent compounded monthly? *Hint:* After t months, $1 is worth $(1.01)^t$ dollars.

C **53.** How long would it take money to double at 15 percent compounded quarterly?

C **54.** A certain substance decays according to the formula $y = 100e^{-.135t}$, where t is in years. Find its half-life.

C **55.** By finding the natural logarithm of the numbers in each pair, determine which is larger.
   (a) $10^5,\ 5^{10}$ (b) $10^9,\ 9^{10}$
   (c) $10^{20},\ 20^{10}$ (d) $10^{1000},\ 1000^{10}$

56. What do your answers in Problem 55 confirm about the growth of 10^x and x^{10} for large x?

57. On the same coordinate plane, graph $y = 3^x$ and $y = x^3$ for $0 \le x \le 4$.

58. By means of a change of variable(s) (as explained in the text), transform each equation below to a linear equation. Find the slope and Y-intercept of the resulting line.
   (a) $y = 3e^{2x}$ (b) $y = 2x^3$ (c) $xy = 12$
   (d) $y = x^e$ (e) $y = 5(3^x)$ (f) $y = ex^{1.1}$

t	N
0	100
1	700
2	5000
3	40,000

Figure 18

EXAMPLE C (Curve Fitting)   The table in Figure 18 shows the number N of bacteria in a certain culture found after t hours. Which is a better description of these data,

$$N = ba^t \quad \text{or} \quad N = bt^a$$

Find a and b.

Solution.   Following the discussion of curve fitting in the text, we begin by plotting $\ln N$ against t. If the resulting points lie along a line, we choose $N = ba^t$ as the appropriate model. If not, we plot $\ln N$ against $\ln t$ to check on the second model.

   Since the fit to a line in the first case is quite good (Figure 19), we accept $N = ba^t$ as our model. To find a and b, we write $N = ba^t$ in the form

$$\ln N = \ln b + t \ln a \quad \text{or} \quad Y = \ln b + (\ln a)t$$

Examination of the line shows that it has a Y-intercept of 4.6 and a slope of about 2; so for its equation we write $Y = 4.6 + 2t$. Comparing this with $Y = \ln b + (\ln a)t$ gives

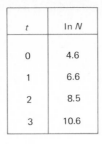

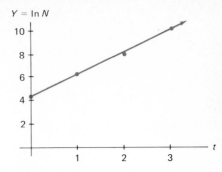

t	$\ln N$
0	4.6
1	6.6
2	8.5
3	10.6

Figure 19

$$\ln b = 4.6 \qquad \ln a = 2$$

Finally, we use the natural logarithm table or a calculator to find

$$b \approx 100 \qquad a \approx 7.4$$

Thus the original data are described reasonably well by the equation

$$N = 100(7.4)^t$$

For the data sets below, decide whether $y = ba^x$ or $y = bx^a$ is the better model. Then determine a and b.

59.

x	1	2	3	4
y	96	145	216	325

60.

x	0	1	2	4
y	243	162	108	48

61.

x	1	2	3	5
y	12	190	975	7490

62.

x	1	4	9
y	16	128	432

MISCELLANEOUS PROBLEMS

63. Evaluate without using a calculator or tables.

    (a) $\ln(e^{4.2})$ (b) $e^{2\ln 2}$ (c) $\dfrac{\ln 3e}{2 + \ln 9}$

64. Solve for x.
    (a) $\ln(5 + x) = 1.2$ (b) $e^{x^2-x} = 2$ (c) $e^{2x} = (.6)8^x$

65. Evaluate without use of a calculator or tables.
    (a) $\ln[(e^{3.5})^2]$ (b) $(\ln e^{3.5})^2$ (c) $\ln(1/\sqrt{e})$
    (d) $(\ln 1)/(\ln \sqrt{e})$ (e) $e^{3\ln 5}$ (f) $e^{\ln(1/2)+\ln(2/3)}$

66. Since $a^x = e^{\ln a^x} = e^{x\ln a}$, the study of exponential functions can be subsumed under the study of the function e^{kx}. Determine k so that each of the following is true. *Hint:* In (a) rewrite the equation as $3^x = (e^k)^x$ which implies that $3 = e^k$. Now take natural logarithms.
    (a) $3^x = e^{kx}$ (b) $\pi^x = e^{kx}$ (c) $(\frac{1}{3})^x = e^{kx}$

67. Solve for x.
    (a) $10^{2x+3} = 200$ (b) $10^{2x} = 8^{x-1}$ (c) $10^{x^2+3x} = 200$
    (d) $e^{-.32x} = 1/2$ (e) $x^{\ln x} = 10$ (f) $(\ln x)^{\ln x} = x$
    (g) $x^{\ln x} = x$ (h) $\ln x = (\ln x)^{\ln x}$ (i) $(x^2 - 5)^{\ln x} = x$

68. Show that each of the following is an identity. Assume $x > 0$.

(a) $(\sqrt{3})^{3\sqrt{3}} = (3\sqrt{3})^{\sqrt{3}}$ (b) $2.25^{3.375} = 3.375^{2.25}$

(c) $a^{\ln(x^b)} = (a^{\ln x})^b$ (d) $x^x = e^{x\ln x}$

(e) $(\ln x)^x = e^{x\ln(\ln x)}$ (f) $\dfrac{\ln\left(\dfrac{x+1}{x}\right)^x}{\ln\left(\dfrac{x+1}{x}\right)^{x+1}} = \dfrac{\left(\dfrac{x+1}{x}\right)^x}{\left(\dfrac{x+1}{x}\right)^{x+1}}$

69. A certain substance decays according to the formula $y = 100e^{-3t}$, where y is the amount present after t years. Find its half-life.

70. Suppose the number of bacteria in a certain culture t hours from now will be $200e^{.468t}$. When will the count reach 10,000?

71. A radioactive substance decays exponentially with a half-life of 240 years. Determine k in the formula $A = A_0e^{-kt}$, where A is the amount present after t years and A_0 is the initial amount.

72. By means of a change of variables using natural logarithms, transform each of the following equations to a linear equation. Then find the slope and the y-intercept of the resulting line.

(a) $xy^2 = 40$ (b) $y = 9e^{2x}$

73. Sketch the graph of

$$y = \frac{1}{\sqrt{2\pi}} e^{-(1/2)x^2}$$

This is the famous *normal curve,* so important in statistics.

74. In calculus it is shown that

$$e^x \approx 1 + x + \frac{x^2}{2} + \frac{x^3}{6} + \frac{x^4}{24} + \frac{x^5}{120}$$

Use this to approximate e and $e^{-1/2}$.

75. From Problem 74 or from looking at the graph of $y = e^x$, you might guess the true result that $e^x > 1 + x$ for all $x > 0$. Use this and the obvious fact that $(\pi/e) - 1 > 0$ to demonstrate algebraically that $e^\pi > \pi^e$.

76. It is important to have a feeling for how various functions grow for large x. Let $\ll$ symbolize the phrase *grows slower than*. Use a calculator to convince yourself that

$$\ln x \ll \sqrt{x} \ll x \ll x^2 \ll e^x \ll x^x$$

77. In calculus, it is shown that $(1 + r/m)^m$ gets closer and closer to e^r as m gets larger and larger. Now if P dollars is invested at rate r (written as a decimal) compounded m times per year, it will grow to $P(1 + r/m)^{mt}$ dollars at the end of t years (Example B of Section 5-2). If interest is compounded continuously (see the box *e* **and interest** on page 194), P dollars will grow to Pe^{rt} dollars at the end of t years. Use these facts to calculate the value of $100 after 10 years if interest is at 12 percent (that is, $r = .12$) and is compounded (a) monthly, (b) daily, (c) hourly, (d) continuously.

78. **TEASER** The *harmonic sum* $S_n = 1 + \frac{1}{2} + \frac{1}{3} + \frac{1}{4} + \cdots + 1/n$ has intrigued both amateur and professional mathematicians since the time of the Greeks. In calculus, it is shown that for $n > 1$

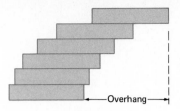

$$\ln n < 1 + \frac{1}{2} + \frac{1}{3} + \frac{1}{4} + \cdots + \frac{1}{n} < 1 + \ln n$$

This shows that S_n grows arbitrarily large but that it grows very very slowly.

(a) About how large would n need to be for $S_n > 100$?

(b) Show how you could stack a pile of identical bricks each of length 1 foot (one brick to a tier as shown in Figure 20) to achieve an overhang of 50 feet. Could you make the overhang 50 million feet? Yes, it does have something to do with part (a).

Figure 20

Henry Briggs 1561–1631

When 10 is used as the base for logarithms, some very nice things happen as the display at the right shows. Realizing this, John Napier and his friend Henry Briggs decided to produce a table of these logarithms. The job was accomplished by Briggs after Napier's death and published in 1624 in the famous book *Arithmetica Logarithmica*.

$\log .0853 = .9309 - 2$

$\log .853 \ = .9309 - 1$

$\log 8.53 \ = .9309$

$\log 85.3 \ = .9309 + 1$

$\log 853 \ = .9309 + 2$

$\log 8530 = .9309 + 3$

5-5   Common Logarithms (Optional)

In Section 5-3, we defined the logarithm of a positive number N to the base a as follows.

$$\log_a N = x \quad \text{if and only if} \quad a^x = N$$

Then in Section 5-4, we introduced base e, calling the result natural logarithms. These are the logarithms that are most important in advanced branches of mathematics such as calculus.

In this section, we shall consider logarithms to the base 10, often called **common logarithms,** or Briggsian logarithms. They have been studied by high school and college students for centuries as an aid to computation, a subject we take up in the next section. We shall write $\log N$ instead of $\log_{10} N$ (just as we write $\ln N$ instead of $\log_e N$). Note that

$$\log N = x \quad \text{if and only if} \quad 10^x = N$$

In other words, the common logarithm of any power of 10 is simply its exponent. Thus

$$\log 100 = \log 10^2 = 2$$

$$\log 1 = \log 10^0 = 0$$

$$\log .001 = \log 10^{-3} = -3$$

But how do we find common logarithms of numbers that are not integral powers of 10, such as 8.53 or 14,600? That is the next topic.

FINDING COMMON LOGARITHMS

We know that $\log 1 = 0$ and $\log 10 = 1$. If $1 < N < 10$, we correctly expect $\log N$ to be between 0 and 1. Table D (in the appendix) gives us four-place approximations of the common logarithms of all three-digit numbers between 1 and 10. For example,

$$\log 8.53 = .9309$$

We find this value by locating 8.5 in the left column and then moving right to the entry with 3 as heading. Similarly,

$$\log 1.08 = .0334$$

and

$$\log 9.69 = .9863$$

You should check these values.

You may have noticed in the opening box that log 8.53, log 8530, and log .0853 all have the same positive fractional part, .9309, called the **mantissa** of the logarithm. They differ only in the integer part, called the **characteristic** of the logarithm. To see why this is so, recall that

$$\log(M \cdot N) = \log M + \log N$$

Thus

$$\log 8530 = \log(8.53 \times 10^3) = \log 8.53 + \log 10^3$$

$$= .9309 + 3$$

$$\log .0853 = \log(8.53 \times 10^{-2}) = \log 8.53 + \log 10^{-2}$$

$$= .9309 - 2$$

Clearly the mantissa .9309 is determined by the sequence of digits 8, 5, 3, while the characteristic is determined by the position of the decimal point. Let us say that the decimal point is in **standard position** when it occurs immediately after the first nonzero digit. The characteristic for the logarithm of a number with the decimal point in standard position is 0. The characteristic is k if the decimal point is k places to the right of standard position; it is $-k$ if it is k places to the left of standard position.

CAUTION

$\log .00345 = -3.5378$
$\log .00345 = .5378 - 3$
$= -2.4622$

8530000. $\log 8,530,000 = .9309 + 6$

6 places right

.0000853   log .0000853 = .9309 − 5

5 places left

Here is another way of describing the mantissa and characteristic. If $c \times 10^n$ is the scientific notation for N, then log c is the mantissa and n is the characteristic of log N.

FINDING ANTILOGARITHMS

If we are to make significant use of common logarithms, we must know how to find a number when its logarithm is given. This process is called finding the inverse logarithm, or the **antilogarithm.** The process is simple: Use the mantissa to find the sequence of digits and then let the characteristic tell you where to put the decimal point.

Suppose, for example, that you are given

$$\log N = .4031 − 4$$

Locate .4031 in the body of Table D. You will find it across from 2.5 and below 3. Thus the number N must have 2, 5, 3 as its sequence of digits. Since the characteristic is −4, put the decimal point 4 places to the left of standard position. The result is

$$N = .000253$$

As a second example, let us find antilog 5.9547. The mantissa .9547 gives us the digits 9, 0, 1. Since the characteristic is 5,

$$\text{antilog } 5.9547 = 901,000$$

LINEAR INTERPOLATION

Suppose that for some function f we know $f(a)$ and $f(b)$, but we want $f(c)$, where c is between a and b (see the diagrams in Figure 21). As a reasonable approximation, we may pretend that the graph of f is a straight line between a and b. Then

$$f(c) \approx f(a) + d$$

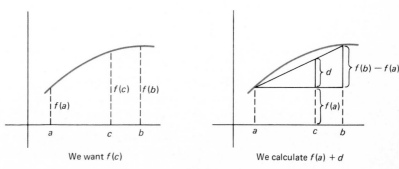

We want $f(c)$ We calculate $f(a) + d$

Figure 21

where, by similarity of triangles (see Figure 22).

$$\frac{d}{f(b) - f(a)} = \frac{c - a}{b - a}$$

That is,

$$d = \frac{f(b) - f(a)}{b - a}(c - a)$$

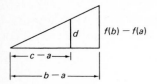

Figure 22

The process just described is called **linear interpolation.** The process of linear interpolation for the logarithm function is explained in Examples A and B of the problem set, as well as at the beginning of the appendix.

Problem Set 5-5

Find the common logarithm of each of the following numbers.

1. 10,000
2. 1,000,000
3. .01
4. .0001
5. $10^4 10^{3/2}$
6. $10^3 10^{4/3}$
7. $(10^3)^{-5}$
8. $(10^5)^{-1/10}$

Find N in each case.

9. $\log N = 4$
10. $\log N = 6$
11. $\log N = -2$
12. $\log N = -5$
13. $\log N = \frac{3}{2}$
14. $\log N = \frac{1}{3}$
15. $\log N = -\frac{3}{4}$
16. $\log N = -\frac{1}{6}$

Use Table B to find the following logarithms.

17. log 4.32
18. log 3.09
19. log 158
20. log 47.3
21. log .0329
22. log .0715
23. log 563,000
24. log 420,000
25. log (9.23×10^8)
26. log (2.83×10^{-11})

Find N in each case.

27. $\log N = 1.5159$
28. $\log N = 3.9015$
29. $\log N = .0043 - 2$
30. $\log N = .8627 - 4$
31. $\log N = 8.5999$
32. $\log N = 4.7427$

Find the antilogarithm of each number.

33. 2.2201
34. 3.8639
35. $.9232 - 1$
36. $.8500 - 5$

EXAMPLE A (Linear Interpolation in Finding Logarithms)   Find log 34.67.

Solution.   Our table gives the logarithms of 34.6 and 34.7, so we use linear interpolation to get an intermediate value. Here is how we arrange our work.

$$.10\left[.07\left[\begin{array}{l}\log 34.60 = 1.5391 \\ \log 34.67 = \quad ? \\ \log 34.70 = 1.5403\end{array}\right]d\right].0012$$

$$\frac{d}{.0012} = \frac{.07}{.10} = \frac{7}{10}$$

$$d = \frac{7}{10}(.0012) \approx .0008$$

$$\log 34.67 \approx \log 34.60 + d \approx 1.5391 + .0008 = 1.5399$$

Use linear interpolation in Table D to find each value.

37. log 5.237   38. log 9.826   39. log 7234

40. log 68.04   41. log .001234   42. log .09876

EXAMPLE B (Interpolation in Finding Antilogarithms)   Find antilog 2.5285.

Solution.   We find .5285 sandwiched between .5276 and .5289 in the body of Table D.

$$.0013\left[.0009\left[\begin{array}{l}\text{antilog } 2.5276 = 337.0 \\ \text{antilog } 2.5285 = \quad ? \\ \text{antilog } 2.5289 = 338.0\end{array}\right]d\right]1.0$$

$$\frac{d}{1.0} = \frac{.0009}{.0013} = \frac{9}{13}$$

$$d = \frac{9}{13}(1.0) \approx .7$$

$$\text{antilog } 2.5285 \approx 337.0 + .7 = 337.7$$

Find the antilogarithm of each of the following using linear interpolation.

43. 0.8497   44. 0.8516   45. 3.9130

46. 1.9849   47. .6004 − 2   48. .4946 − 4

MISCELLANEOUS PROBLEMS

49. Without using a calculator, find the common logarithm of each of the following.
   (a) $10^{3/2}\, 10^{-1/4}$ (b) $(.0001)^{1/3}$
   (c) $\sqrt[3]{10}\,\sqrt{10}$ (d) $10^{\log(.001)}$

50. Find N in each case.
   (a) $\log N = 0$ (b) $\log N = -2$
   (c) $\log N = \frac{1}{2}$ (d) $\log N = 10$

51. Use Table D with interpolation to find each of the following.
   (a) log 492.7   (b) log .04705
   (c) antilog 2.9327   (d) antilog(.2698 − 3)

52. Use Table D to find N. *Hint:* $-2.4473 = .5527 - 3$.
   (a) $\log N = -2.4473$ (b) $\log N = -4.0729$

[c] 53. Do Problem 52 using your calculator.

54. Find the characteristic of log N if:
    (a) $10^{11} < N < 10^{12}$;   (b) $.00001 < N < .0001$.

55. Use Table D to find $\log \dfrac{982 - 467}{(982)(267)}$.

56. If $b = .001\,a$ and $\log a = 5.5$, find $\log(a^3/b^4)$.

57. Evaluate antilog$\left[\log\left(\dfrac{999}{4.71 \times 328}\right) - \log 999 + \log 328 \right]$.

58. Find x if $(\log x)^2 + \log(x^2) = 10^{\log 3}$

59. Let $\log N = 15.992$. In the decimal notation for N, how many digits are there before the decimal point?

60. Convince yourself that the number of digits in a positive integer N is $[\log N] + 1$. Then find the number of digits in 50^{50} when expanded out in the usual way in decimal notation.

[c] 61. If one can write 6 digits to the inch, about how many miles long would the number $9^{(9^9)}$ be when written in decimal notation?

62. **TEASER**   Let $N = 9^{(9^9)}$, A = sum of digits in N, B = sum of digits in A, and C = sum of digits in B. Find C. *Hint:* If a number is divisible by 9, so is the sum of its digits (why?). Since N is divisible by 9, it follows that A, B, and C are all divisible by 9.

The Laws of Logs

If $M > 0$ and $N > 0$, then

1. $\log M \cdot N = \log M + \log N$

2. $\log \dfrac{M}{N} = \log M - \log N$

3. $\log M^n = n \log M$

"The miraculous powers of modern calculation are due to three inventions: the Arabic Notation, Decimal Fractions, and Logarithms."

F. Cajori, 1897

"Electronic calculators make calculations with logarithms as obsolete as whale oil lamps."

Anonymous Reviewer, 1982

5-6   Calculations with Logarithms (Optional)

For 300 years, scientists depended on logarithms to reduce the drudgery associated with long computations. The invention of electronic computers and calculators has diminished the importance of this long-established technique. Still, we think that any student of algebra should know how products, quotients, powers, and roots can be calculated by means of common logarithms.

About all you need are the three laws stated above and Appendix Table D. A little common sense and the ability to organize your work will help.

PRODUCTS

Suppose you want to calculate $(.00872)(95,300)$. Call this number x. Then by Law 1 and Table D,

$$\log x = \log .00872 + \log 95,300$$
$$= (.9405 - 3) + 4.9791$$
$$= 5.9196 - 3$$
$$= 2.9196$$

Now use Table D backward to find that antilog $.9196 = 8.31$, so $x =$ antilog $2.9196 = 831$.

Here is a good way to organize your work in a compact systematic way.

$$x = (.00872)(95,300)$$

$$\begin{array}{r} \log .00872 = \quad .9405 - 3 \\ (+) \ \underline{\log 95300 = 4.9791} \\ \log x = \overline{5.9196 - 3} \end{array}$$

$$x = 831$$

QUOTIENTS

Suppose we want to calculate $x = .4362/91.84$. Then by Law 2,

$$\log x = \log .4362 - \log 91.84$$
$$= (.6397 - 1) - 1.9630$$
$$= .6397 - 2.9630$$
$$= -2.3233$$

What we have done is correct; however, it is poor strategy. The result we found for $\log x$ is not in characteristic-mantissa form and therefore is not usable. Remember that the mantissa must be positive. We can bring this about by adding and subtracting 3.

$$-2.3233 = (-2.3233 + 3) - 3 = .6767 - 3$$

Actually it is better to anticipate the need for doing this and arrange the work as follows.

$$x = \frac{.4362}{91.84}$$

$$\begin{array}{r} \log .4362 = .6397 - 1 = 2.6397 - 3 \\ (-) \ \underline{\log 91.48 = \qquad\qquad 1.9630} \\ \log x = \qquad\qquad .6767 - 3 \end{array}$$

$$x = .00475$$

POWERS OR ROOTS

Here the main tool is Law 3. We illustrate with two examples.

$$x = (31.4)^{11}$$

$$\log x = 11 \log 31.4$$

$$\log 31.4 = 1.4969$$

$$11 \log 31.4 = 16.4659$$

$$\log x = 16.4659$$

$$x = 29{,}230{,}000{,}000{,}000{,}000$$

$$= 2.923 \times 10^{16}$$

$$x = \sqrt[4]{.427} = (.427)^{1/4}$$

$$\log x = \frac{1}{4} \log .427$$

$$\log .427 = .6304 - 1$$

$$\frac{1}{4} \log .427 = \frac{1}{4}(.6304 - 1) = \frac{1}{4}(3.6304 - 4) = .9076 - 1$$

$$\log x = .9076 - 1$$

$$x = .8084$$

Notice in the second example that we wrote $3.6304 - 4$ in place of $.6304 - 1$, so that multiplication by $\frac{1}{4}$ gave the logarithm in characteristic-mantissa form.

Problem Set 5-6

Use logarithms and Table D without interpolation to find approximate values for each of the following.

1. $(46.3)(2.76)$ 　　　2. $(378)(9.63)$ 　　　3. $\dfrac{46.3}{483}$

4. $\dfrac{437}{92300}$ 　　　5. $\dfrac{.00912}{.439}$ 　　　6. $\dfrac{.0429}{15.7}$

7. $(37.2)^5$ 　　　8. $(113)^3$ 　　　9. $\sqrt[3]{42.9}$

10. $\sqrt[4]{312}$ 　　　11. $\sqrt[5]{.918}$ 　　　12. $\sqrt[3]{.0307}$

13. $(14.9)^{2/3}$ 　　　14. $(98.6)^{3/4}$

Use logarithms and Table D with interpolation to approximate each of the following.

15. $(31.96)(149)$ 　　　16. $(6236)(.00108)$ 　　　17. $\dfrac{43.98}{7.16}$

18. $\dfrac{115}{4.623}$ 19. $(.1234)^6$ 20. $(92.83)^3$

EXAMPLE A (More Complicated Calculations)   Use logarithms, without interpolation, to calculate

$$\frac{(31.4)^3(.982)}{(.0463)(824)}$$

Solution.   Let N denote the entire numerator, D the entire denominator, and x the fraction N/D. Then

$$\log x = \log N - \log D$$

where

$$\log N = 3 \log 31.4 + \log .982$$

$$\log D = \log .0463 + \log 824$$

Here is a good way to organize the work.

$$\log 31.4 = 1.4969$$

$$3 \log 31.4 = 4.4907 \qquad\qquad \log .0463 = .6656 - 2$$
$$(+)\quad \log .982 = \underline{.9921 - 1} \qquad (+)\quad \underline{\log 824 = 2.9159}$$
$$\log N = 5.4828 - 1 \qquad\qquad \log D = \overline{3.5815 - 2}$$
$$= 4.4828 \qquad\qquad\qquad\qquad = 1.5815$$

$$\log N = 4.4828$$
$$(-) \log D = \underline{1.5815}$$
$$\log x = \overline{2.9013}$$
$$x = 797$$

Carry out the following calculations using logarithms without interpolation.

21. $\dfrac{(.56)^2(619)}{21.8}$ 22. $\dfrac{.413}{(4.9)^2(.724)}$

23. $\dfrac{(14.3)\sqrt{92.3}}{\sqrt[3]{432}}$ 24. $\dfrac{(91)(41.3)^{2/3}}{42.6}$

EXAMPLE B (Solving Exponential Equations)   Solve the equation

$$2^{2x-1} = 13$$

Solution.   We begin by taking logarithms of both sides and then solving for x.

$$(2x - 1)\log 2 = \log 13$$

$$2x - 1 = \frac{\log 13}{\log 2} = \frac{1.1139}{.3010}$$

$$\log(2x - 1) = \log 1.1139 - \log .3010$$

$$= (1.0469 - 1) - (.4786 - 1)$$

$$= .5683$$

$$2x - 1 = \text{antilog} \ .5683 = 3.70$$

$$2x = 4.70$$

$$x = 2.35$$

Notice that we did the division of (log 13)/(log 2) by means of logarithms. We could have done it by long division (or on a calculator) if we preferred.

Use logarithms to solve the following exponential equations. You need not interpolate.

25. $3^x = 300$ 26. $5^x = 14$ 27. $10^{2-3x} = 6240$

28. $10^{5x-1} = .00425$ 29. $2^{3x} = 3^{x+2}$ 30. $2^{x^2} = 3^x$

MISCELLANEOUS PROBLEMS

31. Use common logarithms and Table D to calculate each of the following.
    (a) $\sqrt[3]{.0427}$ (b) $\dfrac{(42.9)^2(.983)}{\sqrt{323}}$ (c) $\dfrac{10^{6.42}}{8^{7.2}}$

32. Solve for x by using common logarithms.
    (a) $4^{2x} = 150$ (b) $(.975)^x = .5$

33. Solve for x.
    (a) $\log(x + 2) - \log x = 1$ (b) $\log(2x + 1) - \log(x + 3) = 0$
    (c) $\log(x + 3) - \log x + \log 2x^2 = \log 8$
    (d) $\log(x + 3) + \log(x - 1) = \log 4x$

34. Suppose that the amount Q of a radioactive substance (in grams) remaining t years from now will be $Q = (42)2^{-.017t}$. After how many years will the amount remaining be .42 grams?

35. Suppose that the bacteria count in a certain culture t hours from now is $(800)3^t$. When will the count reach 100,000?

36. Assume the population of the earth was 4.19 billion people in 1977 and that the growth rate is 2 percent per year. Then the population t years after 1977 should be $4.19(1.02)^t$ billion.
    (a) What will the population be in the year 2000?
    (b) When will the population reach 8.3 billion?

37. Answer the two questions of Problem 36 assuming the growth rate is only 1.7 percent.

38. **TEASER** Show that log 2 is irrational. For what positive integers n is log n irrational?

Chapter Summary

The symbol $\sqrt[n]{a}$, the principal nth root of a, denotes one of the numbers whose nth power is a. For odd n, that is all that needs to be said. For n even and $a > 0$, we specify that $\sqrt[n]{a}$ signifies the positive nth root. Thus $\sqrt[3]{-8} = -2$ and $\sqrt{16} = \sqrt[2]{16} = 4$. (It is wrong to write $\sqrt{16} = -4$.)

The key to understanding **rational exponents** is the definition $a^{1/n} = \sqrt[n]{a}$, which implies $a^{m/n} = (\sqrt[n]{a})^m$. Thus $16^{5/4} = (\sqrt[4]{16})^5 = 2^5 = 32$. The meaning of **real exponents** is determined by considering rational approximations. For example, 2^π is the number that the sequence 2^3, $2^{3.1}$, $2^{3.14}$, ... approaches. The function $f(x) = a^x$ (and more generally $f(x) = b \cdot a^x$) is called an **exponential function.**

A variable y is **growing exponentially** or **decaying exponentially** according as $a > 1$ or $0 < a < 1$ in the equation $y = b \cdot a^x$. Typical of the former are biological populations; of the latter, radioactive elements. Corresponding key ideas are **doubling times** and **half-lives.**

Logarithms are exponents. In fact, $\log_a N = x$ means $a^x = N$, that is, $a^{\log_a N} = N$. The functions $f(x) = \log_a x$ and $g(x) = a^x$ are **inverses** of each other. Logarithms have three primary properties (page 187). **Natural logarithms** correspond to the choice of base $a = e = 2.71828 \ldots$ and play a fundamental role in advanced courses. **Common logarithms** correspond to base 10 and have historically been used to simplify arithmetic calculations.

Chapter Review Problem Set

1. Simplify, writing your answer in exponential form.

   (a) $\sqrt[3]{-8y^6z^2}$ (b) $\sqrt[4]{32x^2y^8}$ (c) $\sqrt{4\sqrt[3]{5}}$

2. Solve the equations.

   (a) $(x - 3)^{1/2} = 3$ (b) $x^{1/2} = 6 - x$

3. Simplify, writing your answer in exponential form with all positive exponents.

   (a) $(25a^2)^{3/2}$ (b) $(a^{-1/2}aa^{-3/4})^2$ (c) $\dfrac{1}{5\sqrt[4]{5^3}}$

   (d) $\dfrac{(3x^{-2}y^{3/4})^2}{3x^2y^{-2/3}}$ (e) $(x^{1/2} - y^{1/2})^2$ (f) $\sqrt[3]{4}\sqrt[4]{4}$

4. Sketch the graph of $y = (\tfrac{3}{2})^x$ and use your graph to estimate the value of $(\tfrac{3}{2})^\pi$.

5. A certain radioactive substance has a half-life of 3 days. What fraction of an initial amount will be left after 243 days?

6. A population grows so that its doubling time is 40 years. If this population is 1 million today, what will it be after 160 years?

7. If $100 is put in the bank at 8 percent interest compounded quarterly, what will it be worth at the end of 10 years?

8. Find x in each of the following.

   (a) $\log_4 64 = x$ (b) $\log_2 x = -3$
   (c) $\log_x 49 = 2$ (d) $\log_4 x = 0$
   (e) $\log_9 27 = x$ (f) $\log_{10} x + \log_{10}(x - 3) = 1$
   (g) $a^{\log_a 10} = x$ (h) $x = \log_a(a^{1.14})$

9. Write as a single logarithm.

$$2 \log_4(3x + 1) - \frac{1}{2} \log_4 x + \log_4(x - 1)$$

10. Evaluate.
    (a) $\log_{10}\sqrt{1000}$ (b) $\log_{27} 81$
    (c) $\ln\sqrt{e}$ (d) $\ln(1/2^4)$

11. Use the table of natural logarithms to determine each of the following.
    (a) $\ln(\frac{7}{4})^3$ (b) N if $\ln N = 2.230$
    (c) $\ln[(3.4)(9.9)]$ (d) N if $\ln N = -0.105$

12. By taking ln of both sides, solve $2^{x+1} = 7$.

13. A certain substance decays according to the formula $y = y_0 e^{-.05t}$, where t is measured in years. Find its half-life.

14. A substance initially weighing 100 grams decays exponentially according to the formula $y = 100e^{-kt}$. If 30 grams are left after 10 days, determine k.

15. Sketch the graphs of $y = \log_3 x$ and $y = 3^x$ using the same coordinate axes.

16. Use common logarithms to calculate

$$\frac{(13.2)^4\sqrt{15.2}}{29.6}$$

Appollonius' metric treatment of the conic sections—ellipses, hyperbolas, and parabolas— was one of the great mathematical achievements of antiquity. The importance of conic sections for pure and applied mathematics (for example, the orbits of the planets and of electrons in the hydrogen atom are conic sections) can hardly be overestimated.

Richard Courant and Herbert Robbins

CHAPTER 6

Analytic Geometry

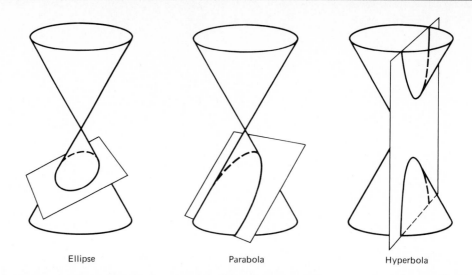

Ellipse   Parabola   Hyperbola

6-1 Parabolas

Analytic geometry could well be called algebraic geometry for it is the study of geometric concepts such as curves and surfaces by means of algebra. Among the most important curves are the conic sections, curves that are obtained by intersecting a cone of two nappes with a plane (see our opening display). We are especially interested in the three general cases of an ellipse, parabola, and hyperbola though we shall also consider certain limiting forms like a circle, two intersecting lines, and so on. We begin with the parabola, a curve already discussed in Section 1-3. Here we give a very general treatment based on the geometric definition given to us by the Greeks.

THE GEOMETRIC DEFINITION OF A PARABOLA

A parabola is the set of points P that are equidistant from a fixed line l (the directrix) and a fixed point F (the focus). In other words, a parabola is the set of points P in Figure 1 satisfying

$$d(P, L) = d(P, F)$$

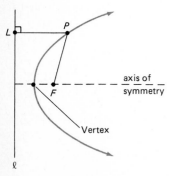

Figure 1

Here, L is the point of l closest to P and, as usual, $d(A, B)$ denotes the distance between the points A and B.

A little thought convinces us that a parabola is a two-armed curve opening ever wider and symmetric with respect to the line through the focus perpendicular to the directrix. This line is called the **axis of symmetry** and the point where this line intersects the parabola is called the **vertex**. Note that the vertex is the point of the parabola closest to the directrix.

THE EQUATION OF A PARABOLA

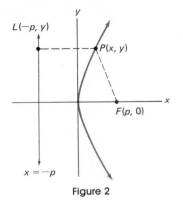
Figure 2

Place the parabola in the coordinate system so that its axis is the x-axis and its vertex is the origin (Figure 2). Let the focus be to the right of the origin, for example, at $(p, 0)$; then the directrix is the line $x = -p$. If $P(x, y)$ is any point on the curve, it must satisfy

$$d(P, F) = d(P, L)$$

which, because of the distance formula, assumes the form

$$\sqrt{(x - p)^2 + (y - 0)^2} = \sqrt{(x + p)^2 + (y - y)^2}$$

Since both sides are positive, this is equivalent to the result when both sides are squared.

$$x^2 - 2px + p^2 + y^2 = x^2 + 2px + p^2$$

This, in turn, simplifies to

$$y^2 = 4px$$

The final equation is called the **standard equation of the parabola.** It is easy to write, simple to graph, and has a form which is straightforward to interpret. For example, we may replace y by $-y$ without affecting the equation, which means that the graph is symmetric with respect to the x-axis. It crosses the x-axis at the origin which is the vertex. The positive number p measures the distance from the focus to the vertex.

The equation just derived has three other variants. If we interchange the roles of x and y (giving $x^2 = 4py$), we have the equation of a parabola that opens upward with the y-axis as its axis. The corresponding parabolas which open to the left and down have equations $y^2 = -4px$ and $x^2 = -4py$, respectively. All of this is summarized in Figure 3 on the next page.

What would be the equation of the parabola with vertex at the origin and focus at the point $(0, -3)$? This is a parabola of the fourth type in Figure 3; it turns down with $p = 3$. We conclude that its equation is $x^2 = -4(3)y = -12y$.

Conversely, suppose that we want to find the focus of the parabola with equation $y^2 = -2x$. Write the equation as $y^2 = -4 \cdot \frac{1}{2} \cdot x$, which is of the third type in Figure 3. We conclude that the parabola opens left, that $p = \frac{1}{2}$, and that the focus is at $(-\frac{1}{2}, 0)$.

In Section 1-4, we claimed that the graph of an equation of the form $y = ax^2$ was a parabola. Thus, for example, the graph of $y = \frac{1}{3}x^2$ should be a parabola. To confirm this, write the equation in the form $x^2 = 3y = 4 \cdot \frac{3}{4}y$. This is the equation of a parabola with vertex at the origin and opening up. Since $p = \frac{3}{4}$, its focus is at $(0, \frac{3}{4})$.

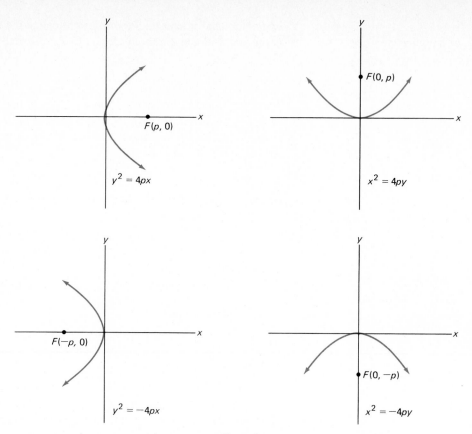

Figure 3

APPLICATIONS OF THE PARABOLA

Perhaps the most important property of the parabola is its *optical property*. Consider a cup-shaped mirror with a parabolic cross section (Figure 4). If a light source is placed at the focus, the resulting rays of light are reflected from the mirror in a beam in which all the rays are parallel to the axis (Figure 5). This fact is used in designing search lights. Conversely, if parallel light rays (as from a star) hit a parabolic mirror, they will be "focused" at the focus. This is the basis for the design of one type of reflecting telescope. The optical property of the parabola is usually demonstrated by means of calculus, but

Cross section of a parabolic mirror
with light source at focus

Figure 4

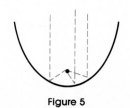

Figure 5

there is a way to do it that involves only geometry and algebra (see Problems 34 and 35).

In calculus, it is shown that the path of a projectile is a parabola and that the cables of a suspension bridge have a parabolic shape. We touch on these and other applications in the problem set.

Problem Set 6-1

Each equation below determines a parabola with vertex at the origin. In what direction does the parabola open?

1. $x^2 = 8y$
2. $y^2 = -2x$
3. $y^2 = 6x$
4. $x^2 = -3y$
5. $3y^2 = -5x$
6. $y = -2x^2$

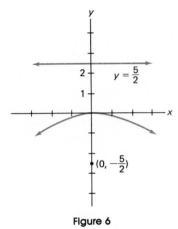

Figure 6

EXAMPLE A (Finding the Focus and Directrix) Determine the focus and directrix of the parabola with equation $y = -\frac{1}{10}x^2$. Then sketch its graph.

Solution. We first write the equation in standard form by solving for the quadratic term and then factoring out 4.

$$x^2 = -10y = -4(\tfrac{5}{2})y$$

This is the equation of a parabola with $p = \frac{5}{2}$; it opens downward. The focus is $(0, -\frac{5}{2})$ and the directrix is the line $y = \frac{5}{2}$. All this and the graph are shown in Figure 6.

In Problems 7–14, find p and then sketch the graph, showing the focus and directrix.

7. $x^2 = -8y$
8. $y^2 = 3x$
9. $y^2 = \frac{1}{2}x$
10. $y = -3x^2$
11. $y = \frac{1}{2}x^2$
12. $6x = -4y^2$
13. $9x = 4y^2$
14. $x = .125y^2$

15. Determine the coordinates of two points on the parabola $y = 4x^2$ with y-coordinate 1.

16. Determine two points on the parabola $y^2 = 8x$ with the same x-coordinate as the focus.

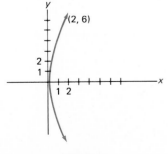

Figure 7

EXAMPLE B (Parabolas with Side Conditions) Find the equation of the parabola that has vertex at the origin, opens to the right, and passes through $(2, 6)$. See Figure 7.

Solution. The equation has the form $y^2 = 4px$. Since $(2, 6)$ lies on the parabola, $6^2 = 4p(2)$, or $p = \frac{36}{8} = \frac{9}{2}$. Thus $y^2 = (4)(\frac{9}{2})x$, which simplifies to $y^2 = 18x$.

Find the equation of the parabola with vertex at (0, 0) that satisfies the given conditions.

17. Opens up; goes through (2, 6).
18. Opens down; goes through (−2, −4).
19. Directrix is $y = 3$.
20. Focus is (−4, 0).
21. Goes through (1, 2) and (1, −2).
22. Goes through (1, 4) and (2, 16).

MISCELLANEOUS PROBLEMS

23. Find the focus and directrix of the parabola with equation $4x = -5y^2$.
24. Find the equation of the parabola with vertex at (0, 0) which in addition satisfies the following condition.
    (a) Its focus is at ($\frac{5}{2}$, 0)
    (b) It opens down and passes through (3, −10)
25. The chord of a parabola through the focus and perpendicular to the axis is called **the latus rectum** of the parabola. Find the length of the latus rectum for the parabola $4px = y^2$.
26. The chord of a parabola that is perpendicular to the axis and 1 unit from the vertex has length 3 units. How long is its latus rectum?
27. A door in the shape of a parabolic arch (Figure 8) is 12 feet high at the center and 5 feet wide at the base. A rectangular box 9 feet tall is to be slid through the door. What is the widest the box can be?
28. The path of a projectile fired from ground level is a parabola opening down. If the greatest height reached by the projectile is 100 meters and if its range (horizontal reach) is 800 meters, what is the horizontal distance from the point of firing to the point where the projectile first reaches a height of 64 meters?
29. The cables for the central span of a suspension bridge take the shape of a parabola as shown in Figure 9. If the towers are 800 meters apart and the cables are attached to them at points 400 meters above the floor of the bridge, how long is the vertical strut that is 100 meters from the tower? Assume the vertex of the parabola is on the floor of the bridge.
30. In Figure 10, AP is parallel to the x-axis and Q is the midpoint of AP. Find the equation of the path of $Q(x_1, y_1)$ as $P(x, y)$ moves along the path $x^2 = 4py$.

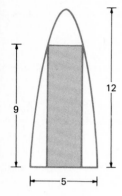

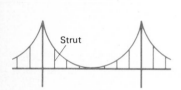

Figure 8

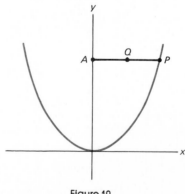

Figure 9

Figure 10

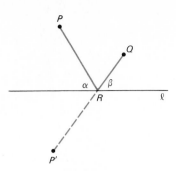

Figure 11

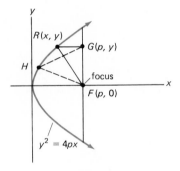

Figure 12

31. Suppose that a submarine has been ordered to follow a path that keeps it equidistant from a circular island of radius r and a straight shoreline that is $2p$ units from the edge of the island. Derive the equation of the submarine's path, assuming that the shoreline has equation $x = -p$ and that the center of the island is on the x-axis.

32. Show that there is no point $P(x, y)$ on the parabola $x^2 = 8y$ for which OP is perpendicular to PF, F being the focus of the parabola. *Hint:* Two lines with slopes m_1 and m_2 are perpendicular if and only if $m_1 m_2 = -1$.

33. An equilateral triangle is inscribed in the parabola $y^2 = 4px$ with one vertex at the origin. Find the length of a side of the triangle.

34. Consider a line l, two fixed points P and Q on the same side of l, and a (variable) point R on l. Use Figure 11 to show that the distance $\overline{PR} + \overline{RQ}$ is minimized precisely when $\alpha = \beta$. *Note:* Since a light ray is known to be reflected from a mirror l so that the angle of incidence equals the angle of reflection, we see that a light ray from P to l to Q picks the shortest path.

35. (Optical property of the parabola)   Imagine the parabola $y^2 = 4px$ of Figure 12 to be a mirror with points F, R, G, and H as indicated and with RG parallel to the x-axis.
    (a) Show that $\overline{FR} + \overline{RG} = 2p$.
    (b) Show that $\overline{FH} + \overline{HG} > 2p$.

    Conclude from Problem 34 that a light ray from the focus to a parabolic mirror is reflected parallel to the axis of the parabola.

36. **TEASER**   Consider the parabola $y = x^2$ (Figure 13). Let T_1 be the triangle with vertices on this parabola at a, c, and b with c midway between a and b. Let T_2 be the union of the two triangles with vertices on the parabola at a, d, c and c, e, b, respectively, with d midway between a and c and e midway between c and b. In a similar manner, let T_3 be the union of four triangles with vertices on the parabola, and so on.
    (a) Show that the area of T_1 is given by $A(T_1) = (b - a)^3/8$.
    (b) Show that $A(T_2) = A(T_1)/4$.
    (c) Find the area of the curved parabolic segment below the line PQ.

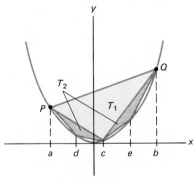

Figure 13

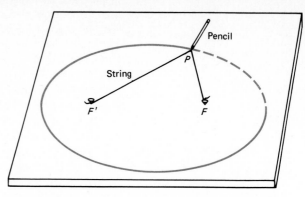

This shows the drawing of an ellipse. A string is tacked down at its ends by thumbtacks. A pencil pulls the string taut.

6-2 Ellipses

Our opening display suggests the geometric definition of the ellipse. An **ellipse** is the set of points P for which the sum of the distances from two fixed points F' and F is a constant. In other words, an ellipse is the set of points P in Figure 14 satisfying

$$d(P, F') + d(P, F) = 2a$$

for some positive constant a.

The two fixed points F' and F are called **foci** (plural of focus) and the point midway between the foci is the **center** of the ellipse. We call the line through the two foci the **major axis** of the ellipse; the line through the center and perpendicular to the major axis is its **minor axis**. Note that the ellipse is symmetric with respect to both its major and minor axes. Finally, the intersection of the ellipse with the major axis determines the two points A' and A, which are called **vertices**. All this is shown in Figure 14.

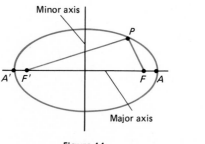

Figure 14

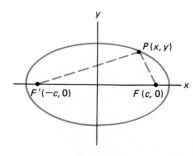

Figure 15

THE EQUATION OF AN ELLIPSE

Place the ellipse in the coordinate system so that its center is at the origin with the major axis along the x-axis. We may suppose the two foci F' and F to be located at $(-c, 0)$ and $(c, 0)$, where c is a positive constant (Figure 15). Then $d(P, F') + d(P, F) = 2a$ combined with the distance formula yields

$$\sqrt{(x + c)^2 + (y - 0)^2} + \sqrt{(x - c)^2 + (y - 0)^2} = 2a$$

or, equivalently,

$$\sqrt{(x + c)^2 + y^2} = 2a - \sqrt{(x - c)^2 + y^2}$$

After squaring both sides, we obtain

$$(x + c)^2 + y^2 = 4a^2 - 4a\sqrt{(x - c)^2 + y^2} + (x - c)^2 + y^2$$

and this in turn simplifies to

$$4cx - 4a^2 = -4a\sqrt{(x - c)^2 + y^2}$$

If we now divide both sides by 4 and square again, we get

$$(cx - a^2)^2 = a^2[(x - c)^2 + y^2]$$

$$c^2x^2 - 2a^2cx + a^4 = a^2[x^2 - 2cx + c^2 + y^2]$$

$$a^4 - a^2c^2 = a^2x^2 - c^2x^2 + a^2y^2$$

$$a^2(a^2 - c^2) = (a^2 - c^2)x^2 + a^2y^2$$

Finally, divide both sides by $a^2(a^2 - c^2)$ and interchange the two sides of the equation to obtain

$$\frac{x^2}{a^2} + \frac{y^2}{a^2 - c^2} = 1$$

It is clear from Figure 15 that $a > c$, so we may let $b^2 = a^2 - c^2$. This results in what we shall call the **standard equation of the ellipse**, namely,

$$\frac{x^2}{a^2} + \frac{y^2}{b^2} = 1$$

INTERPRETING a, b, AND c

We have used c to denote the distance from the center to a focus. If we apply the defining condition for the ellipse to the vertex A, we obtain

$$d(A, F') + d(A, F) = 2a$$

which implies (see Figure 14) that the distance between the two vertices is $2a$. For this reason, we refer to the number $2a$ as the **major diameter** of the ellipse. Since $b^2 + c^2 = a^2$, b and c are the legs of a right triangle with hypotenuse a and it follows that $2b$ is the **minor diameter** of the ellipse. All this is

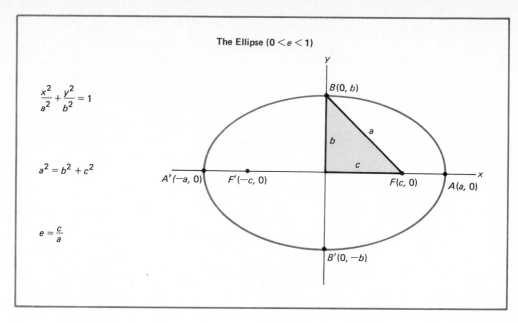

The Ellipse $(0 < e < 1)$

$$\frac{x^2}{a^2} + \frac{y^2}{b^2} = 1$$

$$a^2 = b^2 + c^2$$

$$e = \frac{c}{a}$$

Figure 16

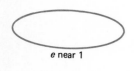

e near 1

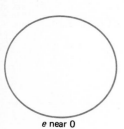

e near 0

Figure 17

summarized in Figure 16, where you should note especially the significance of the triangle with sides a, b, and c. We call it the fundamental triangle for the ellipse.

The number $e = c/a$, which varies between 0 and 1, measures the **eccentricity** of the ellipse. If e is near 1, the ellipse is very eccentric (long and narrow); if e is near 0, the ellipse is almost circular (Figure 17). Sometimes, a circle is referred to as an ellipse of eccentricity 0; for in this case $c = 0$ and $a = b$ and the standard equation takes the form

$$\frac{x^2}{a^2} + \frac{y^2}{a^2} = 1$$

which is equivalent to the familiar circle equation $x^2 + y^2 = a^2$.

As an example of the equation of an ellipse, consider

$$\frac{x^2}{36} + \frac{y^2}{4} = 1$$

Note that $a = 6$ and $b = 2$, so this is the equation of an ellipse with center at the origin, major diameter $2a = 12$ and minor diameter $2b = 4$. Since $c = \sqrt{a^2 - b^2} = \sqrt{32} = 4\sqrt{2}$, the foci are at $(\pm 4\sqrt{2}, 0)$ and the eccentricity is $4\sqrt{2}/6 \approx .94$ (see Figure 18).

One other observation should be made. If we interchange the roles of x and y, then the standard equation takes the form

$$\frac{x^2}{b^2} + \frac{y^2}{a^2} = 1$$

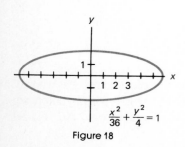

$$\frac{x^2}{36} + \frac{y^2}{4} = 1$$

Figure 18

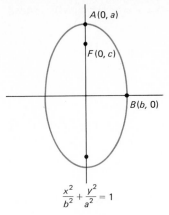

$$\frac{x^2}{b^2} + \frac{y^2}{a^2} = 1$$

Figure 19

The major axis is now the y-axis; the vertices and foci lie on it (Figure 19). For example, the equation

$$\frac{x^2}{16} + \frac{y^2}{25} = 1$$

represents a *vertical* ellipse (the major axis is vertical). Its vertices are at $(0, \pm 5)$, its foci are at $(0, \pm 3)$, its x-intercepts are at $(\pm 4, 0)$, and its eccentricity is $3/5 = .6$. One can always tell from the equation whether the corresponding ellipse is horizontal or vertical by noting whether the larger square in the denominator is in the x-term or the y-term.

APPLICATIONS

Like the parabola, the ellipse has an important optical property. If we imagine the ellipse to represent a mirror, then a light ray emanating from one focus will be reflected from the ellipse back through the other focus (see Problem 33 for a demonstration of this fact). This is the basis for the whispering gallery effect resulting from the elliptical shaped domes of St. Paul's Cathedral in London and the National Statuary Hall in the United States Capitol.

A much more significant application is the observation by Kepler that the planets move around the sun in elliptical orbits. Later, Newton established that this is a consequence of the fact that the gravitational force of attraction between two bodies is inversely proportional to the square of the distance between them. For the same reason, the electrons in the Bohr model of the hydrogen atom travel in elliptical orbits.

Problem Set 6-2

In Problems 1-8, decide whether the ellipse with the given equation is horizontal or vertical and then determine the major and minor diameters. In Problems 5–8, you will first have to rewrite the equation in standard form.

1. $\frac{x^2}{7} + \frac{y^2}{16} = 1$ 2. $\frac{x^2}{9} + \frac{y^2}{8} = 1$

3. $\frac{x^2}{36} + \frac{y^2}{20} = 1$ 4. $\frac{x^2}{12} + \frac{y^2}{25} = 1$

5. $4x^2 + 9y^2 = 4$ 6. $9x^2 + 8y^2 = 18$

7. $4k^2x^2 + k^2y^2 = 1$ 8. $k^2x^2 + (k^2 + 1)y^2 = k^2$

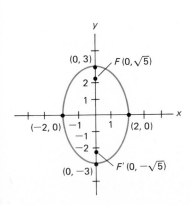

Figure 20

EXAMPLE A (Graphing the Equation of an Ellipse) Graph the equation $x^2/4 + y^2/9 = 1$, showing all the key features.

Solution. We identify this as the equation of a vertical ellipse (because the larger denominator is in the y term.) Also $a = 3$, $b = 2$, and $c = \sqrt{9 - 4} = \sqrt{5}$, from which we determine the four intercepts and the foci as shown in Figure 20.

In Problems 9–12, decide whether the corresponding ellipse is horizontal or vertical, determine a, b, and c, and sketch the graph.

9. $\dfrac{x^2}{25} + \dfrac{y^2}{9} = 1$

10. $\dfrac{x^2}{16} + \dfrac{y^2}{9} = 1$

11. $\dfrac{x^2}{1} + \dfrac{y^2}{4} = 1$

12. $\dfrac{x^2}{25} + \dfrac{y^2}{169} = 1$

EXAMPLE B (Finding Ellipses with Given Properties) Write the equations of the three ellipses with vertices at $(0, \pm 8)$ and foci at (a) $(0, \pm 7)$; (b) $(0, \pm 4)$; (c) $(0, \pm 1)$. Determine the eccentricity e in each case. Sketch the graphs.

Solution. In each case, the ellipse is vertical. From the formulas $b = \sqrt{a^2 - c^2}$ and $e = c/a$, we determine the following.
(a) $a = 8$, $c = 7$, $b = \sqrt{15}$, $e = \frac{7}{8}$
(b) $a = 8$, $c = 4$, $b = \sqrt{48}$, $e = \frac{1}{2}$
(c) $a = 8$, $c = 1$, $b = \sqrt{63}$, $e = \frac{1}{8}$
The three graphs and the corresponding equations are shown in Figure 21. Note that the smaller e is, the more circular the ellipse.

(a) $\dfrac{x^2}{15} + \dfrac{y^2}{64} = 1$

(b) $\dfrac{x^2}{48} + \dfrac{y^2}{64} = 1$

(c) $\dfrac{x^2}{63} + \dfrac{y^2}{64} = 1$

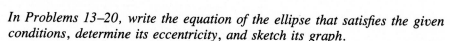

Figure 21

In Problems 13–20, write the equation of the ellipse that satisfies the given conditions, determine its eccentricity, and sketch its graph.

13. Vertices at $(0, \pm 5)$, foci at $(0, \pm 3)$.
14. Vertices at $(0, \pm 10)$, foci at $(0, \pm 8)$.
15. Center at $(0, 0)$, a vertex at $(-7, 0)$, a focus at $(3, 0)$.
16. Center at $(0, 0)$, a focus at $(6, 0)$, major diameter 20.
17. Horizontal, center at $(0, 0)$, major diameter 14, minor diameter 4.
18. Foci at $(\pm 10, 0)$, minor diameter 10.
19. Vertices at $(\pm 9, 0)$, curve passes through $(3, \sqrt{8})$.
20. Ends of minor diameter at $(\pm 4, 0)$, curve passes through $(\sqrt{2}, 4\sqrt{3})$.

MISCELLANEOUS PROBLEMS

21. Determine a, b, c, and e for the ellipse $4x^2 + 25y^2 = 100$.
22. Find the eccentricity of the ellipse $8x^2 + 2y^2 = 8$.
23. Find the equation of the ellipse with eccentricity $\frac{1}{3}$ and foci at $(0, \pm 4)$.

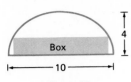

Figure 22

24. Find the equation of the ellipse that goes through $(\frac{1}{4}, \sqrt{3}/2)$ and has vertices at $(0, \pm 1)$.

25. A door has the shape of an elliptical arch (a half-ellipse) that is 10 feet wide and 4 feet high at the center (Figure 22). A box 2 feet high is to be pushed through the door. How wide can it be?

26. How long is the **latus rectum** (chord through the focus perpendicular to the major axis) for the ellipse $x^2/a^2 + y^2/b^2 = 1$?

27. Assume that the center of the earth (a sphere of radius 4000 miles) is at one focus of the elliptical path of a satellite. If the satellite's nearest approach to the surface of the earth is 2000 miles and its farthest distance away is 10,000 miles, what are the major and minor diameters of the elliptical path?

28. ABC is a right triangle with the right angle at B. A and B are the foci of an ellipse and C is on the ellipse. Determine the major and minor diameters of the ellipse given that $\overline{AB} = 8$ and $\overline{BC} = 6$.

29. The area of the ellipse $x^2/a^2 + y^2/b^2 = 1$ is πab. Find the area of the ellipse $11x^2 + 7y^2 = 77$.

30. A square with sides parallel to the coordinate axes is inscribed in the ellipse $b^2x^2 + a^2y^2 = a^2b^2$. Determine the area of the square.

31. A dog's collar is attached by a ring to a loop of rope 32 feet long. The loop of rope is thrown over two stakes 12 feet apart.
    (a) How much area can the dog cover?
    (b) If the dog should manage to nudge the rope over the top of one of the stakes, how much would this increase the area it can cover?

32. Let P be a point on a 16-foot ladder 7 feet from the top end. As the ladder slides with its top end against a wall (the y-axis) and its bottom end along the ground (the x-axis), P traces a curve. Find the equation of this curve.

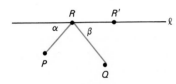

Figure 23

33. (Optical property of the ellipse) In Figure 23, let P and Q be the foci of an ellipse, R be a point on the ellipse, l be the tangent line at R, and R' be any other point of l.
    (a) Show that $\overline{PR'} + \overline{R'Q} > \overline{PR} + \overline{RQ}$.
    (b) Show that $\alpha = \beta$. (See Problem 34 of Section 6-1.)

From this we conclude that a light ray from one focus P of an elliptic mirror is reflected back through the other focus.

34. **TEASER** Two ellipses with the same eccentricity e are such that the major diameter of the smaller ellipse coincides with the minor diameter of the larger ellipse and the area of the smaller ellipse is 19 percent of the area of the larger one. Use this information to determine e.

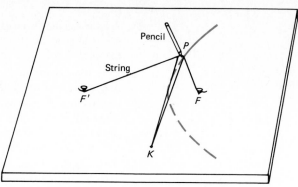

This shows how to draw a hyperbola. Take a string with a knot at K and tack its ends down with thumbtacks. Insert pencil as shown and pull string taut. Pull on knot at K.

6-3 Hyperbolas

Our opening display hints at the geometric definition of a hyperbola. A **hyperbola** is the set of points P for which the difference of the distances from two fixed points F' and F is a constant. More precisely, a hyperbola is the set of points P in Figure 24 satisfying

$$\left| d(P, F') - d(P, F) \right| = 2a$$

for some constant a.

As with the ellipse, the two fixed points F' and F are called **foci** and the point midway between the foci is the **center** of the hyperbola. The line through the foci is the **major axis** (or transverse axis) of the hyperbola and the line through the center and perpendicular to the major axis is the **minor axis** (or conjugate axis). Also, the points A' and A where the hyperbola intersects the major axis are the **vertices** of the hyperbola. By applying the defining condition with $P = A$, we see that $2a$ is the distance between the vertices. Also, if $2c$ denotes the distance between the foci, then $2c$ must be greater than $2a$, that is $c > a$.

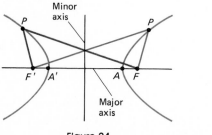

Figure 24

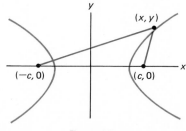

Figure 25

THE EQUATION OF A HYPERBOLA

Place the hyperbola in the coordinate system so that its center is at the orgin with the major axis along the x-axis and the foci at $(-c, 0)$ and $(c, 0)$ as in Figure 25. Then the condition $\left| d(P, F') - d(P, F) \right| = 2a$ combined with the distance formula yields

$$\left| \sqrt{(x + c)^2 + (y - 0)^2} - \sqrt{(x - c)^2 + (y - 0)^2} \right| = 2a$$

If we now employ the same kind of procedure used in obtaining the equation of the ellipse (square both sides, simplify, square again, and simplify), we get

$$(c^2 - a^2)x^2 - a^2 y^2 = a^2(c^2 - a^2)$$

or, equivalently, after dividing both sides by $a^2(c^2 - a^2)$,

$$\frac{x^2}{a^2} - \frac{y^2}{c^2 - a^2} = 1$$

Finally, we let $b^2 = c^2 - a^2$ to obtain what is called the **standard equation of the hyperbola**, namely,

$$\frac{x^2}{a^2} - \frac{y^2}{b^2} = 1$$

Note that both x and y occur to the second power, which corresponds to the fact that the graph of this equation is symmetric with respect to both the x- and y-axes as well as the origin.

INTERPRETING a, b, AND c

We have already noted that $2c$ is the distance between the foci and $2a$ is the distance between the vertices of the hyperbola. To interpret b, observe that if we solve the standard equation for y in terms of x, we get

$$y = \pm \frac{b}{a} \sqrt{x^2 - a^2}$$

For large x, $\sqrt{x^2 - a^2}$ behaves much like x; in fact, as $|x|$ gets larger and larger, $x - \sqrt{x^2 - a^2}$ approaches zero (Problem 26) and hence so does b/a times this quantity. Thus the two branches of the hyperbola $y = \pm \frac{b}{a} \sqrt{x^2 - a^2}$ approach the two lines $y = \pm \frac{b}{a} x$. We say that the hyperbola has these lines as (oblique) asymptotes (see Section 1-5).

Since $c^2 = a^2 + b^2$, the numbers a, b, and c determine a right triangle, which we call the fundamental triangle for the hyperbola. Its role in determining the asymptotes of the hyperbola is clear from Figure 26 on the next page. This figure summarizes all the key facts for a hyperbola.

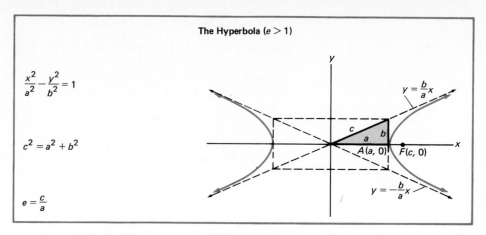

The Hyperbola ($e > 1$)

$$\frac{x^2}{a^2} - \frac{y^2}{b^2} = 1$$

$$c^2 = a^2 + b^2$$

$$e = \frac{c}{a}$$

Figure 26

The number $e = c/a$, which in this case is greater than 1, is called the **eccentricity** of the hyperbola. If e is near 1, then b is small relative to a and the hyperbola is very thin; if e is large the hyperbola is fat.

As a first example, consider the equation

$$\frac{x^2}{9} - \frac{y^2}{16} = 1$$

In this case, $a = 3$, $b = 4$, and $c = \sqrt{a^2 + b^2} = 5$. Thus the vertices are at $(\pm 3, 0)$, the foci are at $(\pm 5, 0)$, and the asymptotes have equations $y = \pm\frac{4}{3}x$. All this is shown in Figure 27.

Again we should consider what happens if we interchange the roles of x and y. The standard equation then takes the form

$$\frac{y^2}{a^2} - \frac{x^2}{b^2} = 1$$

This equation represents a hyperbola with major axis along the y-axis (we call it a *vertical hyperbola*). Its vertices are at $(0, \pm a)$ and its foci are at $(0, \pm c)$. As an example, consider the equation

$$\frac{y^2}{9} - \frac{x^2}{16} = 1$$

which again has $a = 3$ and $b = 4$. Its graph has vertices at $(0, \pm 3)$ and foci at $(0, \pm 5)$. Note how the fundamental triangle determines the asymptotes and helps us draw the graph (Figure 28). The hyperbolas of Figures 27 and 28 have the same eccentricity, namely, $e = c/a = 5/3$.

We make one final important observation. It is not the relative sizes of the denominators in the x- and y-terms that determine whether the hyperbola is vertical or horizontal (as it was with the ellipse). Rather, this is determined by whether in the standard form the minus is associated with the x- or the y-term.

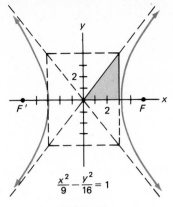

$$\frac{x^2}{9} - \frac{y^2}{16} = 1$$

Figure 27

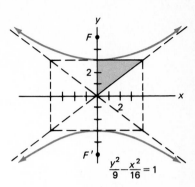

$$\frac{y^2}{9} - \frac{x^2}{16} = 1$$

Figure 28

APPLICATIONS

The hyperbola, too, has an optical property, as is illustrated in Figure 29. If we imagine one branch of the hyperbola to be a mirror, then a light ray from the opposite focus upon hitting the mirror will be reflected away along the line which passes through the nearby focus. The optical properties of the parabola and the hyperbola are combined in one design for a reflecting telescope (Figure 30). Other applications are treated in the problem set.

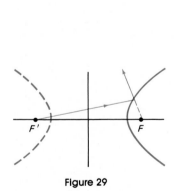

Figure 29

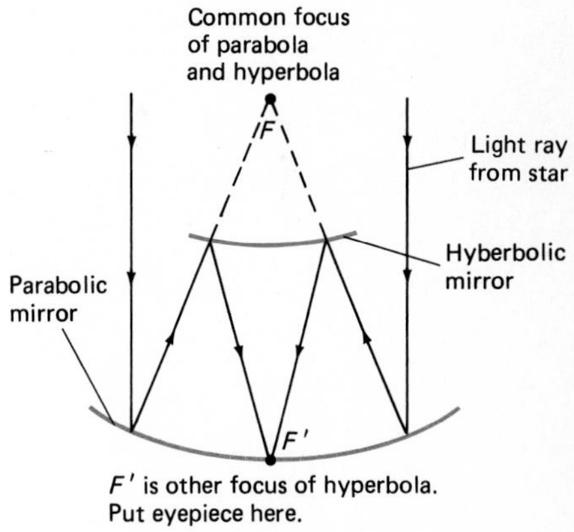

Figure 30

Common focus
of parabola
and hyperbola

Light ray
from star

Hyperbolic
mirror

Parabolic
mirror

F' is other focus of hyperbola.
Put eyepiece here.

Problem Set 6-3

In Problems 1–8, decide whether the given equation determines a horizontal or vertical hyperbola. Also find a (the distance from the center to a vertex), b, and c. Be sure the equation is in standard form before you try to give the answers.

1. $\dfrac{x^2}{16} - \dfrac{y^2}{36} = 1$ 2. $-\dfrac{x^2}{1} + \dfrac{y^2}{8} = 1$ 3. $\dfrac{x^2}{16} - \dfrac{y^2}{9} = -1$

4. $\dfrac{x^2}{4} - \dfrac{y^2}{9} = 1$ 5. $4x^2 - 16y^2 = 1$ 6. $25y^2 - 9x^2 = 1$

7. $4x^2 - y^2 = 16$ 8. $k^2y^2 - 4k^2x^2 = 1$

EXAMPLE A (Graphing the Equation of a Hyperbola)   Sketch the graph of $x^2/4 - y^2/9 = 1$, showing all the important features.

Solution.   The graph is a hyperbola and, since the minus sign is associated with the y-term, the major axis is horizontal. We conclude that $a = 2$, $b = 3$, and $c = \sqrt{4 + 9} = \sqrt{13}$. The asymptotes are the lines $y = \pm\frac{3}{2}x$. With this information, we may sketch the graph shown in Figure 31.

In Problems 9–12, decide whether the corresponding hyperbola is horizontal or vertical, give the values of a, b, and c, and sketch the graph. Be sure to show the asymptotes.

9. $\dfrac{x^2}{25} - \dfrac{y^2}{9} = 1$ 10. $\dfrac{x^2}{16} - \dfrac{y^2}{9} = 1$

11. $\dfrac{y^2}{64} - \dfrac{x^2}{36} = 1$ 12. $\dfrac{x^2}{4} - \dfrac{y^2}{4} = 1$

EXAMPLE B   **(Finding Hyperbolas Satisfying Given Conditions)**   Find the equation of the hyperbola with vertices at $(\pm 2, 0)$ that passes through $(2\sqrt{2}, 4)$. Sketch its graph.

Solution.   Since the vertices are on the x-axis, the hyperbola is horizontal with $a = 2$. The equation has the form

$$\frac{x^2}{4} - \frac{y^2}{b^2} = 1$$

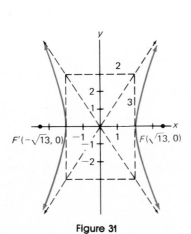

Figure 31   Figure 32

Since the point $(2\sqrt{2}, 4)$ is on the graph,

$$\frac{(2\sqrt{2})^2}{4} - \frac{4^2}{b^2} = 1$$

which gives $b = 4$. Thus the equation is

$$\frac{x^2}{4} - \frac{y^2}{16} = 1$$

The graph is shown in Figure 32.

In Problems 13–18, find the equation of the hyperbola satisfying the given conditions and sketch its graph, displaying the asymptotes.

13. Vertices at $(0, \pm 3)$ and going through $(2, 5)$.
14. Vertices at $(\pm 3, 0)$ and going through $(2\sqrt{3}, 9)$.
15. Foci at $(\pm 4, 0)$, vertices at $(\pm 1, 0)$.
16. Vertices at $(\pm 5, 0)$, equations of asymptotes $y = \pm x$.
17. Vertices at $(\pm 3, 0)$, equations of asymptotes $y = \pm 2x$.
18. Vertices at $(0, \pm 3)$, eccentricity $e = \frac{4}{3}$.

MISCELLANEOUS PROBLEMS

19. Find the equation of the hyperbola centered at the origin with a focus at $(0, 8)$ and a vertex at $(0, -6)$.
20. Determine the eccentricity of the hyperbola with equation $16x^2 - 20y^2 = 320$.
21. A conic has eccentricity 3 and foci at $(\pm 12, 0)$. Find its equation.
22. Find the equations of the asymptotes of the vertical hyperbola with eccentricity 2 and center at the origin.
23. How long is the **focal chord** (chord through a focus perpendicular to the major axis) of the hyperbola $x^2/9 - y^2/16 = 1$.
24. Generalize Problem 23 by finding the length of the focal chord for the hyperbola $x^2/a^2 - y^2/b^2 = 1$.
25. Find the eccentricity of the hyperbola with asymptotes $y = \pm x$.
26. Show that $x - \sqrt{x^2 - a^2}$ approaches 0 as $|x|$ gets larger and larger. *Hint:* Multiply and divide by $x + \sqrt{x^2 - a^2}$.
27. A ball shot from $(-5, 0)$ hit the right branch of the hyperbolic bangboard $x^2/16 - y^2/9 = 1$ at the point $(8, 3\sqrt{3})$. What was the ball's y-coordinate when its x-coordinate was 10?
28. The rectangle $PQRS$ with sides parallel to the coordinate axes is inscribed in the hyperbola $x^2/4 - y^2/9 = 1$ as shown in Figure 33 on the next page. Find the coordinates of P if the area of the rectangle is $6\sqrt{5}$.
29. Andrew, located at $(0, -2200)$, fired a rifle. The sound echoed off a cliff at $(0, 2200)$ to Brian, located at the point (x, y). Brian heard this echo 6 seconds after he heard the original shot. Find the xy-equation of the curve on which Brian is located. Assume that distances are in feet and that sound travels 1100 feet per second.

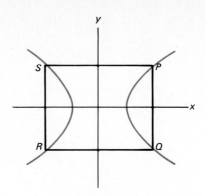

Figure 33

30. **TEASER** Amy, Betty, and Cindy, located at $(-8, 0)$, $(8, 0)$, and $(8, 10)$, respectively, recorded the exact times when they heard an explosion. On comparing notes, they discovered that Betty and Cindy heard the explosion at the same time but that Amy heard it 12 seconds later. Assuming that distances are in kilometers and that sound travels $\frac{1}{3}$ kilometer per second, determine the point of the explosion.

Conic Sections		Limiting Forms	

4. Parallel lines: $y^2 = 3$

1. Parabola: $y^2 = 3x$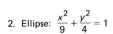

5. Single line: $y^2 = 0$

2. Ellipse: $\dfrac{x^2}{9} + \dfrac{y^2}{4} = 1$

6. Circle: $x^2 + y^2 = 4$

7. Point: $2x^2 + y^2 = 0$

8. Empty set: $3x^2 + y^2 = -1$

3. Hyperbola: $\dfrac{x^2}{9} - \dfrac{y^2}{4} = 1$

9. Intersecting lines: $2x^2 - y^2 = 0$

6-4 Translation of Axes

An astute—or perhaps even a casual—observer will note that the standard equations of the three conic sections involve the second power of x or y. This observation suggests a question. Suppose we graph a polynomial equation that is of second degree in x and y. Will it always be a conic section? The answer is no; that is, it is no unless we admit the six limiting forms of the conic sections illustrated in the right half of our opening display. But if we do admit them, the answer is yes. In particular, we claim that the graph of any equation of the form

$$Ax^2 + Cy^2 + Dx + Ey + F = 0 \qquad (A, C \text{ not both } 0)$$

is a conic section or one of its limiting forms. We will show you why by moving the coordinate axes in just the right way.

TRANSLATIONS

Consider the equation

$$x^2 + y^2 - 4x - 6y - 12 = 0$$

If we let

$$x = u + 2 \qquad y = v + 3$$

this equation becomes

$$(u + 2)^2 + (v + 3)^2 - 4(u + 2) - 6(v + 3) - 12 = 0$$

or

$$u^2 + 4u + 4 + v^2 + 6v + 9 - 4u - 8 - 6v - 18 - 12 = 0$$

This simplifies to

$$u^2 + v^2 = 25$$

which we recognize as the equation of a circle of radius 5.

    To understand what we have just done, introduce new coordinate axes in the plane (the u- and v-axes) parallel to the old x- and y-axes, but with the new origin at $x = 2$ and $y = 3$ (see Figure 34). Each point now has two sets of coordinates: (x, y) and (u, v). They are related by the equations $x = u + 2$ and $y = v + 3$. In the new coordinate system, which is called a **translation** of the old one, the uv-equation represents a circle of radius 5 centered at the origin. In the old coordinate system, the xy-equation must also have represented a circle of radius 5, but centered at $(2, 3)$.

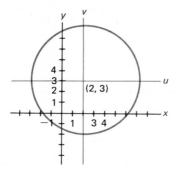

$$x^2 + y^2 - 4x - 6y - 12 = 0$$

$$\downarrow$$

$$\begin{cases} x = u + 2 \\ y = v + 3 \end{cases}$$

$$\downarrow$$

$$u^2 + v^2 = 25$$

Figure 34

    Let us see what happens in general to coordinates of points under a translation of axes. If u- and v-axes are introduced with the same directions as the old x- and y-axes so that the new origin has coordinates (h, k) relative to the old axes (Figure 35), then the two sets of coordinates for a point P are connected by

$$u = x - h \qquad v = y - k$$

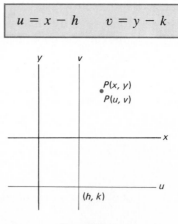

Figure 35

or, equivalently, by

$$x = u + h \qquad y = v + k$$

The shape of a curve is not changed by a translation of axes since it is the axes, not the curve, that are moved. But as we saw in the example above, the resulting change in the equation may enable us to recognize the curve.

The concept of a translation was discussed from a slightly different perspective in Example C of Problem Set 1-5. There we translated the graph; here we are translating the axes.

COMPLETING THE SQUARE

Given an equation, how do we know what translation to make? Here an old algebraic friend, completing the square, comes to our aid. As a typical example, consider

$$x^2 + y^2 - 6x + 8y + 10 = 0$$

We first rewrite the equation and then complete each square.

$$(x^2 - 6x \quad) + (y^2 + 8y \quad) = -10$$
$$(x^2 - 6x + 9) + (y^2 + 8y + 16) = -10 + 9 + 16$$
$$(x - 3)^2 + (y + 4)^2 \quad = 15$$

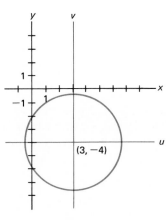

Figure 36

We can recognize this as the equation of a circle of radius $\sqrt{15}$, centered at $(3, -4)$. In terms of translations (Figure 36), we note that the substitutions $u = x - 3$ and $v = y + 4$ transform the equation into

$$u^2 + v^2 = 15$$

As a second example, consider

$$4x^2 + 9y^2 - 24x + 18y + 9 = 0$$

We may rewrite this equation successively as

$$4(x^2 - 6x \quad) + 9(y^2 + 2y \quad) = -9$$
$$4(x^2 - 6x + 9) + 9(y^2 + 2y + 1) = -9 + 36 + 9$$
$$4(x - 3)^2 + 9(y + 1)^2 \quad = 36$$
$$\frac{(x - 3)^2}{9} + \frac{(y + 1)^2}{4} \quad = 1$$

The translation $u = x - 3$ and $v = y + 1$ transforms this to

$$\frac{u^2}{9} + \frac{v^2}{4} = 1$$

which we recognize as an ellipse with $a = 3$ and $b = 2$ (Figure 37).

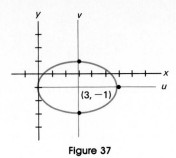

$$4x^2 + 9y^2 - 24x + 18y + 9 = 0$$

$$\downarrow$$

$$\begin{cases} x = u + 3 \\ y = v - 1 \end{cases}$$

$$\downarrow$$

$$\frac{u^2}{9} + \frac{v^2}{4} = 1$$

Figure 37

THE GENERAL CASE

Consider the general equation

$$Ax^2 + Cy^2 + Dx + Ey + F = 0$$

For the moment, we assume that at least one of the coefficients A or C is different from zero. When we apply the process of completing of the square, we transform this equation into one of several forms, the most typical being the following.

1. $(y - k)^2 = 4p(x - h)$

2. $\dfrac{(x - h)^2}{a^2} + \dfrac{(y - k)^2}{b^2} = 1$

3. $\dfrac{(x - h)^2}{a^2} - \dfrac{(y - k)^2}{b^2} = 1$

Perhaps these are already recognizable as the equations of a parabola with vertex at (h, k), an ellipse with center at (h, k), and a hyperbola with center at (h, k). But to remove any doubt, we may translate the axes by the substitutions $u = x - h$ and $v = y - k$, thereby obtaining

1. $v^2 = 4pu$

2. $\dfrac{u^2}{a^2} + \dfrac{v^2}{b^2} = 1$

3. $\dfrac{u^2}{a^2} - \dfrac{v^2}{b^2} = 1$

Our work may also yield these equations with u and v interchanged or we may get one of the six limiting forms illustrated in our opening display. There are no other possibilities.

Problem Set 6-4

In Problems 1–6, make the indicated change of variables (a translation of axes) and then name the conic section or limiting form represented by the equation.

1. $x^2 + 2y^2 - 4y = 0; x = u, y = v + 1$
2. $x^2 - 4y^2 - 4x - 5 = 0; x = u + 2, y = v$
3. $x^2 + y^2 - 4x + 2y = -4; x = u + 2, y = v - 1$
4. $x^2 + y^2 - 4x + 2y = -5; x = u + 2, y = v - 1$
5. $x^2 - 6x - 4y + 13 = 0; x = u + 3, y = v + 1$
6. $x^2 - 4x + 1 = 0; x = u + 2, y = v$

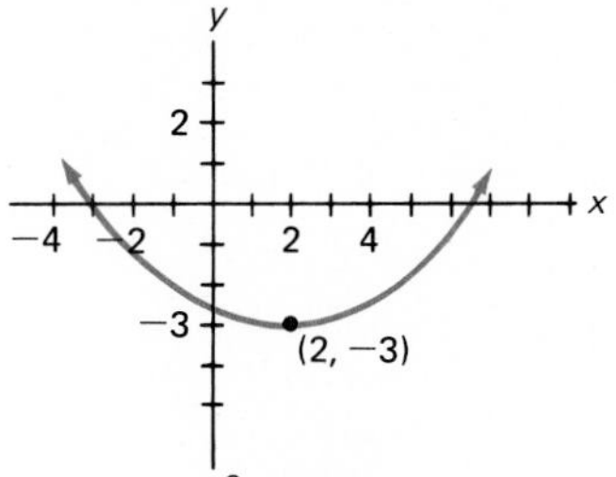

EXAMPLE A (Graphing Conics) In the xy-plane, sketch the graphs of
(a) $(x - 2)^2 = 8(y + 3)$; (b) $(x - 2)^2 - (y + 3)^2 = 0$.

Solution. (a) We could formally make the translation of axes corresponding to $u = x - 2$ and $v = y + 3$, which would yield $u^2 = 8v$. But perhaps we can do this mentally and thereby recognize that $(x - 2)^2 = 8(y + 3)$ is the equation of a vertical parabola with vertex at $(2, -3)$ and $p = 2$. With that information, we can make the sketch shown in Figure 38.
(b) The mental substitutions $u = x - 2$ and $v = y + 3$ transform the second equation into $u^2 - v^2 = 0$, which is equivalent to $(u - v) \times (u + v) = 0$. We recognize this as the equation of two intersecting lines, $v = u$ and $v = -u$. In terms of x and y, this gives $y + 3 = x - 2$ and $y + 3 = -x + 2$ (see Figure 39).

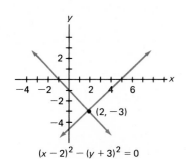

$(x - 2)^2 = 8(y + 3)$

Figure 38

$(x - 2)^2 - (y + 3)^2 = 0$

Figure 39

Sketch the graphs of each of the following equations.

7. $\dfrac{(x + 3)^2}{4} + \dfrac{(y + 2)^2}{16} = 1$

8. $(x + 3)^2 + (y - 4)^2 = 25$

9. $\dfrac{(x + 3)^2}{4} - \dfrac{(y + 2)^2}{16} = 1$

10. $4(x + 3) = (y + 2)^2$

11. $(x + 2)^2 = 8(y - 1)$

12. $(x + 2)^2 = 4$

13. $(y - 1)^2 = 16$

14. $\dfrac{(x + 3)^2}{4} + \dfrac{(y - 2)^2}{8} = 0$

EXAMPLE B (Identifying Conics by Completing Squares) Identify the conic whose equation is

$$4x^2 - 8x - 2y^2 + 16y = 0$$

Solution. We complete the squares.

$$4(x^2 - 2x + \quad) - 2(y^2 - 8y + \quad) = 0$$
$$4(x^2 - 2x + 1) - 2(y^2 - 8y + 16) = 4 - 32$$
$$4(x - 1)^2 - 2(y - 4)^2 = -28$$
$$-\frac{(x - 1)^2}{7} + \frac{(y - 4)^2}{14} = 1$$

We recognize this as the equation of a vertical hyperbola with center at $(1, 4)$.

Identify the conics determined by the equations in Problems 15–24.

15. $4x^2 + 16x + 4y^2 - 8y = 0$
16. $x^2 + 2x + 4y^2 - 8y = 0$
17. $4x^2 - 16x + y^2 - 8y = -6$
18. $4x^2 - 16x - y^2 - 8y = 2$
19. $4x^2 - 16x + y^2 - 8y = -32$
20. $4x^2 - 16x + y^2 - 8y = -40$
21. $4x^2 - 16x + y - 8 = 0$
22. $4x^2 - 16x + 12 = 0$
23. $4x^2 - 16x - 9y^2 + 18y + 7 = 0$
24. $4x^2 - 16x - 9y^2 + 18y + 8 = 0$
25. Sketch the graph of $9x^2 - 18x + 4y^2 + 16y = 11$.
26. Sketch the graph of $4x^2 + 16x - 16y + 32 = 0$.
27. Determine the distance between the vertices of the graph of $-9x^2 + 18x + 4y^2 + 24y = 9$.
28. Find the focus and the directrix of the parabola with equation $x^2 - 4x + 8y = 0$.
29. Find the focus and directrix of the parabola with equation $2y^2 - 4y - 10x = 0$.
30. Find the foci of the ellipse with equation $16(x - 1)^2 + 25(y + 2)^2 = 400$.

MISCELLANEOUS PROBLEMS

31. Sketch the graph of each of the following.

    (a) $\dfrac{(x + 5)^2}{16} + \dfrac{(y - 3)^2}{9} = 1$ (b) $\dfrac{(x + 5)^2}{16} - \dfrac{(y - 3)^2}{9} = 1$

32. Identify the curve with the given equation.

    (a) $4x^2 + 9y^2 - 16x + 54y + 61 = 0$
    (b) $x^2 + 8x + 8y = 0$
    (c) $x^2 - 4y^2 + 6x + 16y = 16$
    (d) $4x^2 + 9y^2 + 16x - 18y + 25 = 0$

33. Name the conic with equation $y^2 + ax^2 = x$ for the various values of a.
34. Find the equation of the parabola with the line $y = 4$ as directrix and the point $(2, -1)$ as focus.
35. Write the equation of the parabola with vertex $(4, 5)$ and focus $(3, 5)$.
36. Write the equation of the ellipse with vertices at $(2, -2)$ and $(2, 10)$ as ends of the major diameter and $(2, 6)$ as a focus.
37. Write the equation of the ellipse with foci $(\pm 2, 2)$ that goes through the origin.
38. Write the equation of the hyperbola with foci $(0, 0)$ and $(4, 0)$ that passes through $(9, 12)$
39. Find the equation of the hyperbola with the lines $y = 2x - 10$ and $y = -2x + 2$ as asymptotes and one focus at $(3, 2)$.
40. Transform the equation $xy - 2x + 3y = 18$ by translation of axes so that the new equation has no first degree terms. Sketch the graph showing both sets of axes.
41. A curve C goes through the three points $(-1, 2)$, $(0, 0)$, and $(3, 6)$. Write the equation for C if C is;

(a) A vertical parabola;

(b) A horizontal parabola;

(c) A circle.

42. Find the equation of the hyperbola with eccentricity 2 that has the y-axis as one directrix and the corresponding focus at $(6, 0)$.

43. Find the equation of the circle that goes through the two foci and the upper y-intercept of the ellipse $36x^2 + 100y^2 = 3600$.

44. **TEASER** Let C be an arbitrary horizontal ellipse that intersects the parabola $y = x^2$ in four points (x_1, y_1), (x_2, y_2), (x_3, y_3), and (x_4, y_4). Prove that $x_1 + x_2 + x_3 + x_4 = 0$.

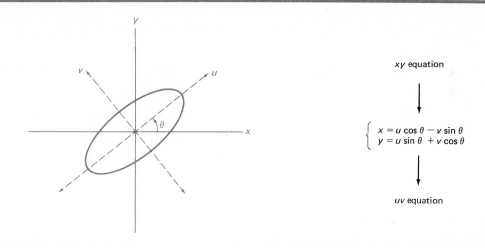

xy equation

$$\begin{cases} x = u \cos \theta - v \sin \theta \\ y = u \sin \theta + v \cos \theta \end{cases}$$

uv equation

6-5 Rotation of Axes

We want you to observe two facts about what we have done in the first four sections of this chapter. First, all the conic sections we have considered so far were oriented in the coordinate system with their major axis parallel to either the x-axis or the y-axis. Second, none of the equations of these conics had an xy-term. We would not mention these facts unless there were a connection between them. To see the connection, we need to discuss rotation of axes.

ROTATIONS

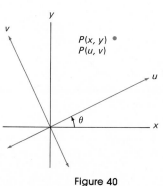

Figure 40

Introduce a new pair of axes, called the u- and v-axes, into the xy-plane. These axes have the same origin as the old x- and y-axes, but they are rotated so that the positive u-axis makes an angle θ with the positive x-axis (see the diagram in the opening panel and also the one in Figure 40). A point P then has two sets of coordinates: (x, y) and (u, v). How are they related?

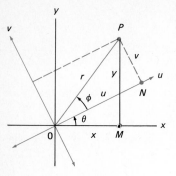

Figure 41

Draw a line segment from the origin O to P, let r denote the length of OP, and let φ denote the angle from the u-axis to OP. Then x, y, u, and v will have the geometric interpretations indicated in Figure 41.

Looking at the right triangle OPM, we see that

$$\cos(\varphi + \theta) = \frac{x}{r}$$

so

$$x = r \cos(\varphi + \theta) = r(\cos \varphi \cos \theta - \sin \varphi \sin \theta)$$
$$= (r \cos \varphi) \cos \theta - (r \sin \varphi) \sin \theta$$

Next, the right triangle OPN tells us that $u = r \cos \varphi$ and $v = r \sin \varphi$. Thus

$$\boxed{x = u \cos \theta - v \sin \theta}$$

Similarly

$$y = r \sin(\varphi + \theta) = r(\sin \varphi \cos \theta + \cos \varphi \sin \theta)$$
$$= (r \sin \varphi) \cos \theta + (r \cos \varphi) \sin \theta$$

so

$$\boxed{y = u \sin \theta + v \cos \theta}$$

We call the boxed results **rotation formulas.**

A SIMPLE EXAMPLE

Consider the equation

$$xy = 1$$

Let us make a rotation of axes through 45° to see what happens to this equation. The required substitutions are

$$x = u \cos 45° - v \sin 45° = \frac{\sqrt{2}}{2}(u - v)$$

$$y = u \sin 45° + v \cos 45° = \frac{\sqrt{2}}{2}(u + v)$$

When we make these substitutions in $xy = 1$, we obtain

$$\frac{\sqrt{2}}{2}(u - v)\frac{\sqrt{2}}{2}(u + v) = 1$$

which simplifies to

$$\frac{u^2}{2} - \frac{v^2}{2} = 1$$

We recognize this as the equation of a hyperbola with $a = b = \sqrt{2}$. Note how the cross-product term xy disappeared as a result of the rotation. The choice of a 45° angle was just right to make this happen (see Figure 42).

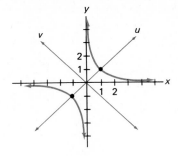

$$xy = 1$$
$$\downarrow$$
$$\begin{cases} x = u \cos 45° - v \sin 45° \\ y = u \sin 45° + v \cos 45° \end{cases}$$
$$\downarrow$$
$$\frac{u^2}{2} - \frac{v^2}{2} = 1$$

Figure 42

THE GENERAL SECOND DEGREE EQUATION

How do we know what rotation to make? Consider the most general second degree equation in x and y:

$$Ax^2 + Bxy + Cy^2 + Dx + Ey + F = 0$$

If we make the substitutions

$$x = u \cos \theta - v \sin \theta$$
$$y = u \sin \theta + v \cos \theta$$

this equation takes the form

$$au^2 + buv + cv^2 + du + ev + f = 0$$

where a, b, c, d, e and f are numbers which depend upon θ. We could find values for all of them, but we really care only about b. When we do the necessary algebra, we find

$$b = B(\cos^2 \theta - \sin^2 \theta) - 2(A - C) \sin \theta \cos \theta$$
$$= B \cos 2\theta - (A - C) \sin 2\theta$$

We would like to have $b = 0$; that is,

$$B \cos 2\theta = (A - C) \sin 2\theta$$

This will occur if

$$\cot 2\theta = \frac{A - C}{B}$$

This formula is the answer to our question: to eliminate the cross-product (xy) term, choose θ so it satisfies this formula. In the example $xy = 1$, we have $A = 0$, $B = 1$, and $C = 0$ so we choose θ to satisfy

$$\cot 2\theta = \frac{0 - 0}{1} = 0$$

One angle that works is $\theta = 45°$. We could also use $\theta = 135°$ or $\theta = -225°$, but it is customary to choose a first quadrant angle.

ANOTHER EXAMPLE

Consider the equation

$$4x^2 + 2\sqrt{3}\,xy + 2y^2 + 10\sqrt{3}x + 10y = 5$$

To remove the cross-product term, we rotate the axes through an angle θ satisfying

$$\cot 2\theta = \frac{A - C}{B} = \frac{4 - 2}{2\sqrt{3}} = \frac{1}{\sqrt{3}}$$

this means that $2\theta = 60°$ and so $\theta = 30°$. When we use this value of θ in the rotation formulas, we obtain

$$x = u \cdot \frac{\sqrt{3}}{2} - v \cdot \frac{1}{2} = \frac{\sqrt{3}\,u - v}{2}$$

$$y = u \cdot \frac{1}{2} + v \cdot \frac{\sqrt{3}}{2} = \frac{u + \sqrt{3}\,v}{2}$$

Substituting these in the original equation gives

$$4\frac{(\sqrt{3}\,u - v)^2}{4} + 2\sqrt{3}\frac{(\sqrt{3}\,u - v)(u + \sqrt{3}v)}{4}$$

$$+ 2\frac{(u + \sqrt{3}\,v)^2}{4} + 10\sqrt{3}\frac{\sqrt{3}\,u - v}{2} + 10\frac{u + \sqrt{3}\,v}{2} = 5$$

After collecting terms and simplifying, we have

$$5u^2 + v^2 + 20u = 5$$

Next we complete the squares.

$$5(u^2 + 4u + 4) + v^2 = 5 + 20$$

$$\frac{(u + 2)^2}{5} + \frac{v^2}{25} = 1$$

As a final step, we make the translation determined by $r = u + 2$ and $s = v$, which gives

$$\frac{r^2}{5} + \frac{s^2}{25} = 1$$

This is the equation of a vertical ellipse in the rs-coordinate system. It has major diameter of length 10 and minor diameter of length $2\sqrt{5}$. All of this is shown in Figure 43.

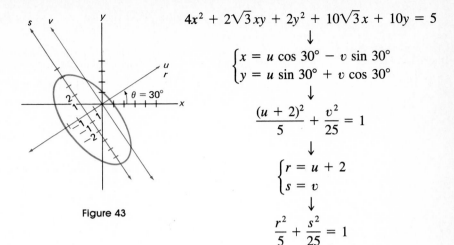

$$4x^2 + 2\sqrt{3}\,xy + 2y^2 + 10\sqrt{3}\,x + 10y = 5$$

$$\downarrow$$

$$\begin{cases} x = u \cos 30° - v \sin 30° \\ y = u \sin 30° + v \cos 30° \end{cases}$$

$$\downarrow$$

$$\frac{(u + 2)^2}{5} + \frac{v^2}{25} = 1$$

$$\downarrow$$

$$\begin{cases} r = u + 2 \\ s = v \end{cases}$$

$$\downarrow$$

$$\frac{r^2}{5} + \frac{s^2}{25} = 1$$

Figure 43

Problem Set 6-5

In the text, we derived the following rotation formulas.

$$\boxed{\begin{aligned} x &= u \cos \theta - v \sin \theta \\ y &= u \sin \theta + v \cos \theta \end{aligned}}$$

In Problems 1–10, transform the given xy-equation to a uv-equation by a rotation through the specified angle θ.

1. $y = \sqrt{3}\,x$; $\theta = 60°$
2. $y = x$; $\theta = 45°$
3. $x^2 + 4y^2 = 16$; $\theta = 90°$
4. $4y^2 - x^2 = 4$; $\theta = 90°$
5. $y^2 = 4\sqrt{2}\,x$; $\theta = 45°$
6. $x^2 = -\sqrt{2}\,y + 3$; $\theta = 45°$
7. $x^2 - xy + y^2 = 4$; $\theta = 45°$
8. $x^2 + 3xy + y^2 = 10$; $\theta = 45°$
9. $6x^2 - 24xy - y^2 = 30$; $\theta = \cos^{-1}(\frac{3}{5})$
10. $3x^2 - \sqrt{3}\,xy + 2y^2 = 39$; $\theta = 60°$

EXAMPLE A (Eliminating the *xy*-term) By rotation of axes, eliminate the *xy*-term from

$$x^2 + 24xy + 8y^2 = 136$$

and then draw its graph.

Solution. We review the example in the text, noting that we must choose θ to satisfy

$$\cot 2\theta = \frac{A - C}{B} = \frac{1 - 8}{24} = -\frac{7}{24}$$

Here our problem is complicated by the fact that 2θ is not a special angle.

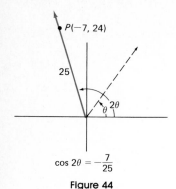

P(-7, 24)

25

θ 2θ

$\cos 2\theta = -\dfrac{7}{25}$

Figure 44

How shall we find the values of $\sin \theta$ and $\cos \theta$ needed for the rotation formulas?

First, we place 2θ in standard position (see Figure 44), noting that $P(-7, 24)$ is on its terminal side. Since P is a distance $r = \sqrt{(-7)^2 + (24)^2} = 25$ from the origin, $\cos 2\theta = -\frac{7}{25}$.

Second, we recall the half-angle formulas.

$$\sin \theta = \pm\sqrt{\frac{1 - \cos 2\theta}{2}} \qquad \cos \theta = \pm\sqrt{\frac{1 + \cos 2\theta}{2}}$$

Since our θ is in the first quadrant, we use the plus sign in both cases. We obtain

$$\sin \theta = \sqrt{\frac{1 + \frac{7}{25}}{2}} = \frac{4}{5} \qquad \cos \theta = \sqrt{\frac{1 - \frac{7}{25}}{2}} = \frac{3}{5}$$

These, in turn, give the rotation formulas

$$x = \frac{3u - 4v}{5} \qquad y = \frac{4u + 3v}{5}$$

All this was preliminary; our main task is to substitute these expressions for x and y in the original equation and simplify.

$$\left(\frac{3u - 4v}{5}\right)^2 + 24\left(\frac{3u - 4v}{5}\right)\left(\frac{4u + 3v}{5}\right) + 8\left(\frac{4u + 3v}{5}\right)^2 = 136$$

After multiplying by 25 and collecting terms, we have

$$425u^2 - 200v^2 = 136 \cdot 25$$

or

$$\frac{u^2}{8} - \frac{v^2}{17} = 1$$

We summarize the process below and show the graph in Figure 45.

$$x^2 + 24xy + 8y^2 = 136$$
$$\downarrow$$
$$\begin{cases} x = \frac{3}{5}u - \frac{4}{5}v \\ y = \frac{4}{5}u + \frac{3}{5}v \end{cases}$$
$$\downarrow$$
$$\frac{u^2}{8} - \frac{v^2}{17} = 1$$

In Problems 11–20, eliminate the xy-term by a suitable rotation of axes and then, if necessary, translate axes (complete the squares) to put the equation in standard form. Finally, graph the equation showing all axes used. (Note: Some problems involve special angles, but several do not.)

11. $3x^2 + 10xy + 3y^2 + 8 = 0$ 12. $2x^2 + xy + 2y^2 = 90$
13. $4x^2 - 3xy = 18$ 14. $4xy - 3y^2 = 64$

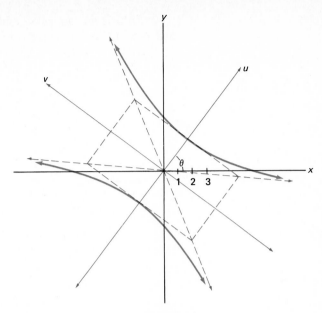

Figure 45

15. $x^2 - 2\sqrt{3}xy + 3y^2 - 12\sqrt{3}x - 12y = 0$

16. $x^2 + 2\sqrt{3}xy + 3y^2 + 8\sqrt{3}x - 8y = 0$

17. $13x^2 + 6\sqrt{3}xy + 7y^2 - 32 = 0$

18. $17x^2 + 12xy + 8y^2 + 17 = 0$

19. $9x^2 - 24xy + 16y^2 - 60x + 80y + 75 = 0$

20. $16x^2 + 24xy + 9y^2 - 20x - 15y - 150 = 0$

EXAMPLE B (The Inverse Rotation Formulas)   For a rotation of axes through angle θ, obtain the formulas that express u and v in terms of x and y. Then use the result to obtain the uv-coordinates of the point that has xy-coordinates $(4, 2\sqrt{3})$ if the angle θ is 30°.

Solution.   Consider the rotation formulas at the beginning of this problem set. Multiply the first one by $\cos \theta$ and the second by $\sin \theta$, then add.

$$x \cos \theta = u \cos^2 \theta - v \sin \theta \cos \theta$$
$$y \sin \theta = u \sin^2 \theta + v \sin \theta \cos \theta$$

$$\overline{x \cos \theta + y \sin \theta = u(\cos^2 \theta + \sin^2 \theta)} \qquad = u$$

Similarly, multiply the first formula by $-\sin \theta$ and the second by $\cos \theta$, and add. The two resulting formulas are

$$u = x \cos \theta + y \sin \theta$$
$$v = -x \sin \theta + y \cos \theta$$

To find the values of u and v, simply substitute $x = 4$, $y = 2\sqrt{3}$, and $\theta = 30°$ in the above formulas. This gives

$$u = 4 \cdot \frac{\sqrt{3}}{2} + 2\sqrt{3} \cdot \frac{1}{2} = 3\sqrt{3}$$

$$v = -4 \cdot \frac{1}{2} + 2\sqrt{3} \cdot \frac{\sqrt{3}}{2} = 1$$

The geometric interpretation of these numbers is shown in Figure 46.

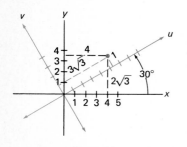

Figure 46

In Problems 21–26, find u and v for the given values of x, y, and θ. Then make a diagram to check that your answers make sense.

21. $(5, -3)$; $60°$

22. $(-2, 5)$; $60°$

23. $(3\sqrt{2}, \sqrt{2})$; $45°$

24. $(5/\sqrt{2}, -5/\sqrt{2})$; $45°$

25. $(3, 4)$; $\tan^{-1}(\frac{4}{3})$

26. $(5, -12)$; $\arctan(\frac{5}{12})$

27. Find the xy-equation that simplifies to $u^2 = 4v$ when the axes are rotated through an angle of $60°$.

28. Find the xy-equation that simplifies to $u^2 - 4v^2 = 4$ when the axes are rotated through an angle of $30°$.

MISCELLANEOUS PROBLEMS

29. Transform the equation $2x^2 + \sqrt{3}xy + y^2 = 5$ to a uv-equation by rotating the axes through $30°$. Use the result to identify the corresponding curve.

30. Without any algebra, determine the uv-equation corresponding to the equation $x^2/16 + y^2/9 = 1$ when the axes are rotated through $90°$. Then do the algebra to corroborate your answer.

31. Without any algebra, determine the uv-equation corresponding to $(x - 2\sqrt{2})^2 + (y - 2\sqrt{2})^2 = 16$ when the axes are rotated through $45°$.

32. Find the angle θ through which the axes must be rotated so that the circle $(x - 4)^2 + (y - 3)^2 = 4$ lies above the u-axis and is tangent to it. Write the uv-equation of this circle.

33. By a rotation of axes, remove the cross-product term from $13x^2 + 24xy + 3y^2 = 105$ and identify the corresponding conic section.

34. Transform the equation $(y^2 - x^2)(y + x) = 8\sqrt{2}$ to a uv-equation by rotating the axes through $45°$. Sketch the graph showing both sets of axes.

35. The graph of $x \cos \alpha + y \sin \alpha = d$ is a line. Show that the perpendicular distance from the origin to this line is $|d|$ by making a rotation of axes through the angle α.

36. Use Problem 35 to show that the perpendicular distance from the origin to the line $ax + by = c$ is $|c|/\sqrt{a^2 + b^2}$.

37. Use the result of Problem 36 to find the perpendicular distance from the origin to the line $5x + 12y = 39$.

38. When $Ax^2 + Bxy + Cy^2 = K$ is transformed to $au^2 + buv + cv^2 = K$ by a rotation of axes, it turns out that $A + C = a + c$ and $B^2 - 4AC = b^2 - 4ac$. (Ambitious students will find showing this to be a straightforward but somewhat

lengthy algebraic exercise.) Use these results to transform $x^2 - 8xy + 7y^2 = 9$ to $au^2 + cv^2 = 9$ without actually carrying out the rotation.

39. Recall that the area of an ellipse with major diameter $2a$ and minor diameter $2b$ is πab. Use the first sentence of Problem 38 to show that if $A + C$ and $4AC - B^2$ are both positive, then $Ax^2 + Bxy + Cy^2 = 1$ is the equation of an ellipse with area $2\pi/\sqrt{4AC - B^2}$.

40. **TEASER** The graph of $x^2 - 2xy + 3y^2 = 32$ is an ellipse and therefore can be circumscribed by a rectangle with sides parallel to the x- and y-axes. Find the vertices of this rectangle.

Which Curve Has the Simpler Equation?

Parabola

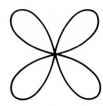

Four-leaved Rose

6-6 The Polar Coordinate System

The question we have asked above makes no sense unless coordinate axes are present. Most people would then choose the parabola as having the simpler equation; but the question is still more subtle than one might think. You already know that the complexity of the equation of a curve depends on the placement of the coordinate axes. Placed just right, the equation of the parabola might be as simple as $y = x^2$. Placed less wisely, the equation might be as complicated as $x - 3 = -(y + 7)^2$, or even worse. However, the four- leaved rose has a very messy equation no matter where the x- and y-axes are placed.

But there is another aspect to the question, one that Fermat and Descartes did not think about when they gave us Cartesian coordinates. There are many different kinds of coordinate systems, that is, different ways of specifying the position of a point. One of these systems, when placed the best possible way, gives the four-leaved rose a delightfully simple equation (see Example B). This system is called the **polar coordinate system;** it simplifies many problems that arise in calculus.

POLAR COORDINATES

In place of two perpendicular axes as in Cartesian coordinates, we introduce in the plane a single horizontal ray, called the **polar axis,** emanating from a fixed point O, called the **pole**. On the polar axis, we mark off the positive half of a number scale with zero at the pole. Any point P other than the pole is the intersection of a unique circle with center O and a unique ray emanating from O, (Figure 47). If r is the radius of the circle and θ is the angle the ray makes with the polar axis, then (r, θ) are the polar coordinates of P.

Points specified by polar coordinates are easiest to plot if we use polar graph paper. The grid on this paper consists of concentric circles and rays emanating from their common center. We have reproduced such a grid in Figure 48 and plotted a few points.

Of course, we can measure the angle θ in degrees as well as radians. More significantly, notice that while a pair of coordinates (r, θ) determines a unique point $P(r, \theta)$, each point has many different pairs of polar coordinates. For example,

$$\left(2, \frac{3\pi}{2}\right) \qquad \left(2, -\frac{\pi}{2}\right) \qquad \left(2, \frac{7\pi}{2}\right)$$

are all coordinates for the same point.

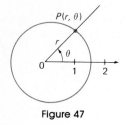

Figure 47

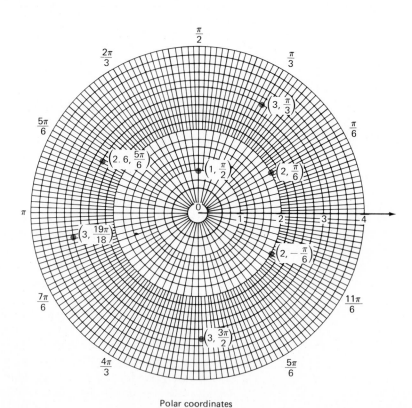

Polar coordinates

Figure 48

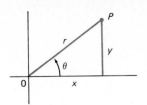

Figure 49

RELATION TO CARTESIAN COORDINATES

Let the positive x-axis of the Cartesian coordinate system serve also as the polar axis of a polar coordinate system, the origin coinciding with the pole (Figure 49). The Cartesian coordinates and polar coordinates are related by two pairs of simple equations.

$$x = r \cos \theta \qquad r^2 = x^2 + y^2$$

$$y = r \sin \theta \qquad \tan \theta = \frac{y}{x}$$

For example, if $(4, \pi/6)$ are the polar coordinates of a point, then its Cartesian coordinates are

$$x = 4 \cos \frac{\pi}{6} = 4 \cdot \frac{\sqrt{3}}{2} = 2\sqrt{3}$$

$$y = 4 \sin \frac{\pi}{6} = 4 \cdot \frac{1}{2} = 2$$

On the other hand, if $(-3, \sqrt{3})$ are the Cartesian coordinates of a point (Figure 50), then

$$r = \sqrt{(-3)^2 + (\sqrt{3})^2} = \sqrt{12} = 2\sqrt{3}$$

$$\tan \theta = \frac{\sqrt{3}}{-3}$$

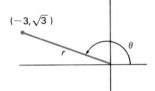

Figure 50

Since the point is in the second quadrant, we choose $5\pi/6$ as an appropriate value of θ. Thus one choice of polar coordinates for the point in question is $(2\sqrt{3}, 5\pi/6)$.

POLAR GRAPHS

The simplest polar equations are $r = k$ and $\theta = k$, where k is a constant. The graph of the first is a circle; the graph of the second is a ray emanating from the origin. Examples are shown in Figure 51. Equations like

$$r = 4 \sin^2\theta \quad \text{and} \quad r = 1 + \cos 2\theta$$

are more complicated. To graph such equations, we suggest making a table of values, plotting the corresponding points, and then connecting those points with a smooth curve. As an example, consider the equation

$$r = \frac{1}{1 - \cos \theta}$$

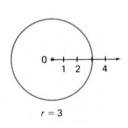

$r = 3$

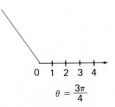

$\theta = \frac{3\pi}{4}$

Figure 51

In Figure 52, we have constructed a table of values and drawn the corresponding graph. It looks suspiciously like a parabola, and in the next section, we will verify that this suspicion is correct.

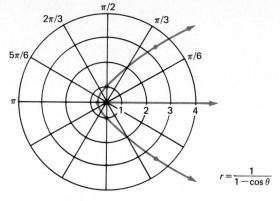

r	θ
—	0
3.4	$\pi/4$
1	$\pi/2$
.6	$3\pi/4$
.5	π
.6	$5\pi/4$
1	$3\pi/2$
3.4	$7\pi/4$
—	2π

$$r = \frac{1}{1-\cos\theta}$$

Figure 52

LIMAÇONS.

We consider next equations of the form

$$r = a \pm b \cos\theta \quad \text{and} \quad r = a \pm b \sin\theta$$

with a and b positive. Their graphs are called **limaçons** with the special case $a = b$ giving a curve called a **cardioid** (a heartlike curve). We assert that these graphs have the shapes shown in Figure 53.

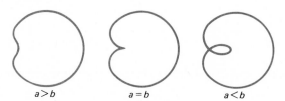

Figure 53

(In the case $a < b$, we allow r to be negative, a matter discussed in Example B in the problem set.)

Consider as an example the equation

$$r = 2(1 + \cos\theta)$$

Its graph (together with a table of values) is the cardioid shown in Figure 54.

Problem Set 6-6

Graph each of the following points given in polar coordinates. Polar graph paper will simplify the graphing process.

1. $\left(3, \frac{\pi}{4}\right)$ 2. $\left(2, \frac{\pi}{3}\right)$ 3. $\left(\frac{3}{2}, \frac{5\pi}{6}\right)$ 4. $\left(1, \frac{5\pi}{3}\right)$

5. $(3, \pi)$ 6. $\left(2, \frac{\pi}{2}\right)$ 7. $(3, -\pi)$ 8. $\left(2, -\frac{3\pi}{2}\right)$

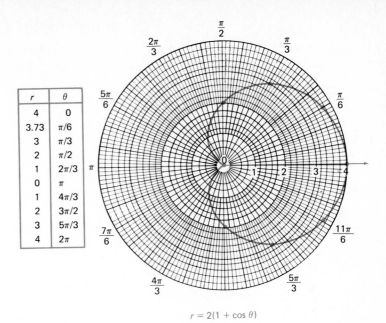

r	θ
4	0
3.73	π/6
3	π/3
2	π/2
1	2π/3
0	π
1	4π/3
2	3π/2
3	5π/3
4	2π

$r = 2(1 + \cos \theta)$

Figure 54

9. $(4, 70°)$ 10. $(3, 190°)$ 11. $\left(\dfrac{5}{2}, \dfrac{7\pi}{3}\right)$ 12. $\left(\dfrac{7}{2}, \dfrac{11\pi}{4}\right)$

Find the Cartesian coordinates of the point having the given polar coordinates.

13. $\left(4, \dfrac{\pi}{4}\right)$ 14. $\left(6, \dfrac{\pi}{6}\right)$ 15. $(3, \pi)$ 16. $\left(2, \dfrac{3\pi}{2}\right)$

17. $\left(10, \dfrac{4\pi}{3}\right)$ 18. $\left(8, \dfrac{11\pi}{6}\right)$ 19. $\left(2, -\dfrac{\pi}{4}\right)$ 20. $\left(3, -\dfrac{2\pi}{3}\right)$

Find polar coordintes for the point with the given Cartesiancoordinates.

21. $(4, 0)$ 22. $(0, 3)$ 23. $(-2, 0)$ 24. $(0, -5)$
25. $(2, 2)$ 26. $(2, -2)$ 27. $(-2, 2)$ 28. $(-2, -2)$
29. $(1, -\sqrt{3})$ 30. $(-2\sqrt{3}, 2)$ 31. $(3, -\sqrt{3})$ 32. $(-\sqrt{3}, -3)$

Graph each of the following equations. Use polar graph paper if it is available.

33. $r = 2$ 34. $r = 5$
35. $\theta = \pi/3$ 36. $\theta = -2\pi/3$
37. $r = |\theta|$ (with θ in radians)   38. $r = \theta^2$
39. $r = 2(1 - \cos \theta)$ 40. $r = 3(1 + \sin \theta)$
41. $r = 2 + \cos \theta$ 42. $r = 2 - \sin \theta$

EXAMPLE A (Transforming Equations)   (a) Change the Cartesian equation $(x^2 + y^2)^2 = x^2 - y^2$ to a polar equation. (b) Change $r = 2 \sin 2\theta$ to a Cartesian equation.

Solution.   (a) Replacing $x^2 + y^2$ by r^2, x by $r \cos \theta$, and y by $r \sin \theta$, we get

$$(r^2)^2 = r^2 \cos^2 \theta - r^2 \sin^2 \theta$$
$$r^4 = r^2(\cos^2 \theta - \sin^2 \theta)$$
$$r^2 = \cos 2\theta$$

Dividing by r^2 at the last step did no harm since the graph of the last equation passes through the pole $r = 0$.

(b)
$$r = 2 \sin \theta$$
$$r = 2 \cdot 2 \sin \theta \cos \theta$$

Multiplying both sides by r^2 gives

$$r^3 = 4(r \sin \theta)(r \cos \theta)$$
$$(x^2 + y^2)^{3/2} = 4yx$$

Transform to a polar equation.

43. $x^2 + y^2 = 4$ 44. $\sqrt{x^2 + y^2} = 6$

45. $y = x^2$ 46. $x^2 + (y - 1)^2 = 1$

Transform to a Cartesian equation.

47. $\tan \theta = 2$ 48. $r = 3 \cos \theta$ 49. $r = \cos 2\theta$ 50. $r^2 = \cos \theta$

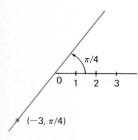

$(-3, \pi/4)$

Figure 55

EXAMPLE B (Allowing Negative Values for r) It is sometimes useful to allow r to be negative. By the point $(-3, \pi/4)$, we shall mean the point 3 units from the pole on the ray in the opposite direction from the ray for $\theta = \pi/4$ (see Figure 55). Allowing r to be negative, graph

$$r = 2 \sin 2\theta$$

Solution. We begin with a table of values (Figure 56), plot the corresponding points, and then sketch the graph (Figure 57).

θ	0	$\frac{\pi}{12}$	$\frac{\pi}{6}$	$\frac{\pi}{4}$	$\frac{\pi}{3}$	$\frac{5\pi}{12}$	$\frac{\pi}{2}$	$\frac{7\pi}{12}$	$\frac{3\pi}{4}$	$\frac{11\pi}{12}$	π	$\frac{5\pi}{4}$	$\frac{3\pi}{2}$	$\frac{7\pi}{4}$	2π
2θ	0	$\frac{\pi}{6}$	$\frac{\pi}{3}$	$\frac{\pi}{2}$	$\frac{2\pi}{3}$	$\frac{5\pi}{6}$	π	$\frac{7\pi}{6}$	$\frac{3\pi}{2}$	$\frac{11\pi}{6}$	2π	$\frac{5\pi}{2}$	3π	$\frac{7\pi}{2}$	4π
r	0	1	$\sqrt{3}$	2	$\sqrt{3}$	1	0	-1	-2	-1	0	2	0	-2	0
			a						*b*				*c*	*d*	

Figure 56

Note: The four leaves correspond to the four parts (a), (b), (c), and (d) of the table of values. For example, leaf (b) results from values of θ between $\pi/2$ and π where r is negative. This graph is the four-leaved rose of our opening display. Its Cartesian equation was obtained in (b) of Example A.

Graph each of the following, allowing r to be negative.

51. $r = 3 \cos 2\theta$ 52. $r = \cos 3\theta$ 53. $r = \sin 3\theta$

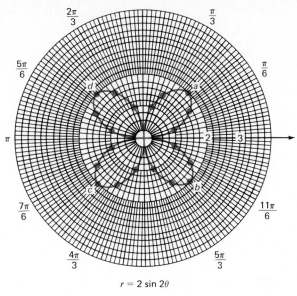

$$r = 2 \sin 2\theta$$

Figure 57

54. $r = 4 \cos \theta$ 　　　 55. $r = \sin 4\theta$ 　　　 56. $r = \cos 4\theta$

MISCELLANEOUS PROBLEMS

57. Transform to a Cartesian equation and identify the corresponding curve.
   (a) $r = 5/(3 \sin \theta - 2 \cos \theta)$ 　　　 (b) $r = 4 \cos \theta - 6 \sin \theta$

58. Transform to a polar equation.
   (a) $x^2 = 4y$ 　　　 (b) $(x - 5)^2 + (y + 2)^2 = 29$

Graph each of the polar equations in Problems 59–64.

59. $r = 4$ 　　　　　　　　　　　 60. $\theta = -\pi/3$

61. $r = 2(1 - \sin \theta)$ 　　　　　 62. $r = 1/\theta, \theta > 0$

63. $r^2 = \sin 2\theta$ 　 *Caution:* Avoid values of θ that make r^2 negative.

64. $r = 2^\theta$ 　 *Note:* Use both negative and positive values for θ.

65. Sketch the graphs of each pair of equations and find their points of intersection.
   (a) $r = 4 \cos \theta, r \cos \theta = 1$ 　　　 (b) $r = 2\sqrt{3} \sin \theta, r = 2(1 + \cos \theta)$

c 66. Find the polar coordinates of the midpoint of the line segment joining the points with polar coordinates $(4, 2\pi/3)$ and $(8, \pi/6)$.

67. Show the distance d between the points with polar coordinates (r_1, θ_1) and (r_2, θ_2) is given by

$$d = \sqrt{r_1^2 + r_2^2 - 2r_1 r_2 \cos(\theta_2 - \theta_1)}$$

and use this result to find the distance between $(4, 2\pi/3)$ and $(8, \pi/6)$.

68. Show that a circle of radius a and center (a, α) has polar equation $r = 2a \cos(\theta - \alpha)$. *Hint:* Law of cosines.

69. Find a formula for the area of the polar rectangle $0 < a < r < b$, $\alpha \le \theta \le \beta, \beta - \alpha < \pi$.

70. A point P moves so that its distance from the pole is always equal to its distance from the horizontal line $r \sin \theta = 4$. Show that the equation of the resulting curve (a parabola) is $r = 4/(1 + \sin \theta)$.

71. A line segment L of length 4 has its two endpoints on the x- and y-axes, respectively. The point P is on L and is such that the line OP from the pole to P is perpendicular to L. Show that the set of points P satisfying this condition is a four-leaved rose by finding its polar equation.

72. **TEASER** Let F and F' be fixed points with polar coordinates $(a, 0)$ and $(-a, 0)$, respectively. A point P moves so that the product of its distances from F and F' is equal to the constant a^2 (that is, $\overline{PF} \cdot \overline{PF'} = a^2$). Find a simple polar equation (of the form $r^2 = f(\theta)$) for the resulting curve and sketch its graph.

If the Greeks had not cultivated conic sections, Kepler could not have superseded Ptolemy.

William Whewell

The planets move around the sun in ellipses; the sun is at one focus of these ellipses.

Kepler's First Law

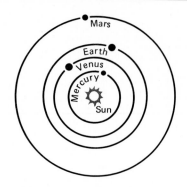

6-7   Polar Equations of Conics

Historians have suggested that the Greeks studied the conic sections simply to satisfy their intellectual cravings after the ideal and that they hardly dreamed that these curves would have important physical applications. Today we know that they describe the motions of the moon, of the planets, of comets, of space probes, and even of tiny electrons as they orbit the nucleus of an atom. In the study of such motions, it is not the Cartesian equations of the conics that prove most useful. It is rather their polar equations that play a central role.

A NEW APPROACH TO THE CONICS

The definitions we have given for the three conics (parabola, ellipse, and hyperbola) are quite dissimilar in form. However, there is another approach to the three curves that treats them in a uniform way; it is this approach that will eventually lead us to the polar equations for these curves.

In the plane, consider a fixed line l (the **directrix**) and a fixed point F (the **focus**). Let a point P move so that its distance $d(P, F)$ from the focus is a constant e times its distance from the directrix $d(P, L)$, that is, so that

$$d(P, F) = ed(P, L)$$

Figure 58

as suggested by Figure 58. Here the constant e (called the **eccentricity**) may be any positive number. Of course, if $e = 1$, we are on familiar ground; the corresponding curve is a parabola. But what if $e \neq 1$?

To get a feeling for this new situation, try graphing the path of P for the two cases $e = \frac{1}{2}$ and $e = 2$. If you do it carefully, you should get curves that look like an ellipse and a hyperbola, respectively. In fact, we claim that the equation $d(P, F) = ed(P, L)$ can serve as the definition of an ellipse when $0 < e < 1$ and of a hyperbola when $e > 1$. We demonstrate this now.

THE NEW DEFINITIONS ARE EQUIVALENT TO THE OLD ONES

Suppose that $e \neq 1$ and consider the curve determined by the defining equation $d(P, F) = ed(P, L)$. It is fairly easy to see that this curve must be symmetric with respect to the line through the focus and perpendicular to the directrix and that the curve must cross this line twice say at A' and A. Place the curve in the coordinate system so that A' and A have coordinates $(-a, 0)$ and $(a, 0)$, respectively. Let the directrix be the line $x = k$ and the focus be the point $(c, 0)$ with a, k, and c all positive. There are two possible arrangements (Figure 59) depending on whether $0 < e < 1$ or $e > 1$. (Do not let the appearance of the curves in the figure lead you to the conclusion that we have proved anything yet.)

Apply the equation $d(P, F) = ed(P, L)$ first with $P = A'$ and then with $P = A$ to obtain the pair of equations

$$a - c = e(k - a) = ek - ea$$
$$a + c = e(k + a) = ek + ea$$

Solve this pair of equations for a and c to get

$$c = ea \quad \text{and} \quad k = \frac{a}{e}$$

and note for later reference that this implies $e = c/a$. Now let $P(x, y)$ be any

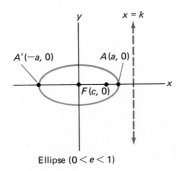

Figure 59

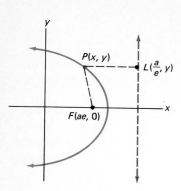

Figure 60

point on the curve. Then $L(a/e, y)$ is its projection on the directrix (see Figure 60 for the case $0 < e < 1$) and the condition $d(P, F) = ed(P, L)$ translates to

$$\sqrt{(x - ae)^2 + y^2} = e\sqrt{\left(x - \frac{a}{e}\right)^2}$$

Squaring both sides and collecting like terms yields

$$x^2 - 2aex + a^2e^2 + y^2 = e^2\left(x^2 - 2\frac{a}{e}x + \frac{a^2}{e^2}\right)$$

or

$$(1 - e^2)x^2 + y^2 = a^2(1 - e^2)$$

or

$$\frac{x^2}{a^2} + \frac{y^2}{(1 - e^2)a^2} = 1$$

If $0 < e < 1$, this is the standard equation of an ellipse; if $e > 1$, it is the standard equation of a hyperbola. Morever since $e = c/a$, our use of e in this section is consistent with our usage in Sections 6-2 and 6-3.

POLAR EQUATIONS OF THE CONICS

To simplify matters, we will place the conic in the polar coordinate system so that the focus is at the pole (origin) and the directrix is d units away as in Figure 61. The defining equation $d(P, F) = ed(P, L)$ takes the form

$$r = e(d - r \cos(\theta - \theta_0))$$

which is equivalent to

$$\boxed{r = \frac{ed}{1 + e \cos(\theta - \theta_0)}}$$

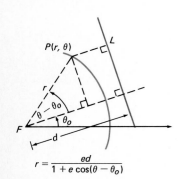

$$r = \frac{ed}{1 + e \cos(\theta - \theta_0)}$$

Figure 61

As an example, consider a case where $\theta_0 = 0$, namely,

$$r = \frac{2}{1 + \frac{1}{2} \cos \theta} = \frac{\frac{1}{2} \cdot 4}{1 + \frac{1}{2} \cos \theta}$$

Since $e = \frac{1}{2}$, the graph is an ellipse, the one shown in Figure 62. On the other hand,

$$r = \frac{12}{3 + 4 \cos \theta} = \frac{4}{1 + \frac{4}{3} \cos \theta} = \frac{\frac{4}{3} \cdot 3}{1 + \frac{4}{3} \cos \theta}$$

is a hyperbola with $e = \frac{4}{3}$ and $d = 3$. Examples A–C in the problem set give a complete discussion of the cases $\theta_0 = 0$, $\pi/2$, π, and $3\pi/2$.

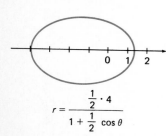

$$r = \frac{\frac{1}{2} \cdot 4}{1 + \frac{1}{2} \cos \theta}$$

Figure 62

POLAR EQUATIONS OF A LINE

If the line passes through the pole, it has the exceedingly simple equation $\theta = \theta_0$, with θ_0 a constant. If the line does not go through the pole, it is some distance d from it. Let θ_0 be the angle from the polar axis to the perpendicular drawn from the pole to the given line (Figure 63). Then if $P(r, \theta)$ is a point on the line,

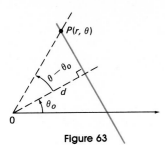

$$\cos(\theta - \theta_0) = \frac{d}{r}$$

or

Figure 63

$$\boxed{r = \frac{d}{\cos(\theta - \theta_0)}}$$

For example,

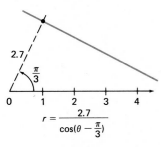

$$r = \frac{2.7}{\cos(\theta - \frac{\pi}{3})}$$

Figure 64

$$r = \frac{2.7}{\cos(\theta - \pi/3)}$$

is the equation of the line shown in Figure 64.

THE POLAR EQUATION OF A CIRCLE

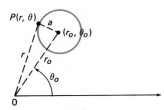

Figure 65

If the circle is centered at the pole, its polar equation is simply $r = a$, where a is the radius of the circle. If the center is at (r_0, θ_0), then by the law of cosines (see Figure 65)

$$a^2 = r^2 + r_0^2 - 2rr_0 \cos(\theta - \theta_0)$$

which is too complicated to be of much use. However, if the circle passes through the pole so $r_0 = a$, the equation simplifies to

$$r^2 = 2ra \cos(\theta - \theta_0)$$

or, after dividing by r,

$$\boxed{r = 2a \cos(\theta - \theta_0)}$$

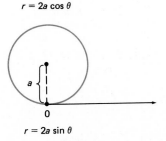

$r = 2a \cos \theta$

$r = 2a \sin \theta$

Figure 66

The cases $\theta_0 = 0$ and $\theta_0 = \pi/2$ are particularly nice. The first gives $r = 2a \cos \theta$; the second gives $r = 2a \cos(\theta - \pi/2)$, which is equivalent to $r = 2a \sin \theta$. The graphs for these two cases are shown in Figure 66.

In Figure 67, we have summarized most of the results of this section.

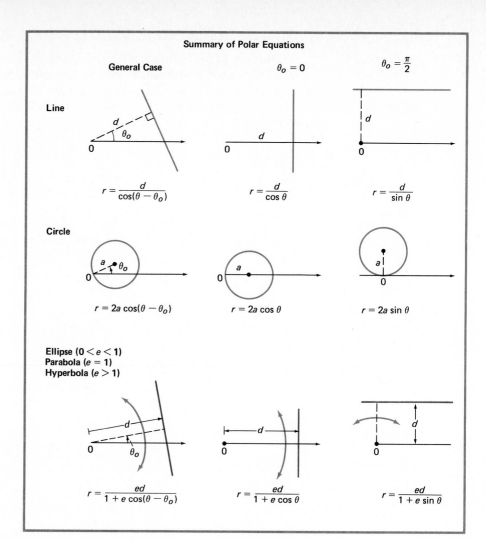

Figure 67

Problem Set 6-7

In doing the following problems, it will be helpful to keep three things in mind.

1. *The summary of polar equations in Figure 67.*
2. *The addition formulas* $\cos(\theta \mp \theta_0) = \cos\theta \cos\theta_0 \pm \sin\theta \sin\theta_0$.
3. *The relations* $x = r\cos\theta$ *and* $y = r\sin\theta$.

In Problems 1–4, write a simple polar equation for each equation.

1. $x = 4$ 2. $y = 3$ 3. $x = -3$ 4. $y = -2$

Each of the polar equations in Problems 5–10 represents a line. Sketch its graph and then write its xy-equation in the form Ax + By = C.

5. $r = \dfrac{6}{\cos \theta}$

6. $r = \dfrac{3}{\sin \theta}$

7. $r = \dfrac{4}{\cos(\theta - \pi/3)}$

8. $r = \dfrac{10}{\cos(\theta - \pi/4)}$

9. $r = \dfrac{5}{\cos(\theta + \pi/4)}$

10. $r = \dfrac{4}{\cos(\theta + \pi/3)}$

In Problems 11–14, write the polar equation of the circle which passes through the pole and whose center has the given polar coordinates. Then find the corresponding xy-equations.

11. $(4, 0)$ 12. $(3, 90°)$ 13. $(5, \pi/3)$ 14. $(8, 150°)$

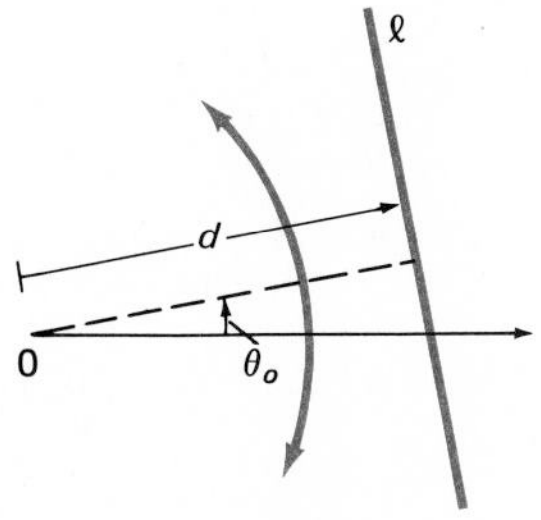

EXAMPLE A (Conics with Directrix Perpendicular to the Polar Axis)   Refer to Figure 68. If the directrix l is perpendicular to the polar axis, then $\theta_0 = 0$ or $\theta_0 = \pi$, and the polar equation of the conic takes one of the two forms.

$$\theta_0 = 0 \qquad\qquad \theta_0 = \pi$$

$$r = \frac{ed}{1 + e\cos\theta} \qquad r = \frac{ed}{1 - e\cos\theta}$$

In the first case, the directrix is to the right of the focus; in the second, it is to the left.

   Identify each of the following conics by name, find its eccentricity, and write the xy-equation of its directrix.

(a) $r = \dfrac{4}{3 + 3\cos\theta}$

(b) $r = \dfrac{5}{2 - 3\cos\theta}$

$r = \dfrac{ed}{1 + e\cos(\theta - \theta_0)}$

Figure 68

Solution.

(a) Divide numerator and denominator by 3, obtaining

$$r = \frac{\frac{4}{3}}{1 + \cos\theta}$$

The conic is a parabola, since the eccentricity $e = 1$. The equation of the directrix is $x = \frac{4}{3}$.

(b) The equation can be rewritten as

$$r = \frac{\frac{5}{2}}{1 - \frac{3}{2}\cos\theta} = \frac{\frac{3}{2}\cdot\frac{5}{3}}{1 - \frac{3}{2}\cos\theta}$$

The conic is a hyperbola with $e = \frac{3}{2}$ and directrix $x = -\frac{5}{3}$.

In Problems 15–22, identify the conic by name, give its eccentricity, and write the equation of its directrix (in xy-coordinates).

15. $r = \dfrac{4}{1 + \frac{2}{3}\cos\theta}$

16. $r = \dfrac{\frac{9}{2}}{1 + \frac{3}{4}\cos\theta}$

17. $r = \dfrac{5}{2 + 4\cos\theta}$

18. $r = \dfrac{3}{1 + \cos \theta}$ 19. $r = \dfrac{7}{1 - \cos \theta}$ 20. $r = \dfrac{\frac{1}{2}}{1 - \frac{3}{2} \cos \theta}$

21. $r = \dfrac{\frac{1}{2}}{\frac{3}{2} - \cos \theta}$ 22. $r = \dfrac{3}{6 - 6 \cos \theta}$

EXAMPLE B (Conics with Directrix Parallel to the Polar Axis) Refer to the diagram of Example A. If the directrix is parallel to the polar axis and above it, then $\theta_0 = \pi/2$; if it is below the polar axis, then $\theta_0 = 3\pi/2$. The corresponding equations can be simplified to

$$\theta_0 = \frac{\pi}{2} \qquad\qquad \theta_0 = \frac{3\pi}{2}$$

$$r = \frac{ed}{1 + e \sin \theta} \qquad\qquad r = \frac{ed}{1 - e \sin \theta}$$

(a) Derive the first of these equations.

(b) Identify the conic $r = 5/(2 - \sin \theta)$ by name, give its eccentricity, and write the xy-equation of its directrix.

Solution.

(a) The equation of the conic with $\theta_0 = \pi/2$ is

$$r = \frac{ed}{1 + e \cos (\theta - \frac{\pi}{2})}$$

Since

$$\cos(\theta - \tfrac{\pi}{2}) = \cos \theta \cos \tfrac{\pi}{2} + \sin \theta \sin \tfrac{\pi}{2} = \sin \theta$$

we get

$$r = \frac{ed}{1 + e \sin \theta}$$

(b) Dividing numerator and denominator by 2, we obtain

$$r = \frac{\frac{5}{2}}{1 - \frac{1}{2} \sin \theta} = \frac{\frac{1}{2} \cdot 5}{1 - \frac{1}{2} \sin \theta}$$

The conic is an ellipse with eccentricity $\frac{1}{2}$. The directrix is below the polar axis and has xy-equation $y = -5$.

In Problems 23–28, identify the conic by name, give its eccentricity, and write the xy-equation of its directrix.

23. $r = \dfrac{5}{1 + \sin \theta}$ 24. $r = \dfrac{2}{1 + \frac{2}{3} \sin \theta}$ 25. $r = \dfrac{6}{2 - \sin \theta}$

26. $r = \dfrac{5}{2 - 4 \sin \theta}$ 27. $r = \dfrac{4}{2 + \frac{5}{2} \sin \theta}$ 28. $r = \dfrac{5}{4 - 3 \sin \theta}$

EXAMPLE C (Graphing Conics in Polar Coordinates)   Graph the conic whose polar equation is

$$r = \frac{6}{1 + 2 \cos \theta}$$

Solution.   We recognize this as the polar equation of a hyperbola ($e = 2$) with major axis along the polar axis. Next we make the small table of values shown in Figure 69 and plot the corresponding points (marked with dots). The points marked with a cross are obtained by symmetry ($\cos(-\theta) = \cos \theta$).

Graph each of the following conics.

29. $r = \dfrac{4}{1 + \frac{2}{3} \cos \theta}$ 30. $r = \dfrac{6}{1 + \frac{3}{4} \sin \theta}$ 31. $r = \dfrac{5}{1 + \sin \theta}$

32. $r = \dfrac{4}{2 + 2 \cos \theta}$ 33. $r = \dfrac{18}{2 + 3 \cos \theta}$ 34. $r = \dfrac{3}{1 - 2 \cos \theta}$

r	θ
2	0°
3	60°
6	90°
−8.2	150°
−6	180°

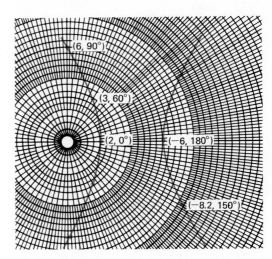

Figure 69

MISCELLANEOUS PROBLEMS

35. Write the polar equation $r = f(\theta)$ that corresponds to each of the following Cartesian equations.
    (a) $y = 3$ (b) $x^2 + y^2 = 9$
    (c) $(x + 9)^2 + y^2 = 81$ (d) $(x - 3)^2 + (y - 3)^2 = 18$
36. Graph each of the following polar equations.
    (a) $r = -4/\cos \theta$ (b) $r = -8 \sin \theta$
    (c) $r = 3$ (d) $r = 6 \cos(\theta - \pi/3)$
    (e) $r = 6/(1 - \cos \theta)$ (f) $r = 6/(2 + \cos \theta)$
37. Determine the polar equations for the parabolas with Cartesian equations given by (a) $4(x + 1) = y^2$ and (b) $-8(y - 2) = x^2$. *Hint:* In each case, the focus is at the origin.

38. Find the polar equation $r = f(\theta)$ corresponding to the parabola with polar equation $y^2 = 16x$. Why is this polar equation unlike any discussed in this section?

39. Write the Cartesian equation of the curve whose polar equation is

$$\cos^2\theta + \sin\theta\cos\theta - 6\sin^2\theta = 0$$

40. Find the points of intersection (in polar coordinates) of the circles with polar equations $r = 3\sin\theta$ and $r = \sqrt{3}\cos\theta$.

41. Find e and d for the ellipse with polar equation $r = 8/(2 + \cos\theta)$. Also find the major and minor diameters of this ellipse.

42. Show that the ellipse $r = ed/(1 + e\cos\theta)$ has major diameter $2ed/(1 - e^2)$ and minor diameter $2ed/\sqrt{1 - e^2}$

43. Find the length of the latus rectum (chord through the focus perpendicular to the major axis) for the ellipse $r = 8/(2 + \cos\theta)$.

44. Express the length of the latus rectum of the conic $r = ed/(1 + e\cos\theta)$ in terms of e and d.

45. The graph of $r = 2a + a\cos\theta$ for $a > 0$ is an example of a curve called a *limaçon* (Section 11-6).
    (a) Sketch this graph for the case $a = 3$.
    (b) Show that every chord through the pole has length $4a$ (a nice property it shares with the circle $r = 2a$).

46. **TEASER** Sketch the graph of the polar equation $r = 1/\theta$ for $\theta \geq \pi/2$ and show that this curve has infinite length.

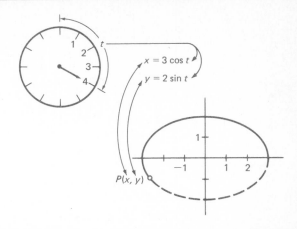

Every physicist knows that a good way to specify the path of a particle is to give its coordinates as functions of the elapsed time t. Suppose a point $P(x, y)$ moves in the plane so that $x = 3 \cos t$ and $y = 2 \sin t$. What is the shape of its path?

6-8 Parametric Equations

So far, we have described the conic sections by their Cartesian equations and by their polar equations. There is still another way to describe the conic sections, as well as many other curves. It arises naturally in the study of motion in physics, but its use goes far beyond that particular application.

Imagine that the xy-coordinates of a point on a curve are specified not by giving a relationship between x and y, but rather by telling how x and y are related to a third variable. For example, it may be that as time t advances from $t = a$ to $t = b$, a point $P(x, y)$ traces out a curve in the xy-plane. Then both x and y are functions of t. That is,

$$x = f(t) \quad \text{and} \quad y = g(t) \qquad a \leq t \leq b$$

We call the boxed equations the **parametric equations** of a curve with t as parameter. A **parameter** is simply an auxiliary variable on which other variables depend.

AN EXAMPLE
Let

$$x = 2t \quad \text{and} \quad y = t^2 - 3 \qquad -1 \leq t \leq 3$$

These equations can be used to make a table of values and then to draw a graph. We illustrate in Figure 70 on the next page.

The curve just drawn looks suspiciously like part of a parabola. We can demonstrate that this is true by **eliminating the parameter** t. Solve the first equation for t, giving $t = x/2$. Then substitute this value of t in the second equation. We obtain

t	x	y
-1	-2	-2
$-1/2$	-1	$-11/4$
0	0	-3
$1/2$	1	$-11/4$
1	2	-2
2	4	1
3	6	6

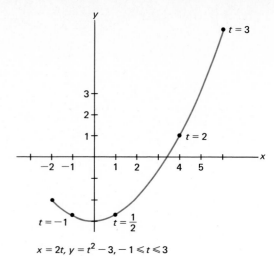

$$x = 2t, \quad y = t^2 - 3, \quad -1 \leqslant t \leqslant 3$$

Figure 70

$$y = \left(\frac{x}{2}\right)^2 - 3 = \frac{1}{4}x^2 - 3$$

which we do recognize as the equation of a parabola.

PARAMETRIC EQUATIONS OF A LINE

Consider first a line l which passes through the points $(0, 0)$ and (a, b). We claim that

$$x = at \quad \text{and} \quad y = bt$$

are a pair of parametric equations for this line. To see why, note that $t = 0$ and $t = 1$ yield the given points. Moreover, if we eliminate t, we get $y = (b/a)x$, which is the equation of a line. This example shows, incidentally, that a parametric representation is not unique since there are many choices for (a, b). (See Problem 29 for more evidence on this point.)

Next we translate the above line by replacing x by $x - x_1$ and y by $y - y_1$. This gives

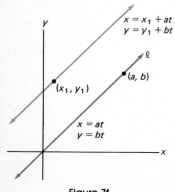

$$\boxed{x = x_1 + at \quad \text{and} \quad y = y_1 + bt}$$

These are the parametric equations of a line through (x_1, y_1) parallel to the line with which we started (Figure 71).

Finally, consider the line through (x_1, y_1) with slope m. This line is parallel to the line through the points $(0, 0)$ and $(1, m)$ and so has parametric equations

Figure 71

$$x = x_1 + t \quad \text{and} \quad y = y_1 + mt$$

Note that if we eliminate t by solving the first equation for t and substituting it in the second, we get $y - y_1 = m(x - x_1)$, the point-slope form for the equation of a line.

THE CIRCLE AND THE ELLIPSE

If you think about the definitions of sine and cosine for a moment, you already know one set of parametric equations for a circle of radius a centered at the origin.

$$x = a \cos t \quad \text{and} \quad y = a \sin t \qquad 0 \leq t \leq 2\pi$$

For other possibilities, see Problem 29.

Consider next an ellipse centered at $(0, 0)$ and passing through $(a, 0)$ and $(0, b)$. We claim that

$$x = a \cos t \quad \text{and} \quad y = b \sin t \qquad 0 \leq t \leq 2\pi$$

are parametric equations for it. To see that this is correct, we shall eliminate the parameter t. One way to do this is to solve for $\cos t$ and $\sin t$, respectively, square the results, and add. This gives

$$\frac{x^2}{a^2} + \frac{y^2}{b^2} = \cos^2 t + \sin^2 t = 1$$

which we recognize as the xy-equation of an ellipse centered at $(0, 0)$ and passing through $(a, 0)$ and $(0, b)$.

Now we can answer the question asked in the opening display of this section. The parametric equations

$$x = 3 \cos t \quad \text{and} \quad y = 2 \sin t$$

determine an ellipse with center at $(0, 0)$ and passing through $(3, 0)$ and $(0, 2)$.

THE HYPERBOLA

By analogy with the situation for the ellipse, we are led to

$$x = a \sec t \quad \text{and} \quad y = b \tan t \qquad 0 \leq t \leq 2\pi, t \neq \frac{\pi}{2}, t \neq \frac{3\pi}{2}$$

as parametric equations for a hyperbola. Note that when we solve for $\sec t$ and $\tan t$ in the two equations, square the results, and subtract, we obtain

$$\frac{x^2}{a^2} - \frac{y^2}{b^2} = \sec^2 t - \tan^2 t = 1$$

There is another set of parametric equations for the hyperbola. It involves an important pair of functions called the *hyperbolic sine* and the *hyperbolic cosine*. These functions are discussed in Problem 36.

THE CYCLOID

So far, our discussion of parametric equations has concentrated on the conic sections. Actually, there are other important curves where parametric representation is almost essential. The cycloid provides an example where the xy-equation is so complicated it is rarely used.

Consider a wheel of radius a which is free to roll along the x-axis. As the wheel turns, a point P on the rim traces out a curve called the **cycloid** (Figure 72).

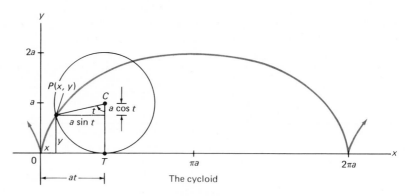

The cycloid

Figure 72

Assume P is initially at the origin and let C and T be as indicated in the diagram, with t denoting the radian measure of angle TCP. Then the arc PT and the segment OT have the same length and so the center C of the rolling circle is at (at, a). Using a little trigonometry, we conclude that

$$x = at - a \sin t = a(t - \sin t)$$

and

$$y = a - a \cos t = a(1 - \cos t)$$

Problem Set 6-8

In Problems 1–4, write parametric equations for the line that passes through the two given points.

1. $(0, 0)$, $(2, -3)$ 2. $(0, 0)$, $(-3, 6)$
3. $(1, 2)$, $(4\ -5)$ 4. $(-2, 3)$, $(3, 7)$
5. Find the slope and y-intercept of the line with parametric equations $x = 3 + 2t$ and $y = -5 - 4t$. *Hint:* Eliminate the parameter t.

6. Write the equation of the line with parametric equations $x = -2 - 3t$ and $y = 4 + 9t$ in the form $Ax + By + C = 0$.

In Problems 7–16, eliminate the parameter to determine the corresponding xy-equation.

7. $x = 3s + 1, y = -2s + 5$ 8. $x = 3s + 1, y = s^2$
9. $x = 2t - 1, y = 2t^2 + t$ 10. $x = 3t, y = t^2 - 3t + 1$
11. $x = 2 \cos t, y = 2 \sin t$ 12. $x = 3 \sin t, y = 3 \cos t$
13. $x = 2 \cos t, y = 3 \sin t$ 14. $x = 6 \sin t, y = \cos t$
15. $x = 3t + 1, y = t^3$ 16. $x = 2 \sec t, y = 3 \tan t$

EXAMPLE A (More Graphing)   Sketch the graph of the curve with parametric equations $x = 8 \cos^3 t$ and $y = 8 \sin^3 t, 0 \le t \le 2\pi$.

Solution.   A table of values and the graph are shown in Figure 73. This curve is called a *hypocycloid* (see Problem 39).

Sketch the graph of each of the following for the indicated interval.

17. $x = 2t - 1, y = t^2 + 2; -2 \le t \le 2$
18. $x = 2t^2, y = 3 - 2t; -2 \le t \le 2$
19. $x = t^3, y = t^2; -2 \le t \le 2$
20. $x = 2^t, y = 3t; -3 \le t \le 2$
21. $x = \dfrac{1 - t^2}{1 + t^2}, y = \dfrac{2t}{1 + t^2};$ all t

22. $x = \dfrac{t^2}{1 + t^2}, y = \dfrac{t^3}{1 + t^2};$ all t

ⓒ 23. $x = 8t - 4 \sin t, y = 8 - 4 \cos t; 0 \le t \le 4\pi$
     Note: This curve is called a *curtate cycloid* (see Problem 37).

t	x	y
0	8	0
$\pi/6$	5.2	1
$\pi/4$	2.8	2.8
$\pi/3$	1	5.2
$\pi/2$	0	8
$3\pi/4$	-2.8	2.8
π	-8	0
$5\pi/4$	-2.8	-2.8
$3\pi/2$	0	-8
$7\pi/4$	2.8	-2.8

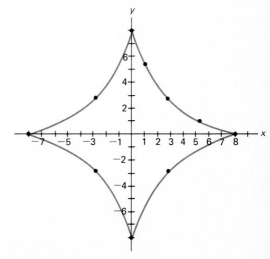

Figure 73

☐ 24. $x = 6t - 8 \sin t$, $y = 6 - 8 \cos t$; $0 \le t \le 4\pi$

   Note: This curve is called a *prolate cycloid* (see Problem 38).

EXAMPLE B (A Projectile Problem)   The path of a projectile fired at 64 feet per second from ground level at an angle of 60° with the horizontal is given by the parametric equations

$$x = 32t \quad \text{and} \quad y = -16t^2 + 32\sqrt{3}\,t$$

where the origin is at the point of firing and the x-axis is along the (horizontal) ground in the plane of the projectile's flight (Figure 74). Find
(a) the xy-equation of the path;
(b) the total time of flight;
(c) the range, that is, the value of x where the projectile strikes the ground;
(d) the greatest height reached.

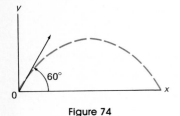

Figure 74

Solution.
   (a) Since $t = x/32$,

$$y = -16\left(\frac{x}{32}\right)^2 + 32\sqrt{3}\left(\frac{x}{32}\right)$$
$$= -\tfrac{1}{64}x^2 + \sqrt{3}\,x$$

which is the equation of a parabola.
   (b) We find the values of t for which $y = 0$.

$$y = -16t^2 + 32\sqrt{3}\,t = -16t(t - 2\sqrt{3})$$

Thus $y = 0$ when $t = 0$ (time of firing) and when $t = 2\sqrt{3}$ (time of landing). The time of flight is $2\sqrt{3}$ seconds.
   (c) The range is $32(2\sqrt{3}) = 64\sqrt{3}$ feet.
   (d) The maximum height occurs when $t = \tfrac{1}{2}(2\sqrt{3}) = \sqrt{3}$. At this time,

$$y = -16(\sqrt{3})^2 + 32\sqrt{3}(\sqrt{3}) = 48 \text{ feet}$$

Answer the four questions of Example B for the data below.

25. $x = 64\sqrt{3}\,t$, $y = -16t^2 + 64t$
26. $x = 48\sqrt{2}\,t$, $y = -16t^2 + 48\sqrt{2}\,t$

MISCELLANEOUS PROBLEMS

27. *Determine the Cartesian equation corresponding to each given pair of parametric equations.*
   (a) $x = 3 - 2t$, $y = 4 + 3t$
   (b) $x = 3t$, $y = 4 \cos t$
   (c) $x = 2 \sec t$, $y = 3 \tan t$
   (d) $x = 1 - t^3$, $y = 2t - 1$
   (e) $x = t^2 + 2t$, $y = \sqrt[3]{t} - t^2 - 2t$
28. Write parametric equations for each of the following.
   (a) The line that passes through the points $(2, -1)$ and $(4, 3)$.

(b) The line through $(4, -2)$ and parallel to the line with parametric equations $x = 3 + 2t$, $y = -2 + t$.

(c) The ellipse with Cartesian equation $x^2/9 + y^2/25 = 1$.

(d) The circle $(x - 4)^2 + (y + 2)^2 = 25$

29. Show that all of the following parametrizations represent the same curve (one quarter of a circle).

(a) $x = 2 \cos t$, $y = 2 \sin t$; $0 \le t \le \pi/2$

(b) $x = \sqrt{t}$, $y = \sqrt{4 - t}$; $0 \le t \le 4$

(c) $x = t + 1$, $y = \sqrt{3 - 2t - t^2}$; $-1 \le t \le 1$

(d) $x = (2 - 2t)/(1 + t)$, $y = 4\sqrt{t}/(1 + t)$; $0 \le t \le 1$

30. Show that the parametric equations $x = \sqrt{2t + 1}$, $y = \sqrt{8t}$, $t \ge 0$, represent part of a hyperbola. Sketch that part.

31. Sketch the graph of $x = 2 + 3 \cos t$, $y = 1 + 4 \sin t$, $0 \le t \le 2\pi$, and determine the corresponding Cartesian equation.

32. Sketch the graph of $x = \sin t$, $y = \tan t$, $-\pi/2 < t < \pi/2$, and determine the corresponding Cartesian equation.

33. The path of a projectile fired from level ground with a speed of v_0 feet per second at an angle α with the ground, is given by the parametric equations

$$x = (v_0 \cos \alpha)t \qquad y = -16t^2 + (v_0 \sin \alpha)t$$

(a) Show that the path is a parabola.

(b) Find the time of flight.

(c) Show that the range is $(v_0^2/32) \sin 2\alpha$.

(d) For a given v_0, what value of α gives the largest possible range?

34. Show that the parametric equations $x = t^{-1/2} \cos t$, $y = t^{-1/2} \sin t$, $t > 0$, represent the same curve as the spiral whose polar equation is $r^2\theta = 1$, $r > 0$, $\theta > 0$.

35. Show that the graph of $x = 5 \sin^2 t - 4 \cos^2 t$, $y = 4 \cos^2 t + 5 \sin^2 t$, $0 \le t \le \pi/2$, is a line segment and find its endpoints. What curve do you get when t varies from $\pi/2$ to π?

36. Define two (important) functions called the **hyperbolic sine** and **hyperbolic cosine** by

$$\sinh t = \frac{e^t - e^{-t}}{2} \qquad \cosh t = \frac{e^t + e^{-t}}{2}$$

(a) Show that $\cosh^2 t - \sinh^2 t = 1$

(b) Show that $x = a \cosh t$, $y = a \sinh t$ give a parameterization for one branch of a hyperbola.

37. Modify the text discussion of the cycloid (and its accompanying diagram) to handle the case where the point P is $b < a$ units from the center of the wheel. You should obtain the parametric equations

$$x = at - b \sin t \qquad y = a - b \cos t$$

The graph of these equations is called a *curtate cycloid* (see Problem 23).

38. Follow the instructions of Problem 37 for the case $b > a$ (a flanged wheel, as on a train) showing that you get the same parametric equations. The graph of these equations is now called a *prolate cycloid* (see Problem 24).

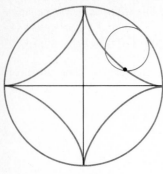

Figure 75

39. Suppose that a wheel of radius b rolls inside a circle of radius $a = 4b$ (Figure 75). Show that the parametric equations for a point P on the rim of the rolling wheel are

$$x = a \cos^3 \theta \qquad y = a \sin^3 \theta$$

provided the fixed wheel is centered at $(0, 0)$ and P is initially at $(a, 0)$. The resulting curve is called a *hypocycloid* (see Example A).

40. **TEASER** A wheel of radius 3 and centered at the origin is rotating counterclockwise at 2 radians per second so that the point P on its rim is at $(3, 0)$ when time $t = 0$. Another wheel of radius 1 and centered at $(8, 0)$ is also rotating but clockwise at 2 radians per second so that the point Q on its rim is at $(9, 0)$ at $t = 0$. A knot K is tied at the middle of an elastic string and then one end of the string is attached to P and the other to Q. The string is short enough to remain taut at all times.
    (a) Obtain the parametric equations for the path of the knot K, using time t as the parameter.
    (b) Write the Cartesian equation for the path of K and use it to describe the path in geometric terms.

Chapter Summary

A **parabola** is the set of points that are equidistant from a fixed line (the **directrix**) and a fixed point (the **focus**). An **ellipse** is the set of points for which the sum of the distances from two fixed points (the **foci**) is a constant. A **hyperbola** is the set of points for which the difference of the distances from two fixed points (the **foci**) is a constant. These three curves (called **conics**) together with their limiting forms (circle, point, empty set, line, two intersecting lines, two parallel lines) are the chief objects of study in this chapter.

When these curves are placed in the Cartesian coordinate plane in an advantageous way, their xy-equations take the **standard forms** below.

$$\text{Parabola:} \qquad y^2 = 4px$$

$$\text{Ellipse:} \qquad \frac{x^2}{a^2} + \frac{y^2}{b^2} = 1$$

$$\text{Hyperbola:} \qquad \frac{x^2}{a^2} - \frac{y^2}{b^2} = 1$$

The curves can, of course, be placed in the plane in other ways; in these cases, their equations are more complicated. However, they can be brought to standard form by a **translation** or a **rotation** of axes. Even the complicated equations are always of the form

$$Ax^2 + Bxy + Cy^2 + Dx + Ey + F = 0$$

Conversely, the graph of any equation of this type is one of the conics or the six limiting forms.

Cartesian coordinates (x, y) are not the only way to specify the position of a point. **Polar coordinates** (r, θ) also determine points, and **polar equations** determine curves. In fact, some very beautiful curves (**limaçons, cardioids, roses**) have simple polar equations but complicated Cartesian equations. Moreover, for purposes of astronomy, the conics are best described by polar equations, since their equations all take the form

$$r = \frac{ed}{1 + e\cos(\theta - \theta_0)}$$

This equation determines an ellipse if e (called the **eccentricity**) satisfies $0 < e < 1$, a parabola if $e = 1$, and a hyperbola if $e > 1$.

Finally, many curves (including the conics) can be described by giving parametric equations $x = f(t)$ and $y = g(t)$ in which both x and y are specified in terms of a **parameter** t.

Chapter Review Problem Set

1. Determine the number of the description that best fits the given equation.

   _____ (a) $(x - 1)^2 + y^2 + 3 = 0$
   _____ (b) $x = y^2 + 2y + 3$
   _____ (c) $x^2 - 4y^2 = 0$
   _____ (d) $4(x - 1)^2 + y^2 = 3$
   _____ (e) $4(x - 1)^2 + 4y^2 = 3$
   _____ (f) $4(x - 1)^2 - 4y^2 = -3$
   _____ (g) $4(x - 1)^2 - 4y^2 = 3$
   _____ (h) $4(x - 1)^2 = 0$
   _____ (i) $4x^2 - 4 = 0$
   _____ (j) $(x - 1)^2 + 4y^2 = 0$
   _____ (k) $4x - 4y = 3$

   (i) horizontal ellipse
   (ii) vertical ellipse
   (iii) horizontal parabola
   (iv) vertical parabola
   (v) horizontal hyperbola
   (vi) vertical hyperbola
   (vii) circle
   (viii) intersecting lines
   (ix) parallel lines
   (x) single line
   (xi) single point
   (xii) empty set

2. Find the focus and vertex for the parabola with equation
$$y = x^2 + 4x + 1$$

3. Find the xy-equation of the parabola that has vertex at the origin, opens up, and passes through $(2, 5)$.

4. Write the xy-equation of a vertical ellipse centered at $(2, 1)$ with major diameter of length 10 and minor diameter of length 6.

5. Sketch the graph of the ellipse with equation $x^2 + 6x + 4y^2 = 7$. Determine its eccentricity.

6. Find the xy-equation of a hyperbola with vertices $(0, \pm 3)$ that passes through $(4, 5)$.

7. Find the vertices of the hyperbola with equation $x^2 - 6x - 2y^2 - 4y = 27$.

8. Eliminate the xy-term from
$$3x^2 + \sqrt{3}\,xy + 2y^2 = 28$$

by making a rotation of axes. Find the distance between the vertices of this conic section.

9. Plot the points with the following polar coordinates.
   (a) $(3, 2\pi/3)$ (b) $(2, -3\pi/2)$ (c) $(6, 210°)$

10. Find the Cartesian coordinates of the points in Problem 9.

11. Find the polar coordinates of the point having the given Cartesian coordinates.
   (a) $(5, 0)$ (b) $(2\sqrt{2}, -2\sqrt{2}$ (c) $(-2\sqrt{3}, 2)$

12. Graph each of the following polar equations.
   (a) $r = 4$ (b) $r = 4 \sin \theta$ (c) $r = 4 \cos 3\theta$

13. Transform $xy = 4$ to a polar equation. Transform $r = \sin 2\theta$ to a Cartesian equation.

14. Determine the number of the description that best fits the given polar equation.

   _____ (a) $r = 6 \sin \theta$ (i) a vertical line
   _____ (b) $r = 6/(1 + 2 \cos \theta)$ (ii) a nonvertical line
   _____ (c) $r = 3/\cos(\theta - \pi/4)$ (iii) a circle
   _____ (d) $r = 12/(2 + \cos \theta)$ (iv) a parabola
   _____ (e) $r = 6$ (v) an ellipse with $e = \frac{1}{2}$
   _____ (f) $r = 6/\sin \theta$ (vi) an ellipse with $e = \frac{1}{4}$
   _____ (g) $r = 6/(2 + 2 \cos \theta)$ (vii) a hyperbola
   _____ (h) $r = 12/(4 + \cos \theta)$ (viii) none of the above

15. Write the equation of a circle of radius 5 centered at the origin (pole) in
   (a) Cartesian coordinates;
   (b) polar coordinates;
   (c) parametric form using the polar angle θ as parameter.

16. Sketch the graph of the Witch of Agnesi, which has parametric equations

$$x = 4 \cot t \qquad y = 4 \sin^2 t$$

Does the pursuit of truth give you as much pleasure as before? Surely it is not the knowing but the learning, not the possessing but the acquiring, not the being-there but the getting-there, that afford the greatest satisfaction. If I have clarified and exhausted something, I leave it in order to go again into the dark. Thus is that insatiable man so strange: when he has completed a structure it is not in order to dwell in it comfortably, but to start another.

Karl Friedrich Gauss

Appendix

Use of Tables

Four tables are included in this appendix. Tables A and B give values of the trigonometric functions for angles measured in degrees and radians, respectively. For a discussion of how to use these tables, see Sections 2-1 and 2-5 of the text. Tables C and D give values for natural logarithms and common logarithms, subjects discussed in Sections 5-4 through 5-6.

In order to find values between those given in any of these tables, we can use a process called **linear interpolation**. Suppose that for some function f, we know $f(a)$ and $f(b)$ but want $f(c)$, where c is between a and b (see the diagrams below). As a reasonable approximation, we may assume that the graph of f is a straight line between a and b. Then

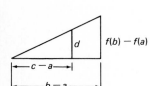

$$f(c) \approx f(a) + d$$

where, by similarity of triangles,

$$\frac{d}{f(b) - f(a)} = \frac{c - a}{b - a}$$

That is,

$$d = \frac{f(b) - f(a)}{b - a}(c - a)$$

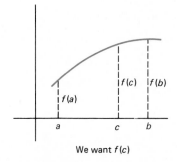 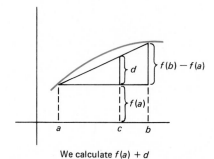

We want $f(c)$ We calculate $f(a) + d$

We illustrate the process of linear interpolation with six examples.

EXAMPLE A (Trigonometric Functions (Degrees)) Find sin 57.44°.

Solution. In Table A, we use the bottom caption and the degree column on the right.

$$.10 \left[.04 \begin{bmatrix} \sin 57.40° = .8425 \\ \sin 57.44° = \quad ? \end{bmatrix} d \\ \sin 57.50° = .8434 \end{bmatrix} .0009$$

$$\frac{d}{.0009} = \frac{.04}{.10} = .4$$

$$d = .4(.0009) \approx .0004$$

$$\sin 57.44° \approx \sin 57.40° + d = .8425 + .0004 = .8429$$

EXAMPLE B (Trigonometric Functions (Radians))   Find $\cos(1.436)$.

Solution.   The cosine is a decreasing function on the interval $0 \le t \le \pi/2$. Thus d is negative. Using Table B, we find the following.

$$.010\left[.006\left[\begin{array}{l}\cos 1.430 = .14033 \\ \cos 1.436 = \quad ? \\ \cos 1.440 = .13042\end{array}\right]d\right]-.00991$$

$$\frac{d}{-.00991} = \frac{.006}{.010} = .6$$

$$d = .6(-.00991) \approx -.00595$$

$$\cos(1.436) \approx \cos(1.430) + d = .14033 - .00595 = .13438$$

EXAMPLE C (Inverse Trigonometric Functions (Degrees))   Find θ if $\sin \theta = .2035$ and $0° \le \theta \le 90°$.

Solution.   We use Table A.

$$.10\left[d\left[\begin{array}{l}\sin 11.70° = .2028 \\ \sin \theta \quad\;\; = .2035 \\ \sin 11.80° = .2045\end{array}\right].0007\right].0017$$

$$\frac{d}{.10} = \frac{.0007}{.0017} \approx .41$$

$$d = .10(.41) \approx .04$$

$$\theta \approx 11.70 + d = 11.70 + .04 = 11.74°$$

EXAMPLE D (Inverse Trigonometric Functions (Radians))   Find t if $\tan t = .43600$ and $0 \le t \le \pi/2$.

Solution.   From Table B, we learn that t is between .41 and .42.

$$0.10\left[d\left[\begin{array}{l}\tan .410° = .43463 \\ \tan t \quad\;\; = .43600 \\ \tan .420° = .44657\end{array}\right].00137\right].01194$$

$$\frac{d}{.010} = \frac{.00137}{.01194} \approx .115$$

$$d = .010(.115) \approx .001$$

$$t \approx .410 + d = .410 + .001 = .411$$

EXAMPLE E (Common Logarithms) Find log 63.26.

Solution. We use Table D.

$$.10\left[.06\left[\begin{array}{l}\log 63.20 = 1.8007\\\log 63.26 = \quad ?\\\log 63.30 = 1.8014\end{array}\right]d\right].0007$$

$$\frac{d}{.0007} = \frac{.06}{.10} = .6$$

$$d = .6(.0007) \approx .0004$$

$$\log 63.26 \approx \log 63.20 + d = 1.8007 + .0004 = 1.8011$$

EXAMPLE F (Antilogarithms) If log x = 1.5085, find x.

Solution. From Table D, we find the following.

$$.10\left[d\left[\begin{array}{l}\log 32.20 = 1.5079\\\log x \quad\ \ = 1.5085\\\log 32.30 = 1.5092\end{array}\right].0006\right].0013$$

$$\frac{d}{.10} = \frac{.0006}{.0013} \approx .5$$

$$d = (.10)(.5) = .05$$

$$x \approx 32.20 + d = 32.20 + .05 = 32.25$$

TABLE A Trigonometric Functions (degrees)

Deg.	Sin	Tan	Cot	Cos		Deg.	Sin	Tan	Cot	Cos	
0.0	0.00000	0.00000	∞	1.0000	**90.0**	**6.0**	0.10453	0.10510	9.514	0.9945	**84.0**
.1	.00175	.00175	573.0	1.0000	89.9	.1	.10626	.10687	9.357	.9943	83.9
.2	.00349	.00349	286.5	1.0000	.8	.2	.10800	.10863	9.205	.9942	.8
.3	.00524	.00524	191.0	1.0000	.7	.3	.10973	.11040	9.058	.9940	.7
.4	.00698	.00698	143.24	1.0000	.6	.4	.11147	.11217	8.915	.9938	.6
.5	.00873	.00873	114.59	1.0000	.5	.5	.11320	.11394	8.777	.9936	.5
.6	.01047	.01047	95.49	0.9999	.4	.6	.11494	.11570	8.643	.9934	.4
.7	.01222	.01222	81.85	.9999	.3	.7	.11667	.11747	8.513	.9932	.3
.8	.01396	.01396	71.62	.9999	.2	.8	.11840	.11924	8.386	.9930	.2
.9	.01571	.01571	63.66	.9999	89.1	.9	.12014	.12101	8.264	.9928	83.1
1.0	0.01745	0.01746	57.29	0.9998	**89.0**	**7.0**	0.12187	0.12278	8.144	0.9925	**83.0**
.1	.01920	.01920	52.08	.9998	88.9	.1	.12360	.12456	8.028	.9923	82.9
.2	.02094	.02095	47.74	.9998	.8	.2	.12533	.12633	7.916	.9921	.8
.3	.02269	.02269	44.07	.9997	.7	.3	.12706	.12810	7.806	.9919	.7
.4	.02443	.02444	40.92	.9997	.6	.4	.12880	.12988	7.700	.9917	.6
.5	.02618	.02619	38.19	.9997	.5	.5	.13053	.13165	7.596	.9914	.5
.6	.02792	.02793	35.80	.9996	.4	.6	.13226	.13343	7.495	.9912	.4
.7	.02967	.02968	33.69	.9996	.3	.7	.13399	.13521	7.396	.9910	.3
.8	.03141	.03143	31.82	.9995	.2	.8	.13572	.13698	7.300	.9907	.2
.9	.03316	.03317	30.14	.9995	88.1	.9	.13744	.13876	7.207	.9905	82.1
2.0	0.03490	0.03492	28.64	0.9994	**88.0**	**8.0**	0.13917	0.14054	7.115	0.9903	**82.0**
.1	.03664	.03667	27.27	.9993	87.9	.1	.14090	.14232	7.026	.9900	81.9
.2	.03839	.03842	26.03	.9993	.8	.2	.14263	.14410	6.940	.9898	.8
.3	.04013	.04016	24.90	.9992	.7	.3	.14436	.14588	6.855	.9895	.7
.4	.04188	.04191	23.86	.9991	.6	.4	.14608	.14767	6.772	.9893	.6
.5	.04362	.04366	22.90	.9990	.5	.5	.14781	.14945	6.691	.9890	.5
.6	.04536	.04541	22.02	.9990	.4	.6	.14954	.15124	6.612	.9888	.4
.7	.04711	.04716	21.20	.9989	.3	.7	.15126	.15302	6.535	.9885	.3
.8	.04885	.04891	20.45	.9988	.2	.8	.15299	.15481	6.460	.9882	.2
.9	.05059	.05066	19.74	.9987	87.1	.9	.15471	.15660	6.386	.9880	81.1
3.0	0.05234	0.05241	19.081	0.9986	**87.0**	**9.0**	0.15643	0.15838	6.314	0.9877	**81.0**
.1	.05408	.05416	18.464	.9985	86.9	.1	.15816	.16017	6.243	.9874	80.9
.2	.05582	.05591	17.886	.9984	.8	.2	.15988	.16196	6.174	.9871	.8
.3	.05756	.05766	17.343	.9983	.7	.3	.16160	.16376	6.107	.9869	.7
.4	.05931	.05941	16.832	.9982	.6	.4	.16333	.16555	6.041	.9866	.6
.5	.06105	.06116	16.350	.9981	.5	.5	.16505	.16734	5.976	.9863	.5
.6	.06279	.06291	15.895	.9980	.4	.6	.16677	.16914	5.912	.9860	.4
.7	.06453	.06467	15.464	.9979	.3	.7	.16849	.17093	5.850	.9857	.3
.8	.06627	.06642	15.056	.9978	.2	.8	.17021	.17273	5.789	.9854	.2
.9	.06802	.06817	14.669	.9977	86.1	.9	.17193	.17453	5.730	.9851	80.1
4.0	0.06976	0.06993	14.301	0.9976	**86.0**	**10.0**	0.1736	0.1763	5.671	0.9848	**80.0**
.1	.07150	.07168	13.951	.9974	85.9	.1	.1754	.1781	5.614	.9845	79.9
.2	.07324	.07344	13.617	.9973	.8	.2	.1771	.1799	5.558	.9842	.8
.3	.07498	.07519	13.300	.9972	.7	.3	.1788	.1817	5.503	.9839	.7
.4	.07672	.07695	12.996	.9971	.6	.4	.1805	.1835	5.449	.9836	.6
.5	.07846	.07870	12.706	.9969	.5	.5	.1822	.1853	5.396	.9833	.5
.6	.08020	.08046	12.429	.9968	.4	.6	.1840	.1871	5.343	.9829	.4
.7	.08194	.08221	12.163	.9966	.3	.7	.1857	.1890	5.292	.9826	.3
.8	.08368	.08397	11.909	.9965	.2	.8	.1874	.1908	5.242	.9823	.2
.9	.08542	.08573	11.664	.9963	85.1	.9	.1891	.1926	5.193	.9820	79.1
5.0	0.08716	0.08749	11.430	0.9962	**85.0**	**11.0**	0.1908	0.1944	5.145	0.9816	**79.0**
.1	.08889	.08925	11.205	.9960	84.9	.1	.1925	.1962	5.079	.9813	78.9
.2	.09063	.09101	10.988	.9959	.8	.2	.1942	.1980	5.050	.9810	.8
.3	.09237	.09277	10.780	.9957	.7	.3	.1959	.1998	5.005	.9806	.7
.4	.09411	.09453	10.579	.9956	.6	.4	.1977	.2016	4.959	.9803	.6
.5	.09585	.09629	10.385	.9954	.5	.5	.1994	.2035	4.915	.9799	.5
.6	.09758	.09805	10.199	.9952	.4	.6	.2011	.2053	4.872	.9796	.4
.7	.09932	.09981	10.019	.9951	.3	.7	.2028	.2071	4.829	.9792	.3
.8	.10106	.10158	9.845	.9949	.2	.8	.2045	.2089	4.787	.9789	.2
.9	.10279	.10334	9.677	.9947	84.1	.9	.2062	.2107	4.745	.9785	78.1
6.0	0.10453	0.10510	9.514	0.9945	**84.0**	**12.0**	0.2079	0.2126	4.705	0.9781	**78.0**
	Cos	Cot	Tan	Sin	Deg.		Cos	Cot	Tan	Sin	Deg.

TABLE A   Trigonometric Functions (degrees)

Deg.	Sin	Tan	Cot	Cos		Deg.	Sin	Tan	Cot	Cos	
12.0	0.2079	0.2126	4.705	0.9781	**78.0**	**18.0**	0.3090	0.3249	3.078	0.9511	**72.0**
.1	.2096	.2144	4.665	.9778	77.9	.1	.3107	.3269	3.060	.9505	71.9
.2	.2113	.2162	4.625	.9774	.8	.2	.3123	.3288	3.042	.9500	.8
.3	.2130	.2180	4.586	.9770	.7	.3	.3140	.3307	3.024	.9494	.7
.4	.2147	.2199	4.548	.9767	.6	.4	.3156	.3327	3.006	.9489	.6
.5	.2164	.2217	4.511	.9763	.5	.5	.3173	.3346	2.989	.9483	.5
.6	.2181	.2235	4.474	.9759	.4	.6	.3190	.3365	2.971	.9478	.4
.7	.2198	.2254	4.437	.9755	.3	.7	.3206	.3385	2.954	.9472	.3
.8	.2215	.2272	4.402	.9751	.2	.8	.3223	.3404	2.937	.9466	.2
.9	.2233	.2290	4.366	.9748	77.1	.9	.3239	.3424	2.921	.9461	71.1
13.0	0.2250	0.2309	4.331	0.9744	**77.0**	**19.0**	0.3256	0.3443	2.904	0.9455	**71.0**
.1	.2267	.2327	4.297	.9740	76.9	.1	.3272	.3463	2.888	.9449	70.9
.2	.2284	.2345	4.264	.9736	.8	.2	.3289	.3482	2.872	.9444	.8
.3	.2300	.2364	4.230	.9732	.7	.3	.3305	.3502	2.856	.9438	.7
.4	.2317	.2382	4.198	.9728	.6	.4	.3322	.3522	2.840	.9432	.6
.5	.2334	.2401	4.165	.9724	.5	.5	.3338	.3541	2.824	.9426	.5
.6	.2351	.2419	4.134	.9720	.4	.6	.3355	.3561	2.808	.9421	.4
.7	.2368	.2438	4.102	.9715	.3	.7	.3371	.3581	2.793	.9415	.3
.8	.2385	.2456	4.071	.9711	.2	.8	.3387	.3600	2.778	.9409	.2
.9	.2402	.2475	4.041	.9707	76.1	.9	.3404	.3620	2.762	.9403	70.1
14.0	0.2419	0.2493	4.011	0.9703	**76.0**	**20.0**	0.3420	0.3640	2.747	0.9397	**70.0**
.1	.2436	.2512	3.981	.9699	75.9	.1	.3437	.3659	2.733	.9391	69.9
.2	.2453	.2530	3.952	.9694	.8	.2	.3453	.3679	2.718	.9385	.8
.3	.2470	.2549	3.923	.9690	.7	.3	.3469	.3699	2.703	.9379	.7
.4	.2487	.2568	3.895	.9686	.6	.4	.3486	.3719	2.689	.9373	.6
.5	.2504	.2586	3.867	.9681	.5	.5	.3502	.3739	2.675	.9367	.5
.6	.2521	.2605	3.839	.9677	.4	.6	.3518	.3759	2.660	.9361	.4
.7	.2538	.2623	3.812	.9673	.3	.7	.3535	.3779	2.646	.9354	.3
.8	.2554	.2642	3.785	.9668	.2	.8	.3551	.3799	2.633	.9348	.2
.9	.2571	.2661	3.758	.9664	75.1	.9	.3567	.3819	2.619	.9342	69.1
15.0	0.2588	0.2679	3.732	0.9659	**75.0**	**21.0**	0.3584	0.3839	2.605	0.9336	**69.0**
.1	.2605	.2698	3.706	.9655	74.9	.1	.3600	.3859	2.592	.9330	68.9
.2	.2622	.2717	3.681	.9650	.8	.2	.3616	.3879	2.578	.9323	.8
.3	.2639	.2736	3.655	.9646	.7	.3	.3633	.3899	2.565	.9317	.7
.4	.2656	.2754	3.630	.9641	.6	.4	.3649	.3919	2.552	.9311	.6
.5	.2672	.2773	3.606	.9636	.5	.5	.3665	.3939	2.539	.9304	.5
.6	.2689	.2792	3.582	.9632	.4	.6	.3681	.3959	2.526	.9298	.4
.7	.2706	.2811	3.558	.9627	.3	.7	.3697	.3979	2.513	.9291	.3
.8	.2723	.2830	3.534	.9622	.2	.8	.3714	.4000	2.500	.9285	.2
.9	.2740	.2849	3.511	.9617	74.1	.9	.3730	.4020	2.488	.9278	68.1
16.0	0.2756	0.2867	3.487	0.9613	**74.0**	**22.0**	0.3746	0.4040	2.475	0.9272	**68.0**
.1	.2773	.2886	3.465	.9608	73.9	.1	.3762	.4061	2.463	.9265	67.9
.2	.2790	.2905	3.442	.9603	.8	.2	.3778	.4081	2.450	.9259	.8
.3	.2807	.2924	3.420	.9598	.7	.3	.3795	.4101	2.438	.9252	.7
.4	.2823	.2943	3.398	.9593	.6	.4	.3811	.4122	2.426	.9245	.6
.5	.2840	.2962	3.376	.9588	.5	.5	.3827	.4142	2.414	.9239	.5
.6	.2857	.2981	3.354	.9583	.4	.6	.3843	.4163	2.402	.9232	.4
.7	.2874	.3000	3.333	.9578	.3	.7	.3859	.4183	2.391	.9225	.3
.8	.2890	.3019	3.312	.9573	.2	.8	.3875	.4204	2.379	.9219	.2
.9	.2907	.3038	3.291	.9568	73.1	.9	.3891	.4224	2.367	.9212	67.1
17.0	0.2924	0.3057	3.271	0.9563	**73.0**	**23.0**	0.3907	0.4245	2.356	0.9205	**67.0**
.1	.2940	.3076	3.251	.9558	72.9	.1	.3923	.4265	2.344	.9198	66.9
.2	.2957	.3096	3.230	.9553	.8	.2	.3939	.4286	2.333	.9191	.8
.3	.2974	.3115	3.211	.9548	.7	.3	.3955	.4307	2.322	.9184	.7
.4	.2990	.3134	3.191	.9542	.6	.4	.3971	.4327	2.311	.9178	.6
.5	.3007	.3153	3.172	.9537	.5	.5	.3987	.4348	2.300	.9171	.5
.6	.3024	.3172	3.152	.9532	.4	.6	.4003	.4369	2.289	.9164	.4
.7	.3040	.3191	3.133	.9527	.3	.7	.4019	.4390	2.278	.9157	.3
.8	.3057	.3211	3.115	.9521	.2	.8	.4035	.4411	2.267	.9150	.2
.9	.3074	.3230	3.096	.9516	72.1	.9	.4051	.4431	2.257	.9143	66.1
18.0	0.3090	0.3249	3.078	0.9511	**72.0**	**24.0**	0.4067	0.4452	2.246	0.9135	**66.0**
	Cos	Cot	Tan	Sin	Deg.		Cos	Cot	Tan	Sin	Deg.

TABLE A Trigonometric Functions (degrees)

Deg.	Sin	Tan	Cot	Cos		Deg.	Sin	Tan	Cot	Cos	
24.0	0.4067	0.4452	2.246	0.9135	**66.0**	**30.0**	0.5000	0.5774	1.7321	0.8660	**60.0**
.1	.4083	.4473	2.236	.9128	65.9	.1	.5015	.5797	1.7251	.8652	59.9
.2	.4099	.4494	2.225	.9121	.8	.2	.5030	.5820	1.7182	.8643	.8
.3	.4115	.4515	2.215	.9114	.7	.3	.5045	.5844	1.7113	.8634	.7
.4	.4131	.4536	2.204	.9107	.6	.4	.5060	.5867	1.7045	.8625	.6
.5	.4147	.4557	2.194	.9100	.5	.5	.5075	.5890	1.6977	.8616	.5
.6	.4163	.4578	2.184	.9092	.4	.6	.5090	.5914	1.6909	.8607	.4
.7	.4179	.4599	2.174	.9085	.3	.7	.5105	.5938	1.6842	.8599	.3
.8	.4195	.4621	2.164	.9078	.2	.8	.5120	.5961	1.6775	.8590	.2
.9	.4210	.4642	2.154	.9070	65.1	.9	.5135	.5985	1.6709	.8581	59.1
25.0	0.4226	0.4663	2.145	0.9063	**65.0**	**31.0**	0.5150	0.6009	1.6643	0.8572	**59.0**
.1	.4242	.4684	2.135	.9056	64.9	.1	.5165	.6032	1.6577	.8563	58.9
.2	.4258	.4706	2.125	.9048	.8	.2	.5180	.6056	1.6512	.8554	.8
.3	.4274	.4727	2.116	.9041	.7	.3	.5195	.6080	1.6447	.8545	.7
.4	.4289	.4748	2.106	.9033	.6	.4	.5210	.6104	1.6383	.8536	.6
.5	.4305	.4770	2.097	.9026	.5	.5	.5225	.6128	1.6319	.8526	.5
.6	.4321	.4791	2.087	.9018	.4	.6	.5240	.6152	1.6255	.8517	.4
.7	.4337	.4813	2.078	.9011	.3	.7	.5255	.6176	1.6191	.8508	.3
.8	.4352	.4834	2.069	.9003	.2	.8	.5270	.6200	1.6128	.8499	.2
.9	.4368	.4856	2.059	.8996	64.1	.9	.5284	.6224	1.6066	.8490	58.1
26.0	0.4384	0.4887	2.050	0.8988	**64.0**	**32.0**	0.5299	0.6249	1.6003	0.8480	**58.0**
.1	.4399	.4899	2.041	.8980	63.9	.1	.5314	.6273	1.5941	.8471	57.9
.2	.4415	.4921	2.032	.8973	.8	.2	.5329	.6297	1.5880	.8462	.8
.3	.4431	.4942	2.023	.8965	.7	.3	.5344	.6322	1.5818	.8453	.7
.4	.4446	.4964	2.014	.8957	.6	.4	.5358	.6346	1.5757	.8443	.6
.5	.4462	.4986	2.006	.8949	.5	.5	.5373	.6371	1.5697	.8434	.5
.6	.4478	.5008	1.997	.8942	.4	.6	.5388	.6395	1.5637	.8425	.4
.7	.4493	.5029	1.988	.8934	.3	.7	.5402	.6420	1.5577	.8415	.3
.8	.4509	.5051	1.980	.8926	.2	.8	.5417	.6445	1.5517	.8406	.2
.9	.4524	.5073	1.971	.8918	63.1	.9	.5432	.6469	1.5458	.8396	57.1
27.0	0.4540	0.5095	1.963	0.8910	**63.0**	**33.0**	0.5446	0.6494	1.5399	0.8387	**57.0**
.1	.4555	.5117	1.954	.8902	62.9	.1	.5461	.6519	1.5340	.8377	56.9
.2	.4571	.5139	1.946	.8894	.8	.2	.5476	.6544	1.5282	.8368	.8
.3	.4586	.5161	1.937	.8886	.7	.3	.5490	.6569	1.5224	.8358	.7
.4	.4602	.5184	1.929	.8878	.6	.4	.5505	.6594	1.5166	.8348	.6
.5	.4617	.5206	1.921	.8870	.5	.5	.5519	.6619	1.5108	.8339	.5
.6	.4633	.5228	1.913	.8862	.4	.6	.5534	.6644	1.5051	.8329	.4
.7	.4648	.5250	1.905	.8854	.3	.7	.5548	.6669	1.4994	.8320	.3
.8	.4664	.5272	1.897	.8846	.2	.8	.5563	.6694	1.4938	.8310	.2
.9	.4679	.5295	1.889	.8838	62.1	.9	.5577	.6720	1.4882	.8300	56.1
28.0	0.4695	0.5317	1.881	0.8829	**62.0**	**34.0**	0.5592	0.6745	1.4826	0.8290	**56.0**
.1	.4710	.5340	1.873	.8821	61.9	.1	.5606	.6771	1.4770	.8281	55.9
.2	.4726	.5362	1.865	.8813	.8	.2	.5621	.6796	1.4715	.8271	.8
.3	.4741	.5384	1.857	.8805	.7	.3	.5635	.6822	1.4659	.8261	.7
.4	.4756	.5407	1.849	.8796	.6	.4	.5650	.6847	1.4605	.8251	.6
.5	.4772	.5430	1.842	.8788	.5	.5	.5664	.6873	1.4550	.8241	.5
.6	.4787	.5452	1.834	.8780	.4	.6	.5678	.6899	1.4496	.8231	.4
.7	.4802	.5475	1.827	.8771	.3	.7	.5693	.6924	1.4442	.8221	.3
.8	.4818	.5498	1.819	.8763	.2	.8	.5707	.6950	1.4388	.8211	.2
.9	.4833	.5520	1.811	.8755	61.1	.9	.5721	.6976	1.4335	.8202	55.1
29.0	0.4848	0.5543	1.804	0.8746	**61.0**	**35.0**	0.5736	0.7002	1.4281	0.8192	**55.0**
.1	.4863	.5566	1.797	.8738	60.9	.1	.5750	.7028	1.4229	.8181	54.9
.2	.4879	.5589	1.789	.8729	.8	.2	.5764	.7054	1.4176	.8171	.8
.3	.4894	.5612	1.782	.8721	.7	.3	.5779	.7080	1.4124	.8161	.7
.4	.4909	.5635	1.775	.8712	.6	.4	.5793	.7107	1.4071	.8151	.6
.5	.4924	.5658	1.767	.8704	.5	.5	.5807	.7133	1.4019	.8141	.5
.6	.4939	.5681	1.760	.8695	.4	.6	.5821	.7159	1.3968	.8131	.4
.7	.4955	.5704	1.753	.8686	.3	.7	.5835	.7186	1.3916	.8121	.3
.8	.4970	.5727	1.746	.8678	.2	.8	.5850	.7212	1.3865	.8111	.2
.9	.4985	.5750	1.739	.8669	60.1	.9	.5864	.7239	1.3814	.8100	54.1
30.0	0.5000	0.5774	1.732	0.8660	**60.0**	**36.0**	0.5878	0.7265	1.3764	0.8090	**54.0**
	Cos	Cot	Tan	Sin	Deg.		Cos	Cot	Tan	Sin	Deg.

TABLE A Trigonometric Functions (degrees)

Deg.	Sin	Tan	Cot	Cos		Deg.	Sin	Tan	Cot	Cos	
36.0	0.5878	0.7265	1.3764	0.8090	**54.0**	**40.5**	0.6494	0.8541	1.1708	0.7604	**49.5**
.1	.5892	.7292	1.3713	.8080	53.9	.6	.6508	.8571	1.1667	.7593	.4
.2	.5906	.7319	1.3663	.8070	.8	.7	.6521	.8601	1.1626	.7581	.3
.3	.5920	.7346	1.3613	.8059	.7	.8	.6534	.8632	1.1585	.7570	.2
.4	.5934	.7373	1.3564	.8049	.6	.9	.6547	.8662	1.1544	.7559	49.1
.5	.5948	.7400	1.3514	.8039	.5	**41.0**	0.6561	0.8693	1.1504	0.7547	**49.0**
.6	.5962	.7427	1.3465	.8028	.4	.1	.6574	.8724	1.1463	.7536	48.9
.7	.5976	.7454	1.3416	.8018	.3	.2	.6587	.8754	1.1423	.7524	.8
.8	.5990	.7481	1.3367	.8007	.2	.3	.6600	.8785	1.1383	.7513	.7
.9	.6004	.7508	1.3319	.7997	53.1	.4	.6613	.8816	1.1343	.7501	.6
37.0	0.6018	0.7536	1.3270	0.7986	**53.0**	.5	.6626	.8847	1.1303	.7490	.5
.1	.6032	.7563	1.3222	.7976	52.9	.6	.6639	.8878	1.1263	.7478	.4
.2	.6046	.7590	1.3175	.7965	.8	.7	.6652	.8910	1.1224	.7466	.3
.3	.6060	.7618	1.3127	.7955	.7	.8	.6665	.8941	1.1184	.7455	.2
.4	.6074	.7646	1.3079	.7944	.6	.9	.6678	.8972	1.1145	.7443	48.1
.5	.6088	.7673	1.3032	.7934	.5	**42.0**	0.6691	0.9004	1.1106	0.7431	**48.0**
.6	.6101	.7701	1.2985	.7923	.4	.1	.6704	.9036	1.1067	.7420	47.9
.7	.6115	.7729	1.2938	.7912	.3	.2	.6717	.9067	1.1028	.7408	.8
.8	.6129	.7757	1.2892	.7902	.2	.3	.6730	.9099	1.0990	.7396	.7
.9	.6143	.7785	1.2846	.7891	52.1	.4	.6743	.9131	1.0951	.7385	.6
38.0	0.6157	0.7813	1.2799	0.7880	**52.0**	.5	.6756	.9163	1.0913	.7373	.5
.1	.6170	.7841	1.2753	.7869	51.9	.6	.6769	.9195	1.0875	.7361	.4
.2	.6184	.7869	1.2708	7859	.8	.7	.6782	.9228	1.0837	.7349	.3
.3	.6198	.7898	1.2662	.7848	.7	.8	.6794	.9260	1.0799	.7337	.2
.4	.6211	.7926	1.2617	.7837	.6	.9	.6807	.9293	1.0761	.7325	47.1
.5	.6225	.7954	1.2572	.7826	.5	**43.0**	0.6820	0.9325	1.0724	0.7314	**47.0**
.6	.6239	.7983	1.2527	.7815	.4	.1	.6833	.9358	1.0686	.7302	46.9
.7	.6252	.8012	1.2482	.7804	.3	.2	.6845	.9391	1.0649	.7290	.8
.8	.6266	.8040	1.2437	.7793	.2	.3	.6858	.9424	1.0612	.7278	.7
.9	.6280	.8069	1.2393	.7782	51.1	.4	.6871	.9457	1.0575	.7266	.6
39.0	0.6293	0.8098	1.2349	0.7771	**51.0**	.5	.6884	.9490	1.0538	.7254	.5
.1	.6307	.8127	1.2305	.7760	50.9	.6	.6896	.9523	1.0501	.7242	.4
.2	.6320	.8156	1.2261	.7749	.8	.7	.6909	.9556	1.0464	.7230	.3
.3	.6334	.8185	1.2218	.7738	.7	.8	.6921	.9590	1.0428	.7218	.2
.4	.6347	.8214	1.2174	.7727	.6	.9	.6934	.9623	1.0392	.7206	46.1
.5	.6361	.8243	1.2131	.7716	.5	**44.0**	0.6947	0.9657	1.0355	0.7193	**46.0**
.6	.6374	.8273	1.2088	.7705	.4	.1	.6959	.9691	1.0319	.7181	45.9
.7	.6388	.8302	1.2045	.7694	.3	.2	.6972	.9725	1.0283	.7169	.8
.8	.6401	.8332	1.2002	.7683	.2	.3	.6984	.9759	1.0247	.7157	.7
.9	.6414	.8361	1.1960	.7672	50.1	.4	.6997	.9793	1.0212	.7145	.6
40.0	0.6428	0.8391	1.1918	0.7660	**50.0**	.5	.7009	.9827	1.0176	.7133	.5
.1	.6441	.8421	1.1875	.7649	49.9	.6	.7022	.9861	1.0141	.7120	.4
.2	.6455	.8451	1.1833	.7638	.8	.7	.7034	.9896	1.0105	.7108	.3
.3	.6468	.8481	1.1792	.7627	.7	.8	.7046	.9930	1.0070	.7096	.2
.4	.6481	.8511	1.1750	.7615	.6	.9	.7059	.9965	1.0035	.7083	45.1
40.5	0.6494	0.8541	1.1708	0.7604	**49.5**	**45.0**	0.7071	1.0000	1.0000	0.7071	**45.0**
	Cos	Cot	Tan	Sin	Deg.		Cos	Cot	Tan	Sin	Deg.

TABLE B Trigonometric Functions (radians)

Rad.	Sin	Tan	Cot	Cos	Rad.	Sin	Tan	Cot	Cos
.00	.00000	.00000	∞	1.00000	**.50**	.47943	.54630	1.8305	.87758
.01	.01000	.01000	99.997	0.99995	.51	.48818	.55936	1.7878	.87274
.02	.02000	.02000	49.993	.99980	.52	.49688	.57256	1.7465	.86782
.03	.03000	.03001	33.323	.99955	.53	.50553	.58592	1.7067	.86281
.04	.03999	.04002	24.987	.99920	.54	.51414	.59943	1.6683	.85771
.05	.04998	.05004	19.983	.99875	.55	.52269	.61311	1.6310	.85252
.06	.05996	.06007	16.647	.99820	.56	.53119	.62695	1.5950	.84726
.07	.06994	.07011	14.262	.99755	.57	.53963	.64097	1.5601	.84190
.08	.07991	.08017	12.473	.99680	.58	.54802	.65517	1.5263	.83646
.09	.08988	.09024	11.081	.99595	.59	.55636	.66956	1.4935	.83094
.10	.09983	.10033	9.9666	.99500	**.60**	.56464	.68414	1.4617	.82534
.11	.10978	.11045	9.0542	.99396	.61	.57287	.69892	1.4308	.81965
.12	.11971	.12058	8.2933	.99281	.62	.58104	.71391	1.4007	.81388
.13	.12963	.13074	7.6489	.99156	.63	.58914	.72911	1.3715	.80803
.14	.13954	.14092	7.0961	.99022	.64	.59720	.74454	1.3431	.80210
.15	.14944	.15114	6.6166	.98877	.65	.60519	.76020	1.3154	.79608
.16	.15932	.16138	6.1966	.98723	.66	.61312	.77610	1.2885	.78999
.17	.16918	.17166	5.8256	.98558	.67	.62099	.79225	1.2622	.78382
.18	.17903	.18197	5.4954	.98384	.68	.62879	.80866	1.2366	.77757
.19	.18886	.19232	5.1997	.98200	.69	.63654	.82534	1.2116	.77125
.20	.19867	.20271	4.9332	.98007	**.70**	.64422	.84229	1.1872	.76484
.21	.20846	.21314	4.6917	.97803	.71	.65183	.85953	1.1634	.75836
.22	.21823	.22362	4.4719	.97590	.72	.65938	.87707	1.1402	.75181
.23	.22798	.23414	4.2709	.97367	.73	.66687	.89492	1.1174	.74517
.24	.23770	.24472	4.0864	.97134	.74	.67429	.91309	1.0952	.73847
.25	.24740	.25534	3.9163	.96891	.75	.68164	.93160	1.0734	.73169
.26	.25708	.26602	3.7591	.96639	.76	.68892	.95045	1.0521	.72484
.27	.26673	.27676	3.6133	.96377	.77	.69614	.96967	1.0313	.71791
.28	.27636	.28755	3.4776	.96106	.78	.70328	.98926	1.0109	.71091
.29	.28595	.29841	3.3511	.95824	.79	.71035	1.0092	.99084	.70385
.30	.29552	.30934	3.2327	.95534	**.80**	.71736	1.0296	.97121	.69671
.31	.30506	.32033	3.1218	.95233	.81	.72429	1.0505	.95197	.68950
.32	.31457	.33139	3.0176	.94924	.82	.73115	1.0717	.93309	.68222
.33	.32404	.34252	2.9195	.94604	.83	.73793	1.0934	.91455	.67488
.34	.33349	.35374	2.8270	.94275	.84	.74464	1.1156	.89635	.66746
.35	.34290	.36503	2.7395	.93937	.85	.75128	1.1383	.87848	.65998
.36	.35227	.37640	2.6567	.93590	.86	.75784	1.1616	.86091	.65244
.37	.36162	.38786	2.5782	.93233	.87	.76433	1.1853	.84365	.64483
.38	.37092	.39941	2.5037	.92866	.88	.77074	1.2097	.82668	.63715
.39	.38019	.41105	2.4328	.92491	.89	.77707	1.2346	.80998	.62941
.40	.38942	.42279	2.3652	.92106	**.90**	.78333	1.2602	.79355	.62161
.41	.39861	.43463	2.3008	.91712	.91	.78950	1.2864	.77738	.61375
.42	.40776	.44657	2.2393	.91309	.92	.79560	1.3133	.76146	.60582
.43	.41687	.45862	2.1804	.90897	.93	.80162	1.3409	.74578	.59783
.44	.42594	.47078	2.1241	.90475	.94	.80756	1.3692	.73034	.58979
.45	.43497	.48306	2.0702	.90045	.95	.81342	1.3984	.71511	.58168
.46	.44395	.49545	2.0184	.89605	.96	.81919	1.4284	.70010	.57352
.47	.45289	.50797	1.9686	.89157	.97	.82489	1.4592	.68531	.56530
.48	.46178	.52061	1.9208	.88699	.98	.83050	1.4910	.67071	.55702
.49	.47063	.53339	1.8748	.88233	.99	.83603	1.5237	.65631	.54869
.50	.47943	.54630	1.8305	.87758	**1.00**	.84147	1.5574	.64209	.54030
Rad.	Sin	Tan	Cot	Cos	Rad.	Sin	Tan	Cot	Cos

TABLE B Trigonometric Functions (radians)

Rad.	Sin.	Tan	Cot	Cos	Rad.	Sin	Tan	Cot	Cos
1.00	.84147	1.5574	.64209	.54030	**1.50**	.99749	14.101	.07091	.07074
1.01	.84683	1.5922	.62806	.53186	1.51	.99815	16.428	.06087	.06076
1.02	.85211	1.6281	.61420	.52337	1.52	.99871	19.670	.05084	.05077
1.03	.85730	1.6652	.60051	.51482	1.53	.99917	24.498	.04082	.04079
1.04	.86240	1.7036	.58699	.50622	1.54	.99953	32.461	.03081	.03079
1.05	.86742	1.7433	.57362	.49757	1.55	.99978	48.078	.02080	.02079
1.06	.87236	1.7844	.56040	.48887	1.56	.99994	92.621	.01080	.01080
1.07	.87720	1.8270	.54734	.48012	1.57	1.00000	1255.8	.00080	.00080
1.08	.88196	1.8712	.53441	.47133	1.58	.99996	− 108.65	−.00920	−.00920
1.09	.88663	1.9171	.52162	.46249	1.59	.99982	− 52.067	−.01921	−.01920
1.10	.89121	1.9648	.50897	.45360	**1.60**	.99957	− 34.233	−.02921	−.02920
1.11	.89570	2.0143	.49644	.44466	1.61	.99923	− 25.495	−.03922	−.03919
1.12	.90010	2.0660	.48404	.43568	1.62	.99879	− 20.307	−.04924	−.04918
1.13	.90441	2.1198	.47175	.42666	1.63	.99825	− 16.871	−.05927	−.05917
1.14	.90863	2.1759	.45959	.41759	1.64	.99761	− 14.427	−.06931	−.06915
1.15	.91276	2.2345	.44753	.40849	1.65	.99687	− 12.599	−.07937	−.07912
1.16	.91680	2.2958	.43558	.39934	1.66	.99602	− 11.181	−.08944	−.08909
1.17	.92075	2.3600	.42373	.39015	1.67	.99508	− 10.047	−.09953	−.09904
1.18	.92461	2.4273	.41199	.38092	1.68	.99404	− 9.1208	−.10964	−.10899
1.19	.92837	2.4979	.40034	.37166	1.69	.99290	− 8.3492	−.11977	−.11892
1.20	.93204	2.5722	.38878	.36236	**1.70**	.99166	− 7.6966	−.12993	−.12884
1.21	.93562	2.6503	.37731	.35302	1.71	.99033	− 7.1373	−.14011	−.13875
1.22	.93910	2.7328	.36593	.34365	1.72	.98889	− 6.6524	−.15032	−.14865
1.23	.94249	2.8198	.35463	.33424	1.73	.98735	− 6.2281	−.16056	−.15853
1.24	.94578	2.9119	.34341	.32480	1.74	.98572	− 5.8535	−.17084	−.16840
1.25	.94898	3.0096	.33227	.31532	1.75	.98399	− 5.5204	−.18115	−.17825
1.26	.95209	3.1133	.32121	.30582	1.76	.98215	− 5.2221	−.19149	−.18808
1.27	.95510	3.2236	.31021	.29628	1.77	.98022	− 4.9534	−.20188	−.19789
1.28	.95802	3.3413	.29928	.28672	1.78	.97820	− 4.7101	−.21231	−.20768
1.29	.96084	3.4672	.28842	.27712	1.79	.97607	− 4.4887	−.22278	−.21745
1.30	.96356	3.6021	.27762	.26750	**1.80**	.97385	− 4.2863	−.23330	−.22720
1.31	.96618	3.7471	.26687	.25785	1.81	.97153	− 4.1005	−.24387	−.23693
1.32	.96872	3.9033	.25619	.24818	1.82	.96911	− 3.9294	−.25449	−.24663
1.33	.97115	4.0723	.24556	.23848	1.83	.96659	− 3.7712	−.26517	−.25631
1.34	.97348	4.2556	.23498	.22875	1.84	.96398	− 3.6245	−.27590	−.26596
1.35	.97572	4.4552	.22446	.21901	1.85	.96128	− 3.4881	−.28669	−.27559
1.36	.97786	4.6734	.21398	.20924	1.86	.95847	− 3.3608	−.29755	−.28519
1.37	.97991	4.9131	.20354	.19945	1.87	.95557	− 2.2419	−.30846	−.29476
1.38	.98185	5.1774	.19315	.18964	1.88	.95258	− 3.1304	−.31945	−.30430
1.39	.98370	5.4707	.18279	.17981	1.89	.94949	− 3.0257	− 33.051	−.31381
1.40	.98545	5.7979	.17248	.16997	**1.90**	.94630	− 2.9271	−.34164	−.32329
1.41	.98710	6.1654	.16220	.16010	1.91	.94302	− 2.8341	−.35284	−.33274
1.42	.98865	6.5811	.15195	.15023	1.92	.93965	− 2.7463	−.36413	−.34215
1.43	.99010	7.0555	.14173	.14033	1.93	.93618	− 2.6632	−.37549	−.35153
1.44	.99146	7.6018	.13155	.13042	1.94	.93262	− 2.5843	−.38695	−.36087
1.45	.99271	8.2381	.12139	.12050	1.95	.92896	− 2.5095	−.39849	−.37018
1.46	.99387	8.9886	.11125	.11057	1.96	.92521	− 2.4383	−.41012	−.37945
1.47	.99492	9.8874	.10114	.10063	1.97	.92137	− 2.3705	−.42185	−.38868
1.48	.99588	10.983	.09105	.09067	1.98	.91744	− 2.3058	−.43368	−.39788
1.49	.99674	12.350	.08097	.08071	1.99	.91341	− 2.2441	−.44562	−.40703
1.50	.99749	14.101	.07091	.07074	**2.00**	.90930	− 2.1850	−.45766	−.41615
Rad.	Sin	Tan	Cot	Cos	Rad.	Sin	Tan	Cot	Cos

TABLE C   Natural Logarithms

	.00	.01	.02	.03	.04	.05	.06	.07	.08	.09
1.0	0.0000	0.0100	0.0198	0.0296	0.0392	0.0488	0.0583	0.0677	0.0770	0.0862
1.1	0.0953	0.1044	0.1133	0.1222	0.1310	0.1398	0.1484	0.1570	0.1655	0.1740
1.2	0.1823	0.1906	0.1989	0.2070	0.2151	0.2231	0.2311	0.2390	0.2469	0.2546
1.3	0.2624	0.2700	0.2776	0.2852	0.2927	0.3001	0.3075	0.3148	0.3221	0.3293
1.4	0.3365	0.3436	0.3507	0.3577	0.3646	0.3716	0.3784	0.3853	0.3920	0.3988
1.5	0.4055	0.4121	0.4187	0.4253	0.4318	0.4383	0.4447	0.4511	0.4574	0.4637
1.6	0.4700	0.4762	0.4824	0.4886	0.4947	0.5008	0.5068	0.5128	0.5188	0.5247
1.7	0.5306	0.5365	0.5423	0.5481	0.5539	0.5596	0.5653	0.5710	0.5766	0.5822
1.8	0.5878	0.5933	0.5988	0.6043	0.6098	0.6152	0.6206	0.6259	0.6313	0.6366
1.9	0.6419	0.6471	0.6523	0.6575	0.6627	0.6678	0.6729	0.6780	0.6831	0.6881
2.0	0.6931	0.6981	0.7031	0.7080	0.7130	0.7178	0.7227	0.7275	0.7324	0.7372
2.1	0.7419	0.7467	0.7514	0.7561	0.7608	0.7655	0.7701	0.7747	0.7793	0.7839
2.2	0.7885	0.7930	0.7975	0.8020	0.8065	0.8109	0.8154	0.8198	0.8242	0.8286
2.3	0.8329	0.8372	0.8416	0.8459	0.8502	0.8544	0.8587	0.8629	0.8671	0.8713
2.4	0.8755	0.8796	0.8838	0.8879	0.8920	0.8961	0.9002	0.9042	0.9083	0.9123
2.5	0.9163	0.9203	0.9243	0.9282	0.9322	0.9361	0.9400	0.9439	0.9478	0.9517
2.6	0.9555	0.9594	0.9632	0.9670	0.9708	0.9746	0.9783	0.9821	0.9858	0.9895
2.7	0.9933	0.9969	1.0006	1.0043	1.0080	1.0116	1.0152	1.0188	1.0225	1.0260
2.8	1.0296	1.0332	1.0367	1.0403	1.0438	1.0473	1.0508	1.0543	1.0578	1.0613
2.9	1.0647	1.0682	1.0716	1.0750	1.0784	1.0818	1.0852	1.0886	1.0919	1.0953
3.0	1.0986	1.1019	1.1053	1.1086	1.1119	1.1151	1.1184	1.1217	1.1249	1.1282
3.1	1.1314	1.1346	1.1378	1.1410	1.1442	1.1474	1.1506	1.1537	1.1569	1.1600
3.2	1.1632	1.1663	1.1694	1.1725	1.1756	1.1787	1.1817	1.1848	1.1878	1.1909
3.3	1.1939	1.1970	1.2000	1.2030	1.2060	1.2090	1.2119	1.2149	1.2179	1.2208
3.4	1.2238	1.2267	1.2296	1.2326	1.2355	1.2384	1.2413	1.2442	1.2470	1.2499
3.5	1.2528	1.2556	1.2585	1.2613	1.2641	1.2669	1.2698	1.2726	1.2754	1.2782
3.6	1.2809	1.2837	1.2865	1.2892	1.2920	1.2947	1.2975	1.3002	1.3029	1.3056
3.7	1.3083	1.3110	1.3137	1.3164	1.3191	1.3218	1.3244	1.3271	1.3297	1.3324
3.8	1.3350	1.3376	1.3403	1.3429	1.3455	1.3481	1.3507	1.3533	1.3558	1.3584
3.9	1.3610	1.3635	1.3661	1.3686	1.3712	1.3737	1.3762	1.3788	1.3813	1.3838
4.0	1.3863	1.3888	1.3913	1.3938	1.3962	1.3987	1.4012	1.4036	1.4061	1.4085
4.1	1.4110	1.4134	1.4159	1.4183	1.4207	1.4231	1.4255	1.4279	1.4303	1.4327
4.2	1.4351	1.4375	1.4398	1.4422	1.4446	1.4469	1.4493	1.4516	1.4540	1.4563
4.3	1.4586	1.4609	1.4633	1.4656	1.4679	1.4702	1.4725	1.4748	1.4770	1.4793
4.4	1.4816	1.4839	1.4861	1.4884	1.4907	1.4929	1.4952	1.4974	1.4996	1.5019
4.5	1.5041	1.5063	1.5085	1.5107	1.5129	1.5151	1.5173	1.5195	1.5217	1.5239
4.6	1.5261	1.5282	1.5304	1.5326	1.5347	1.5369	1.5390	1.5412	1.5433	1.5454
4.7	1.5476	1.5497	1.5518	1.5539	1.5560	1.5581	1.5602	1.5623	1.5644	1.5665
4.8	1.5686	1.5707	1.5728	1.5748	1.5769	1.5790	1.5810	1.5831	1.5851	1.5872
4.9	1.5892	1.5913	1.5933	1.5953	1.5974	1.5994	1.6014	1.6034	1.6054	1.6074
5.0	1.6094	1.6114	1.6134	1.6154	1.6174	1.6194	1.6214	1.6233	1.6253	1.6273
5.1	1.6292	1.6312	1.6332	1.6351	1.6371	1.6390	1.6409	1.6429	1.6448	1.6467
5.2	1.6487	1.6506	1.6525	1.6544	1.6563	1.6582	1.6601	1.6620	1.6639	1.6658
5.3	1.6677	1.6696	1.6715	1.6734	1.6753	1.6771	1.6790	1.6808	1.6827	1.6845
5.4	1.6864	1.6882	1.6901	1.6919	1.6938	1.6956	1.6974	1.6993	1.7011	1.7029

$$\ln(N \cdot 10^m) = \ln N + m \ln 10, \quad \ln 10 = 2.3026$$

TABLE C Natural Logarithms

	.00	.01	.02	.03	.04	.05	.06	.07	.08	.09
5.5	1.7047	1.7066	1.7084	1.7102	1.7120	1.7138	1.7156	1.7174	1.7192	1.7210
5.6	1.7228	1.7246	1.7263	1.7281	1.7299	1.7317	1.7334	1.7352	1.7370	1.7387
5.7	1.7405	1.7422	1.7440	1.7457	1.7475	1.7492	1.7509	1.7527	1.7544	1.7561
5.8	1.7579	1.7596	1.7613	1.7630	1.7647	1.7664	1.7682	1.7699	1.7716	1.7733
5.9	1.7750	1.7766	1.7783	1.7800	1.7817	1.7834	1.7851	1.7867	1.7884	1.7901
6.0	1.7918	1.7934	1.7951	1.7967	1.7984	1.8001	1.8017	1.8034	1.8050	1.8066
6.1	1.8083	1.8099	1.8116	1.8132	1.8148	1.8165	1.8181	1.8197	1.8213	1.8229
6.2	1.8245	1.8262	1.8278	1.8294	1.8310	1.8326	1.8342	1.8358	1.8374	1.8390
6.3	1.8406	1.8421	1.8437	1.8453	1.8469	1.8485	1.8500	1.8516	1.8532	1.8547
6.4	1.8563	1.8579	1.8594	1.8610	1.8625	1.8641	1.8656	1.8672	1.8687	1.8703
6.5	1.8718	1.8733	1.8749	1.8764	1.8779	1.8795	1.8810	1.8825	1.8840	1.8856
6.6	1.8871	1.8886	1.8901	1.8916	1.8931	1.8946	1.8961	1.8976	1.8991	1.9006
6.7	1.9021	1.9036	1.9051	1.9066	1.9081	1.9095	1.9110	1.9125	1.9140	1.9155
6.8	1.9169	1.9184	1.9199	1.9213	1.9228	1.9242	1.9257	1.9272	1.9286	1.9301
6.9	1.9315	1.9330	1.9344	1.9359	1.9373	1.9387	1.9402	1.9416	1.9430	1.9445
7.0	1.9459	1.9473	1.9488	1.9502	1.9516	1.9530	1.9544	1.9559	1.9573	1.9587
7.1	1.9601	1.9615	1.9629	1.9643	1.9657	1.9671	1.9685	1.9699	1.9713	1.9727
7.2	1.9741	1.9755	1.9769	1.9782	1.9796	1.9810	1.9824	1.9838	1.9851	1.9865
7.3	1.9879	1.9892	1.9906	1.9920	1.9933	1.9947	1.9961	1.9974	1.9988	2.0001
7.4	2.0015	2.0028	2.0042	2.0055	2.0069	2.0082	2.0096	2.0109	2.0122	2.0136
7.5	2.0149	2.0162	2.0176	2.0189	2.0202	2.0215	2.0229	2.0242	2.0255	2.0268
7.6	2.0282	2.0295	2.0308	2.0321	2.0334	2.0347	2.0360	2.0373	2.0386	2.0399
7.7	2.0412	2.0425	2.0438	2.0451	2.0464	2.0477	2.0490	2.0503	2.0516	2.0528
7.8	2.0541	2.0554	2.0567	2.0580	2.0592	2.0605	2.0618	2.0631	2.0643	2.0656
7.9	2.0669	2.0681	2.0694	2.0707	2.0719	2.0732	2.0744	2.0757	2.0769	2.0782
8.0	2.0794	2.0807	2.0819	2.0832	2.0844	2.0857	2.0869	2.0882	2.0894	2.0906
8.1	2.0919	2.0931	2.0943	2.0956	2.0968	2.0980	2.0992	2.1005	2.1017	2.1029
8.2	2.1041	2.1054	2.1066	2.1078	2.1090	2.1102	2.1114	2.1126	2.1138	2.1150
8.3	2.1163	2.1175	2.1187	2.1199	2.1211	2.1223	2.1235	2.1247	2.1258	2.1270
8.4	2.1282	2.1294	2.1306	2.1318	2.1330	2.1342	2.1353	2.1365	2.1377	2.1389
8.5	2.1401	2.1412	2.1424	2.1436	2.1448	2.1459	2.1471	2.1483	2.1494	2.1506
8.6	2.1518	2.1529	2.1541	2.1552	2.1564	2.1576	2.1587	2.1599	2.1610	2.1622
8.7	2.1633	2.1645	2.1656	2.1668	2.1679	2.1691	2.1702	2.1713	2.1725	2.1736
8.8	2.1748	2.1759	2.1770	2.1782	2.1793	2.1804	2.1815	2.1827	2.1838	2.1849
8.9	2.1861	2.1872	2.1883	2.1894	2.1905	2.1917	2.1928	2.1939	2.1950	2.1961
9.0	2.1972	2.1983	2.1994	2.2006	2.2017	2.2028	2.2039	2.2050	2.2061	2.2072
9.1	2.2083	2.2094	2.2105	2.2116	2.2127	2.2138	2.2148	2.2159	2.2170	2.2181
9.2	2.2192	2.2203	2.2214	2.2225	2.2235	2.2246	2.2257	2.2268	2.2279	2.2289
9.3	2.2300	2.2311	2.2322	2.2332	2.2343	2.2354	2.2364	2.2375	2.2386	2.2396
9.4	2.2407	2.2418	2.2428	2.2439	2.2450	2.2460	2.2471	2.2481	2.2492	2.2502
9.5	2.2513	2.2523	2.2534	2.2544	2.2555	2.2565	2.2576	2.2586	2.2597	2.2607
9.6	2.2618	2.2628	2.2638	2.2649	2.2659	2.2670	2.2680	2.2690	2.2701	2.2711
9.7	2.2721	2.2732	2.2742	2.2752	2.2762	2.2773	2.2783	2.2793	2.2803	2.2814
9.8	2.2824	2.2834	2.2844	2.2854	2.2865	2.2875	2.2885	2.2895	2.2905	2.2915
9.9	2.2925	2.2935	2.2946	2.2956	2.2966	2.2976	2.2986	2.2996	2.3006	2.3016

TABLE D Common Logarithms

n	0	1	2	3	4	5	6	7	8	9
1.0	.0000	.0043	.0086	.0128	.0170	.0212	.0253	.0294	.0334	.0374
1.1	.0414	.0453	.0492	.0531	.0569	.0607	.0645	.0682	.0719	.0755
1.2	.0792	.0828	.0864	.0899	.0934	.0969	.1004	.1038	.1072	.1106
1.3	.1139	.1173	.1206	.1239	.1271	.1303	.1335	.1367	.1399	.1430
1.4	.1461	.1492	.1523	.1553	.1584	.1614	.1644	.1673	.1703	.1732
1.5	.1761	.1790	.1818	.1847	.1875	.1903	.1931	.1959	.1987	.2014
1.6	.2041	.2068	.2095	.2122	.2148	.2175	.2201	.2227	.2253	.2279
1.7	.2304	.2330	.2355	.2380	.2405	.2430	.2455	.2480	.2504	.2529
1.8	.2553	.2577	.2601	.2625	.2648	.2672	.2695	.2718	.2742	.2765
1.9	.2788	.2810	.2833	.2856	.2878	.2900	.2923	.2945	.2967	.2989
2.0	.3010	.3032	.3054	.3075	.3096	.3118	.3139	.3160	.3181	.3201
2.1	.3222	.3243	.3263	.3284	.3304	.3324	.3345	.3365	.3385	.3404
2.2	.3424	.3444	.3464	.3483	.3502	.3522	.3541	.3560	.3579	.3598
2.3	.3617	.3636	.3655	.3674	.3692	.3711	.3729	.3747	.3766	.3784
2.4	.3802	.3820	.3838	.3856	.3874	.3892	.3909	.3927	.3945	.3962
2.5	.3979	.3997	.4014	.4031	.4048	.4065	.4082	.4099	.4116	.4133
2.6	.4150	.4166	.4183	.4200	.4216	.4232	.4249	.4265	.4281	.4298
2.7	.4314	.4330	.4346	.4362	.4378	.4393	.4409	.4425	.4440	.4456
2.8	.4472	.4487	.4502	.4518	.4533	.4548	.4564	.4579	.4594	.4609
2.9	.4624	.4639	.4654	.4669	.4683	.4698	.4713	.4728	.4742	.4757
3.0	.4771	.4786	.4800	.4814	.4829	.4843	.4857	.4871	.4886	.4900
3.1	.4914	.4928	.4942	.4955	.4969	.4983	.4997	.5011	.5024	.5038
3.2	.5051	.5065	.5079	.5092	.5105	.5119	.5132	.5145	.5159	.5172
3.3	.5185	.5198	.5211	.5224	.5237	.5250	.5263	.5276	.5289	.5302
3.4	.5315	.5328	.5340	.5353	.5366	.5378	.5391	.5403	.5416	.5428
3.5	.5441	.5453	.5465	.5478	.5490	.5502	.5514	.5527	.5539	.5551
3.6	.5563	.5575	.5587	.5599	.5611	.5623	.5635	.5647	.5658	.5670
3.7	.5682	.5694	.5705	.5717	.5729	.5740	.5752	.5763	.5775	.5786
3.8	.5798	.5809	.5821	.5832	.5843	.5855	.5866	.5877	.5888	.5899
3.9	.5911	.5922	.5933	.5944	.5955	.5966	.5977	.5988	.5999	.6010
4.0	.6021	.6031	.6042	.6053	.6064	.6075	.6085	.6096	.6107	.6117
4.1	.6128	.6138	.6149	.6160	.6170	.6180	.6191	.6201	.6212	.6222
4.2	.6232	.6243	.6253	.6263	.6274	.6284	.6294	.6304	.6314	.6325
4.3	.6335	.6345	.6355	.6365	.6375	.6385	.6395	.6405	.6415	.6425
4.4	.6435	.6444	.6454	.6464	.6474	.6484	.6493	.6503	.6513	.6522
4.5	.6532	.6542	.6551	.6561	.6571	.6580	.6590	.6599	.6609	.6618
4.6	.6628	.6637	.6646	.6656	.6665	.6675	.6684	.6693	.6702	.6712
4.7	.6721	.6730	.6739	.6749	.6758	.6767	.6776	.6785	.6794	.6803
4.8	.6812	.6821	.6830	.6839	.6848	.6857	.6866	.6875	.6884	.6893
4.9	.6902	.6911	.6920	.6928	.6937	.6946	.6955	.6964	.6972	.6981
5.0	.6990	.6998	.7007	.7016	.7024	.7033	.7042	.7050	.7059	.7067
5.1	.7076	.7084	.7093	.7101	.7110	.7118	.7126	.7135	.7143	.7152
5.2	.7160	.7168	.7177	.7185	.7193	.7202	.7210	.7218	.7226	.7235
5.3	.7243	.7251	.7259	.7267	.7275	.7284	.7292	.7300	.7308	.7316
5.4	.7324	.7332	.7340	.7348	.7356	.7364	.7372	.7380	.7388	.7396

TABLE D Common Logarithms

n	0	1	2	3	4	5	6	7	8	9
5.5	.7404	.7412	.7419	.7427	.7435	.7443	.7451	.7459	.7466	.7474
5.6	.7482	.7490	.7497	.7505	.7513	.7520	.7528	.7536	.7543	.7551
5.7	.7559	.7566	.7574	.7582	.7589	.7597	.7604	.7612	.7619	.7627
5.8	.7634	.7642	.7649	.7657	.7664	.7672	.7679	.7686	.7694	.7701
5.9	.7709	.7716	.7723	.7731	.7738	.7745	.7752	.7760	.7767	.7774
6.0	.7782	.7789	.7796	.7803	.7810	.7818	.7825	.7832	.7839	.7846
6.1	.7853	.7860	.7868	.7875	.7882	.7889	.7896	.7903	.7910	.7917
6.2	.7924	.7931	.7938	.7945	.7952	.7959	.7966	.7973	.7980	.7987
6.3	.7993	.8000	.8007	.8014	.8021	.8028	.8035	.8041	.8048	.8055
6.4	.8062	.8069	.8075	.8082	.8089	.8096	.8102	.8109	.8116	.8122
6.5	.8129	.8136	.8142	.8149	.8156	.8162	.8169	.8176	.8182	.8189
6.6	.8195	.8202	.8209	.8215	.8222	.8228	.8235	.8241	.8248	.8254
6.7	.8261	.8267	.8274	.8280	.8287	.8293	.8299	.8306	.8312	.8319
6.8	.8325	.8331	.8338	.8344	.8351	.8357	.8363	.8370	.8376	.8382
6.9	.8388	.8395	.8401	.8407	.8414	.8420	.8426	.8432	.8439	.8445
7.0	.8451	.8457	.8463	.8470	.8476	.8482	.8488	.8494	.8500	.8506
7.1	.8513	.8519	.8525	.8531	.8537	.8543	.8549	.8555	.8561	.8567
7.2	.8573	.8579	.8585	.8591	.8597	.8603	.8609	.8615	.8621	.8627
7.3	.8633	.8639	.8645	.8651	.8657	.8663	.8669	.8675	.8681	.8686
7.4	.8692	.8698	.8704	.8710	.8716	.8722	.8727	.8733	.8739	.8745
7.5	.8751	.8756	.8762	.8768	.8774	.8779	.8785	.8791	.8797	.8802
7.6	.8808	.8814	.8820	.8825	.8831	.8837	.8842	.8848	.8854	.8859
7.7	.8865	.8871	.8876	.8882	.8887	.8893	.8899	.8904	.8910	.8915
7.8	.8921	.8927	.8932	.8938	.8943	.8949	.8954	.8960	.8965	.8971
7.9	.8976	.8982	.8987	.8993	.8998	.9004	.9009	.9015	.9020	.9025
8.0	.9031	.9036	.9042	.9047	.9053	.9058	.9063	.9069	.9074	.9079
8.1	.9085	.9090	.9096	.9101	.9106	.9112	.9117	.9122	.9128	.9133
8.2	.9138	.9143	.9149	.9154	.9159	.9165	.9170	.9175	.9180	.9186
8.3	.9191	.9196	.9201	.9206	.9212	.9217	.9222	.9227	.9232	.9238
8.4	.9243	.9248	.9253	.9258	.9263	.9269	.9274	.9279	.9284	.9289
8.5	.9294	.9299	.9304	.9309	.9315	.9320	.9325	.9330	.9335	.9340
8.6	.9345	.9350	.9355	.9360	.9365	.9370	.9375	.9380	.9385	.9390
8.7	.9395	.9400	.9405	.9410	.9415	.9420	.9425	.9430	.9435	.9440
8.8	.9445	.9450	.9455	.9460	.9465	.9469	.9474	.9479	.9484	.9489
8.9	.9494	.9499	.9504	.9509	.9513	.9518	.9523	.9528	.9533	.9538
9.0	.9542	.9547	.9552	.9557	.9562	.9566	.9571	.9576	.9581	.9586
9.1	.9590	.9595	.9600	.9605	.9609	.9614	.9619	.9624	.9628	.9633
9.2	.9638	.9643	.9647	.9652	.9657	.9661	.9666	.9671	.9675	.9680
9.3	.9685	.9689	.9694	.9699	.9703	.9708	.9713	.9717	.9722	.9727
9.4	.9731	.9736	.9741	.9745	.9750	.9754	.9759	.9763	.9768	.9773
9.5	.9777	.9782	.9786	.9791	.9795	.9800	.9805	.9809	.9814	.9818
9.6	.9823	.9827	.9832	.9836	.9841	.9845	.9850	.9854	.9859	.9863
9.7	.9868	.9872	.9877	.9881	.9886	.9890	.9894	.9899	.9903	.9908
9.8	.9912	.9917	.9921	.9926	.9930	.9934	.9939	.9943	.9948	.9952
9.9	.9956	.9961	.9965	.9969	.9974	.9978	.9983	.9987	.9991	.9996

Answers to Odd-Numbered Problems

PROBLEM SET 1-1 (Page 5)

1. 11   **3.** $-3x - 18$ **5.** $\frac{13}{18}$ **7.** $\frac{2}{5}$ **9.** $(x + 5)/(x + 1)$ **11.** $(2x + 3)/(x^2 - 9)$ **13.** 0.31250
15. $0.\overline{27}$ **17.** $1.\overline{1176470588235294}$ **19.** 28/99   **21.** 179/999   **23.** 89/495   **25.** 5   **27.** -30
29. 6   **31.** ± 4 **33.** 3; 4   **35.** -4; 2   **37.** $(-1 \pm \sqrt{13})/6$ **39.** 31 centimeters   **41.** $2 - i$
43. $18 + 26i$ **45.** $\frac{6}{5} - \frac{2}{5}i$ **47.** (a) 1.7; $\frac{17}{10}$ (b) $3.39\overline{3}$; $\frac{509}{150}$ (c) $5.\overline{142857}$; $\frac{36}{7}$ (d) $28.1\overline{6}$; $\frac{169}{6}$
49. (a) $\frac{97}{96}$ (b) $\frac{1}{31}$ (c) $(a - 1)^2$ (d) $-8/(a - 1)$ **51.** (a) $<$ (b) $<$ (c) $<$ (d) $=$ (e) $>$ (f) $=$
53. (a) -3, 7   (b) $-\frac{11}{3}$, 3   (c) ± 2, $\pm 2\sqrt{3}$ **55.** (a) $|x - 17| < 6$ (b) $|x + \frac{9}{8}| < \frac{31}{8}$ (c) $|x - 5.44| \le 2.32$
57. (a) $\frac{13}{9}$ (b) -5, $\frac{1}{2}$ (c) -2, 0, 6   (d) -2, 4   (e) 0, 3   (f)) i, $\frac{3}{2}$ (g) 3   (h) $3 + 2i$ **59.** $\frac{2}{3}$ or $\frac{3}{2}$
61. $\overline{CD} = 4.8$; $\overline{AD} = 6.4$; $\overline{BD} = 3.6$ **63.** $(\sqrt{3} - \frac{\pi}{2})r^2$ **65.** 20 feet by 40 feet

PROBLEM SET 1-2 (Page 15)

Answers may vary slightly depending on calculator used.
1. 48.35   **3.** -2441.7393 **5.** 303.27778   **7.** 2.7721×10^{15} **9.** 1.286×10^{10} **11.** -13.138859
13. 1.7891883   **15.** 1.0498907   **17.** .90569641   **19.** .00000081   **21.** .3827594   **23.** 3.6374×10^4
25. 59.76 centimeters   **27.** 218.8 meters   **29.** 499 seconds   **31.** .0116299   **33.** 3,497,467,766
35. $V = 41.0$ cubic feet; $S = 73.8$ square feet   **37.** 97,600 cubic meters   **39.** $P = 52$ feet; $A = 150$ square feet
41. 108   **43.** $\bar{x} = 70.4$; $s = 2.93$

PROBLEM SET 1-3 (Page 20)

1.

$d(P, Q) = 5$

3.

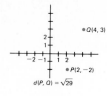

$d(P, Q) = \sqrt{29}$

5.

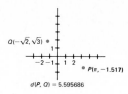

$d(P, Q) = 5.595686$

7.

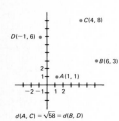

$d(A, C) = \sqrt{58} = d(B, D)$

9. $B(8, -1); D(2, 7)$

11. $(x - 1)^2 + (y + 2)^2 = 9$

13. $(4, -\frac{5}{2}); \sqrt{93}/2$ **15.** $(2, -4); \sqrt{22}$ **17.** $(-\pi/2, 1); \sqrt{\pi^2 + 4}/2$ **19.** $(x - 3)^2 + (y - 6)^2 = 25$
21. $5 \pm 4\sqrt{2}$ **23.** $a = 1 - \sqrt{3}; b = 1 + \sqrt{3}$ **25.** $\sqrt{106}$ **27.** $4\sqrt{13} + 6\pi \approx 33.27$ **29.** 16
31. (a) 12.5   (b) $(20 + 100\pi, 10)$ **33.** $(y - 2)^2 = 16(x + 1)$

35.

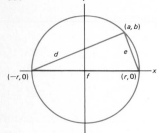

$$d^2 + e^2 = (a + r)^2 + b^2 + (a - r)^2 + b^2 = 2a^2 + 2b^2 + 2r^2$$
$$= 2r^2 + 2r^2 = 4r^2 = f^2.$$

Thus the triangle is a right triangle.

37. $\sqrt{2 + \sqrt{2}}$

PROBLEM SET 1-4 (Page 25)

1.

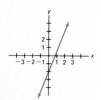

3.

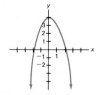

5.

7.

9.

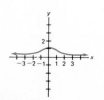

11.

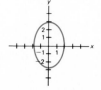

13.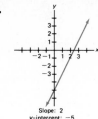

Slope: 2
y-intercept: −5

15. Slope: $\frac{4}{3}$; y-intercept: $\frac{7}{3}$.

17.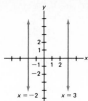

$x=-2$ $x=3$

19.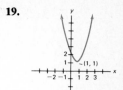

~(1, 1)

21.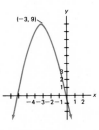

(−3, 9)

23.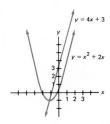

$y = 4x + 3$

$y = x^2 + 2x$

25.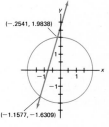

(−.2541, 1.9838)

(−1.1577, −1.6309)

27. −6, 2

29. (a) $y - 3 = \frac{1}{2}(x - 2)$ (b) $y - 1 = -\frac{2}{3}(x + 4)$ (c) $y + 2 = -\frac{3}{4}(x + 1)$

31. (a) 2.5 seconds   (b) 196 feet   (c) 6 seconds

33.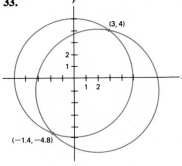

(3, 4)

(−1.4, −4.8)

35.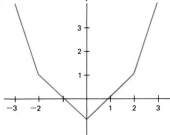

37.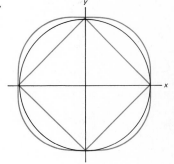

PROBLEM SET 1-5 (Page 32)

1. $\frac{17}{8}$ **3.** 10   **5.** −1000   **7.** $\{x : x \geq 1\}; \{x : x \neq -3\}$ **9.** $f(t) = (t^2 + 2)^3$

11.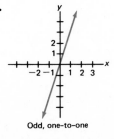

Odd, one–to–one

13.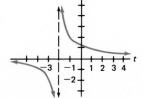

15.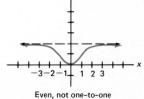

Even, not one-to-one

17. $-1; -1.5$ **19.** $f^{-1}(x) = -2x/(x-2)$

21.

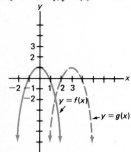

$y = f^{-1}(x) = x^5$

$y = f(x) = \sqrt[5]{x}$

23.

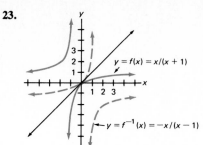

$y = f(x) = x/(x+1)$

$y = f^{-1}(x) = -x/(x-1)$

25.

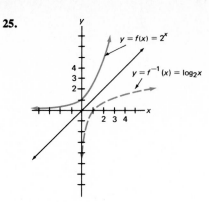

$y = f(x) = 2^x$

$y = f^{-1}(x) = \log_2 x$

27. $\{x : x \geq 0\}; f^{-1}(x) = \sqrt[4]{x}$ **29.** $\{x : x \geq 0\}; f^{-1}(x) = \sqrt{(x+3)/2}$

31.

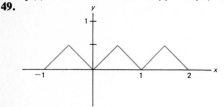

$y = f(x)$

$y = g(x)$

33.

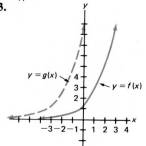

$y = g(x)$

$y = f(x)$

35. (a) -2 (b) -10 (c) $\frac{3}{4}$ (d) $\frac{1}{5}$ (e) $-\frac{1}{3}$ (f) $-\frac{9}{5}$ (g) $4x^4 - 4x^2 + 1$ (h) 3 (i) $\frac{7}{2}$

37. (a) Even. (b) Odd. (c) Odd. (d) Neither. (e) Even. (f) Odd.   **41.** $f^{-1}(x) = -3 + \sqrt{x+9}$

43. $f(x) = 3x^2 + 8$ **45.** $A(x) = x^2/2; P(x) = 2\sqrt{2}x$ **47.** $A(x) = x^2/16 + (120 - x)^2/4\pi$

49.

51.

Period: $\frac{1}{2}$

53. 6

CHAPTER 1. REVIEW PROBLEM SET (Page 36)

1. (a) $\frac{1}{6}$ (b) $\frac{13}{15}$ (c) $\frac{7}{55}$ **2.** (a) $.20402$ (b) $1.\overline{5}$ **3.** (a) $0; \pm\sqrt{5}$ (b) $(3 \pm \sqrt{13})/2$

4. (a) $9 - 7i$ (b) $-\frac{3}{10} + \frac{11}{10}i$ **5.** $-.5 \leq x \leq 4.5$ **6.** 69431.6 **7.** 10 **8.** $(x+5)^2 + (y-5)^2 = 25$

9.

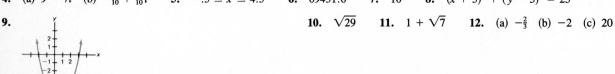

$(\frac{1}{2}, -\frac{25}{4})$

10. $\sqrt{29}$ **11.** $1 + \sqrt{7}$ **12.** (a) $-\frac{2}{3}$ (b) -2 (c) 20

13. $\{x : x \geq 3\}$ **14.** 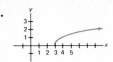   **15.** $f^{-1}(x) = x^2 + 3$

16. (a) Even. (b) Odd. (c) Neither. (d) Even.   **17.** $50\pi(6x - x^2)$

PROBLEM SET 2-1 (Page 43)

1. .6600   **3.** .6534   **5.** 3.133   **7.** 12.5°   **9.** 66.6°   **11.** 69.3°   **13.** 16.97 ≈ 17   **15.** 41.34 ≈ 41
17. 66.60 ≈ 67   **19.** $\beta = 48°$; $a = 23.42 \approx 23$; $b = 26.01 \approx 26$ **21.** $\alpha = 33.8°$; $a = 50.8$; $b = 75.9$
23. $\beta = 50.6°$; $b = 146$; $c = 189$ **25.** $c = 15$; $\alpha = 36.9°$; $\beta = 53.1°$ **27.** $b = 30$; $\alpha = 53.1°$; $\beta = 36.9°$
29. $\alpha = 26.7°$; $\beta = 63.3°$; $b = 29.0$ **31.** $\alpha = 32.9°$; $\beta = 57.1°$; $c = 17.5$ **33.** 14.6°   **35.** 7.0°
37. 31.2 feet   **39.** (a) .25862 (b) 6.6568 (c) 17.445   **41.** 725 feet   **43.** 65.369   **45.** 364 feet
47. 448 meters   **49.** 41.77°   **51.** $P = 24$; $A = 24\sqrt{3}$ **53.** $\sqrt{3}$

PROBLEM SET 2-2 (Page 50)

1. $2\pi/3$ **3.** $4\pi/3$ **5.** $7\pi/6$ **7.** $7\pi/4$ **9.** 3π **11.** $-7\pi/3$ **13.** $8\pi/9$ **15.** $\frac{1}{9}$ **17.** 240°
19. −120°   **21.** 540°   **23.** 259.0°   **25.** 18.2°   **27.** (a) 2 (b) 3.14   **29.** (a) 3 centimeters (b) 5.5 inches
31. II   **33.** III   **35.** II   **37.** IV   **39.** $16\pi/3 \approx 16.76$ feet per second.
41. $320\pi \approx 1005.3$ inches per minute.   **43.** (a) -8π (b) 5π (c) $\frac{1}{3}$
45. (a) 25.5 centimeters (b) .0327 centimeters (c) 37.83 centimeters
47. $9600\pi \approx 30{,}159$ centimeters   **49.** $330\pi \approx 1037$ miles per hour.   **51.** 8.6×10^5 miles
53. $\frac{33\pi}{20} \approx 5.184$ hours   **55.** $264\pi \approx 829$ miles   **57.** $7\pi \approx 21.99$ square inches.   **59.** 130

PROBLEM SET 2-3 (Page 57)

1. $(\sqrt{3}/2, \frac{1}{2})$ **3.** $(-\sqrt{2}/2, \sqrt{2}/2)$ **5.** $(-\sqrt{2}/2, -\sqrt{2}/2)$ **7.** $(-\sqrt{3}/2, \frac{1}{2})$ **9.** $-\sqrt{2}/2$
11. $\sqrt{2}/2$ **13.** $-\sqrt{2}/2$ **15.** $-\frac{1}{2}$ **17.** 1 **19.** 0   **21.** $-\sqrt{3}/2$ **23.** $\frac{1}{2}$ **25.** $-\sqrt{2}/2$
27. $\frac{1}{2}$ **29.** $-\frac{1}{2}$ **31.** $-\sqrt{3}/2$ **33.** −.95557; −.29476   **35.** (a) $(1/\sqrt{5}, 2/\sqrt{5})$ (b) $2/\sqrt{5}, 1/\sqrt{5}$
37. (a) $\sin(\pi + t) = -y = -\sin t$ (b) $\cos(\pi + t) = -x = -\cos t$
39. (a) Negative. (b) Negative. (c) Positive. (d) Positive. (e) Positive. (f) Negative.
41. (a) $\pm\sqrt{3}/2$ (b) $7\pi/6$; $11\pi/6$
43. (a) $\pi/4$; $5\pi/4$ (b) $\pi/6 < t < \pi/3$; $2\pi/3 < t < 5\pi/6$ (c) $0 \leq t \leq \pi/3$; $2\pi/3 \leq t \leq 4\pi/3$;
$5\pi/3 \leq t < 2\pi$ (d) $0 \leq t < \pi/4$; $3\pi/4 < t < 5\pi/4$; $7\pi/4 < t < 2\pi$ **45.** (a) $\frac{3}{5}$ (b) $-\frac{7}{25}$
47. (a) $\frac{3}{5}$ (b) $\frac{4}{5}$ (c) $\frac{4}{3}$ (d) $\frac{4}{5}$ (e) $\frac{3}{5}$ (f) $-\frac{4}{5}$
49. (a) Period 1 (b) Period $\frac{1}{3}$ (c) Not periodic (d) Period 1   **51.** 0

PROBLEM SET 2-4 (Page 63)

1. (a) $-\frac{4}{3}$ (b) $-\frac{3}{4}$ (c) $-\frac{5}{3}$ (d) $\frac{5}{4}$ **3.** $-\sqrt{5}/2$; $\frac{3}{5}\sqrt{5}$ **5.** $\sqrt{3}/3$ **7.** $2\sqrt{3}/3$ **9.** 1
11. $2\sqrt{3}/3$ **13.** $-\sqrt{3}/2$ **15.** $\sqrt{3}$ **17.** 0   **19.** $-\sqrt{3}/3$ **21.** −2
23. (a) $\pi/2$; $3\pi/2$; $5\pi/2$; $7\pi/2$ (b) $\pi/2$; $3\pi/2$; $5\pi/2$; $7\pi/2$ (c) 0; π; 2π; 3π; 4π (d) 0; π; 2π; 3π; 4π
25. −12/13; −12/5; 13/5   **27.** $-2\sqrt{5}/5$; 2; $-\sqrt{5}$ **29.** $\frac{3}{5}$; $\frac{5}{4}$ **31.** $-\frac{12}{13}$; $-\frac{12}{5}$ **33.** $(\frac{5}{13}, -\frac{12}{13})$
35. 111.8°   **37.** (a) $-2\sqrt{3}/3$ (b) $\sqrt{3}$ (c) $\sqrt{2}$ (d) −1 (e) −2 (f) 1
39. (a) $\frac{24}{25}$ (b) $\frac{-7}{25}$ (c) $\frac{-24}{7}$ (d) $\frac{25}{24}$ (e) $\frac{-24}{7}$ (f) $\frac{-25}{7}$ **41.** (a) $3\pi/4$; $7\pi/4$ (b) $\pi/4$; $7\pi/4$ (c) $\pi/2$; $3\pi/2$

43. (a) 1 (b) $\sin \theta - 1/\cos \theta$ (c) $1 + 2 \sin \theta \cos \theta$ (d) $1/\sin \theta$ (e) $\cos \theta + \sin \theta$ (f) $-(\sin^2 \theta + 1)/(\cos^2 \theta)$

45. (a) $\tan(t + \pi) = \sin(t + \pi)/\cos(t + \pi) = (-\sin t)/(-\cos t) = \tan t$
(b) $\cot(t + \pi) = \cos(t + \pi)/\sin(t + \pi) = (-\cos t)/(-\sin t) = \cot t$
(c) $\sec(t + \pi) = 1/\cos(t + \pi) = 1/(-\cos t) = -1/(\cos t) = -\sec t$
(d) $\csc(t + \pi) = 1/\sin(t + \pi) = 1/(-\sin t) = -1/(\sin t) = -\csc t$

47. $\frac{119}{169}$ **49.** -10 **51.** (a) -1.3764 (b) $.3153$ **53.** 428.98 centimeters

PROBLEM SET 2-5 (Page 68)

1. .98185 **3.** .7337 **5.** .93309 **7.** .9291 **9.** 1.30 **11.** .40 **13.** 1.10 **15.** 1.06 **17.** 1.12
19. .50 **21.** $3\pi/8$ **23.** $\pi/3$ **25.** .24 **27.** .24 **29.** $\pi/2$ **31.** .15023 **33.** 5.4707
35. .84147 **37.** -1.2885 **39.** $-.82534$ **41.** 1.25; 1.89 **43.** 1.65; 4.63 **45.** 1.37; 4.51
47. 1.84; 4.98 **49.** 40.4° **51.** 11.3° **53.** 80.2° **55.** .4051 **57.** $-.1962$ **59.** .4051 **61.** .15126
63. .9657 **65.** 21.3°; 158.7° **67.** 26.3°; 206.3° **69.** 155.3°; 204.7°
71. (a) .79608 (b) $-.79560$ (c) -1.5574 (d) $-.7513$ (e) -1.2349 (f) $-.9877$
73. (a) .9999997 (b) .744399 (c) 1.2338651
75. (a) .679996; 2.461597 (b) 1.222007; 5.061178 (c) 1.878966; 5.020558
77. (a) ϕ (b) $90° - \phi$ (c) $90° - \phi$ **79.** $-.514496$ **81.** $\phi \approx 126.9°$

PROBLEM SET 2-6 (Page 76)

1.

t	0	$\frac{\pi}{6}$	$\frac{\pi}{4}$	$\frac{\pi}{3}$	$\frac{\pi}{2}$	$\frac{3\pi}{4}$	π	$\frac{5\pi}{4}$	$\frac{3\pi}{2}$	$\frac{7\pi}{4}$	2π
$\cos t$	1	$\frac{\sqrt{3}}{2}$	$\frac{\sqrt{2}}{2}$	$\frac{1}{2}$	0	$-\frac{\sqrt{2}}{2}$	-1	$-\frac{\sqrt{2}}{2}$	0	$\frac{\sqrt{2}}{2}$	1

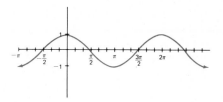

3.

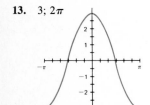

5. $\sec(t + 2\pi) = 1/\cos(t + 2\pi) = 1/\cos t = \sec t$

7. Domain: $\{t : t \neq \pi/2 + k\pi,\ k$ any integer$\}$; range: $\{y : |y| \geq 1\}$. **9.** $\pi; 2\pi$ **11.** $\cot(-t) = -\cot t$

13. 3; 2π **15.** 1; 2π **17.** 1; $\pi/2$ **19.** 2; 4π

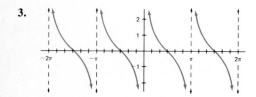

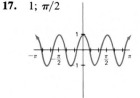

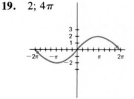

21. 2; $2\pi/3$ **23.** **25.** **27.**

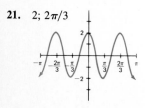

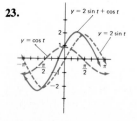

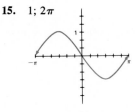

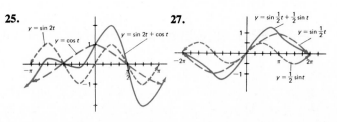

29.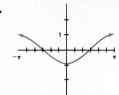

31.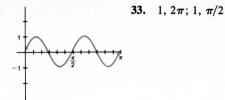

33. $1, 2\pi; 1, \pi/2$

35. (a)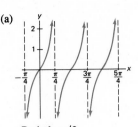

Period: $\pi/2$

(b)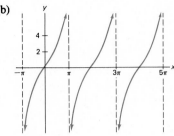

Period: 2π

37.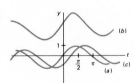

39.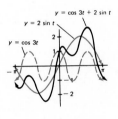

41.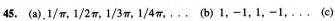

43. (a) $\frac{1}{60}$ seconds (b) 60 (c) 30 amperes

45. (a) $1/\pi, 1/2\pi, 1/3\pi, 1/4\pi, \ldots$ (b) $1, -1, 1, -1, \ldots$ (c)

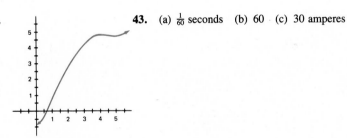

CHAPTER 2. REVIEW PROBLEM SET (Page 80)

1. (a) $\beta = 42.9°$; $a = 27.0$; $b = 25.1$ (b) $\alpha = 46.7°$; $\beta = 43.3°$; $b = 393$ **2.** 5.01 feet **3.** .576; 405°
4. 18,850 centimeters **5.** (a) $-\frac{1}{2}$ (b) $\sqrt{3}/2$ (c) 1 (d) $\frac{1}{2}$ **6.** (a) .7771 (b) $-.6157$ (c) $-.5635$ (d) .5258
7. (a) $-\sin t$ (b) $\sin t$ (c) $-\sin t$ (d) $\sin t$ **8.** (a) $\{t: 0 \le t < \pi/2 \text{ or } 3\pi/2 < t \le 2\pi\}$
(b) $\{t: 0 \le t < \pi/4 \text{ or } 3\pi/4 < t < 5\pi/4 \text{ or } 7\pi/4 < t \le 2\pi\}$ **9.** (a) $\frac{5}{12}$ (b) $-\frac{13}{5}$ **10.** $-2/\sqrt{21}$

11.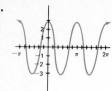

12.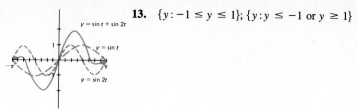

13. $\{y : -1 \le y \le 1\}$; $\{y : y \le -1 \text{ or } y \ge 1\}$

PROBLEM SET 3-1 (Page 86)

1. (a) $1 - \sin^2 t$ (b) $\sin t$ (c) $\sin^2 t$ (d) $(1 - \sin^2 t)/\sin^2 t$ **3.** (a) $1/\tan^2 t$ (b) $1 + \tan^2 t$ (c) $\tan t$ (d) 3

5. $\cos t \sec t = \cos t(1/\cos t) = 1$ **7.** $\tan x \cot x = \tan x(1/\tan x) = 1$ **9.** $\cos y \csc y = \cos y(1/\sin y) = \cot y$

11. $\cot \theta \sin \theta = (\cos \theta/\sin\theta) \sin \theta = \cos \theta$ **13.** $\tan u/\sin u = (\sin u/\cos u)(1/\sin u) = 1/\cos u$

15. $(1 + \sin z)(1 - \sin z) = 1 - \sin^2 z = \cos^2 z = 1/\sec^2 z$

17. $(1 - \sin^2 x)(1 + \tan^2 x) = \cos^2 x \sec^2 x = \cos^2 x(1/\cos^2 x) = 1$

19. $\sec t - \sin t \tan t = 1/(\cos t) - (\sin^2 t)/(\cos t) = (1 - \sin^2 t)/(\cos t) = (\cos^2 t)/(\cos t) = \cos t$

21. $(\sec^2 t - 1)/(\sec^2 t) = 1 - 1/(\sec^2 t) = 1 - \cos^2 t = \sin^2 t$

23. $\cos t(\tan t + \cot t) = \sin t + (\cos^2 t)/(\sin t) = (\sin^2 t + \cos^2 t)/(\sin t) = \csc t$

25. $\sin t = (1 - \cos^2 t)^{1/2}$; $\tan t = (1 - \cos^2 t)^{1/2}/\cos t$; $\cot t = \cos t/(1 - \cos^2 t)^{1/2}$; $\sec t = 1/\cos t$; $\csc t = 1/(1 - \cos^2 t)^{1/2}$

27. $\cos t = -3/5$; $\tan t = -4/3$; $\cot t = -3/4$; $\sec t = -5/3$; $\csc t = 5/4$

29. $\dfrac{\sec t - 1}{\tan t} \cdot \dfrac{\sec t + 1}{\sec t + 1} = \dfrac{\sec^2 t - 1}{\tan t(\sec t + 1)} = \dfrac{\tan^2 t}{\tan t(\sec t + 1)} = \dfrac{\tan t}{\sec t + 1}$

31. $\dfrac{\tan^2 x}{\sec x + 1} = \dfrac{\sec^2 x - 1}{\sec x + 1} = \dfrac{(\sec x - 1)(\sec x + 1)}{\sec x + 1} = \sec x - 1 = \dfrac{1 - \cos x}{\cos x}$

33. $\dfrac{\sin t + \cos t}{\tan^2 t - 1} \cdot \dfrac{\cos^2 t}{\cos^2 t} = \dfrac{(\sin t + \cos t) \cos^2 t}{\sin^2 t - \cos^2 t} = \dfrac{\cos^2 t}{\sin t - \cos t}$

35. (a) $2/\sin^2 x$ (b) $(2 + 2 \tan^2 x)/\tan^2 x$

37. $(1 + \tan^2 t)(\cos t + \sin t) = \sec^2 t \cos t + \sec^2 t \sin t = \sec t + \sec t \tan t = \sec t(1 + \tan t)$

39. $2 \sec^2 y - 1 = \dfrac{2}{\cos^2 y} - 1 = \dfrac{2 - \cos^2 y}{\cos^2 y} = \dfrac{1 + \sin^2 y}{\cos^2 y}$

41. $\dfrac{\sin z}{\sin z + \tan z} = \dfrac{\sin z}{\sin z + \sin z/\cos z} = \dfrac{1}{1 + 1/\cos z} = \dfrac{\cos z}{\cos z + 1}$

43. $(\csc t + \cot t)^2 = \left(\dfrac{1}{\sin t} + \dfrac{\cos t}{\sin t}\right)^2 = \dfrac{(1 + \cos t)^2}{1 - \cos^2 t} = \dfrac{1 + \cos t}{1 - \cos t}$

45. $\dfrac{1 + \tan x}{1 - \tan x} = \dfrac{1 + \sin x/\cos x}{1 - \sin x/\cos x} = \dfrac{\cos x + \sin x}{\cos x - \sin x}$

47. $(\sec t + \tan t)(\csc t - 1) = \left(\dfrac{1 + \sin t}{\cos t}\right)\left(\dfrac{1 - \sin t}{\sin t}\right) = \dfrac{\cos^2 t}{\cos t \sin t} = \cot t$

49. $\dfrac{\cos^3 t + \sin^3 t}{\cos t + \sin t} = \cos^2 t - \cos t \sin t + \sin^2 t = 1 - \sin t \cos t$

51. $\left(\dfrac{1 - \cos \theta}{\sin \theta}\right)^2 = \dfrac{(1 - \cos \theta)^2}{1 - \cos^2 \theta} = \dfrac{1 - \cos \theta}{1 + \cos \theta}$

53. $(\csc t - \cot t)^4(\csc t + \cot t)^4 = (\csc^2 t - \cot^2 t)^4 = 1^4 = 1$

55. $\sin^6 u + \cos^6 u = (1 - \cos^2 u)^3 + \cos^6 u = 1 - 3 \cos^2 u + 3 \cos^4 u$
$$= 1 - 3 \cos^2 u(1 - \cos^2 u) = 1 - 3 \cos^2 u \sin^2 u$$

57. $\cot 3x = \dfrac{1}{\tan 3x} = \dfrac{1 - 3 \tan^2 x}{3 \tan x - \tan^3 x}\left(\dfrac{\cot^3 x}{\cot^3 x}\right) = \dfrac{\cot^3 x - 3 \cot x}{3 \cot^2 x - 1} = \dfrac{3 \cot x - \cot^3 x}{1 - 3 \cot^2 x}$

PROBLEM SET 3-2 (Page 92)

1. (a) $(\sqrt{2} + 1)/2 \approx 1.21$ (b) $(\sqrt{2}\sqrt{3} + \sqrt{2})/4 \approx .97$
3. (a) $(\sqrt{2} - \sqrt{3})/2 \approx -.16$ (b) $(\sqrt{2}\sqrt{3} + \sqrt{2})/4 \approx .97$
5. $\sin(t + \pi) = \sin t \cos \pi + \cos t \sin \pi = -\sin t$
7. $\sin(t + 3\pi/2) = \sin t \cos(3\pi/2) + \cos t \sin(3\pi/2) = -\cos t$
9. $\sin(t - \pi/2) = \sin t \cos(\pi/2) - \cos t \sin(\pi/2) = -\cos t$
11. $\cos(t + \pi/3) = \cos t \cos(\pi/3) - \sin t \sin(\pi/3) = (1/2) \cos t - (\sqrt{3}/2) \sin t$
13. $\cos 2$ 15. $\sin \pi$ 17. $\cos 60°$ 19. $\sin \alpha$ 21. $\frac{56}{65}$; $-\frac{33}{65}$; in quadrant II.
23. $-(1 + 3\sqrt{3})/(2\sqrt{10}) \approx -.9797$; $(-3 + \sqrt{3})/(2\sqrt{10}) \approx -.2005$; quadrant III.
25. $\tan(s - t) = \tan(s + (-t)) = (\tan s + \tan(-t))/(1 - \tan s \tan(-t)) = (\tan s - \tan t)/(1 + \tan s \tan t)$
27. $\tan(t + \pi/4) = (\tan t + \tan \pi/4)/(1 - \tan t \tan \pi/4) = (1 + \tan t)/(1 - \tan t)$
29. (a) $-(\cos t + \sqrt{3} \sin t)/2$ (b) $(\sqrt{3} \cos t + \sin t)/2$
31. (a) $\sqrt{5}/3$ (b) $-2\sqrt{2}/3$ (c) $(4\sqrt{2} - \sqrt{5})/9$ (d) $(2\sqrt{10} - 2)/9$ (e) $(-\frac{2}{3})(\sqrt{2} + \sqrt{5})$ (f) $4\sqrt{2}/9$
33. (a) $\sqrt{3}/2$ (b) $-\sqrt{3}/2$ (c) $\sin 1 \approx .84147$
35. (a) $\sin(x + y) \sin(x - y) = (\sin x \cos y + \cos x \sin y)(\sin x \cos y - \cos x \sin y)$
$= \sin^2 x \cos^2 y - \cos^2 x \sin^2 y = \sin^2 x(1 - \sin^2 y) - \cos^2 x \sin^2 y$
$= \sin^2 x - \sin^2 y(\sin^2 x + \cos^2 x) = \sin^2 x - \sin^2 y$

(b) $\dfrac{\sin(x + y)}{\cos(x - y)} = \dfrac{\sin x \cos y + \cos x \sin y}{\cos x \cos y + \sin x \sin y} = \dfrac{\dfrac{\sin x \cos y}{\cos x \cos y} + \dfrac{\cos x \sin y}{\cos x \cos y}}{\dfrac{\cos x \cos y}{\cos x \cos y} + \dfrac{\sin x \sin y}{\cos x \cos y}} = \dfrac{\tan x + \tan y}{1 + \tan x \tan y}$

(c) $\dfrac{\cos 5t}{\sin t} - \dfrac{\sin 5t}{\cos t} = \dfrac{\cos 5t \cos t - \sin 5t \sin t}{\sin t \cos t} = \dfrac{\cos 6t}{\sin t \cos t}$

37.

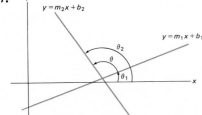

Since $\theta = \theta_2 - \theta_1$,

$$\tan \theta = \frac{\tan \theta_2 - \tan \theta_1}{1 + \tan \theta_2 \tan \theta_1} = \frac{m_2 - m_1}{1 + m_1 m_2}$$

39. (a) $\frac{1}{2}[\cos(s + t) + \cos(s - t)] = \frac{1}{2}[\cos s \cos t - \sin s \sin t + \cos s \cos t + \sin s \sin t] = \cos s \cos t$
(b) $-\frac{1}{2}[\cos s \cos t - \sin s \sin t - \cos s \cos t - \sin s \sin t] = \sin s \sin t$
(c) $\frac{1}{2}[\sin s \cos t + \cos s \sin t + \sin s \cos t - \cos s \sin t] = \sin s \cos t$
(d) $\frac{1}{2}[\sin s \cos t + \cos s \sin t - \sin s \cos t + \cos s \sin t] = \cos s \sin t$

41. (a) $(1 - \sqrt{3})/4$ (b) $-\sqrt{2}/2$ (c) $\dfrac{1 + \sqrt{2} + \sqrt{3} + \sqrt{6}}{2}$

43. $\tan(\alpha + \beta) = (\tan \alpha + \tan \beta)/(1 - \tan \alpha \tan \beta) = (\frac{1}{3} + \frac{1}{2})/(1 - \frac{1}{3} \cdot \frac{1}{2}) = 1 = \tan \gamma$. Thus $\alpha + \beta = \gamma$.

PROBLEM SET 3-3 (Page 99)

1. (a) $\frac{24}{25}$ (b) $\frac{7}{25}$ (c) $3\sqrt{10}/10$ (d) $\sqrt{10}/10$ 3. $\sin 10t$ 5. $\cos 3t$ 7. $\cos(y/2)$
9. $\cos 1.2t$ 11. $-\cos(\pi/4)$ 13. $\cos^2(x/2)$ 15. $\sin^2 2\theta$
17. (a) $\sin(\pi/8) = \sqrt{(1 - \cos \pi/4)/2} \approx .3827$ (b) $\cos 112.5° = -\sqrt{(1 + \cos 225°)/2} \approx -.3827$

19. $\tan 2t = \tan(t + t) = (\tan t + \tan t)/(1 - \tan^2 t) = 2 \tan t/(1 - \tan^2 t)$

21. $\tan\dfrac{t}{2} = \dfrac{\sin t/2}{\cos t/2} = \dfrac{\pm\sqrt{(1 - \cos t)/2}}{\pm\sqrt{(1 + \cos t)/2}} = \pm\sqrt{\dfrac{1 - \cos t}{1 + \cos t}}$

23. $\cos 3t = \cos(2t + t) = \cos 2t \cos t - \sin 2t \sin t = (2 \cos^2 t - 1) \cos t - 2 \sin^2 t \cos t$
$= (2 \cos^2 t - 1) \cos t - 2(1 - \cos^2 t) \cos t = 4 \cos^3 t - 3 \cos t$

25. $\csc 2t + \cot 2t = (1 + \cos 2t)/(\sin 2t) = (2 \cos^2 t)/(2 \sin t \cos t) = \cot t$

27. $\sin \theta/(1 - \cos \theta) = 2 \sin(\theta/2) \cos(\theta/2)/2 \sin^2(\theta/2) = \cot(\theta/2)$

29. $2 \tan \alpha/(1 + \tan^2 \alpha) = 2 \tan \alpha/\sec^2 \alpha = 2(\sin \alpha/\cos \alpha)\cos^2 \alpha = 2 \sin \alpha \cos \alpha = \sin 2\alpha$

31. $\sin 4\theta = 2 \sin 2\theta \cos 2\theta = 2(2 \sin \theta \cos \theta)(2 \cos^2 \theta - 1) = 4 \sin \theta(2 \cos^3 \theta - \cos \theta)$

33. (a) $\sin x$ (b) $\cos 6t$ (c) $-\cos(y/2)$ (d) $-\sin^2 2t$ (e) $\tan^2 2t$ (f) $\tan 3y$

35. (a) $120/169$ (b) $-2\sqrt{13}/13$ (c) $-\frac{3}{2}$ **37.** $\cos^4 z - \sin^4 z = (\cos^2 z + \sin^2 z)(\cos^2 z - \sin^2 z) = 1 \cdot \cos 2z$

39. $1 + (1 - \cos 8t)/(1 + \cos 8t) = 1 + \tan^2 4t = \sec^2 4t$

41. $\tan\frac{\theta}{2} - \sin \theta = (\sin \theta)/(1 + \cos \theta) - \sin \theta = (\sin \theta - \sin \theta - \sin \theta \cos \theta)/(1 + \cos \theta) =$
$(-\sin \theta \cos \theta)/(1 + \cos \theta) = -(\sin \theta)/(\sec \theta + 1)$

43. $(3 \cos t - \sin t)(\cos t + 3 \sin t) = 3 \cos^2 t - 3 \sin^2 t + 8 \sin t \cos t = 3 \cos 2t + 4 \sin 2t$

45. $2(\cos 3x \cos x + \sin 3x \sin 3x)^2 = 2 \cos^2 2x = 1 + \cos 4x$

47. $\tan 3t = \tan(2t + t) = \dfrac{\tan 2t + \tan t}{1 - \tan 2t \tan t} = \dfrac{\dfrac{2 \tan t}{1 - \tan^2 t} + \tan t}{1 - \dfrac{2 \tan^2 t}{1 - \tan^2 t}} = \dfrac{3 \tan t - \tan^3 t}{1 - 3 \tan^2 t}$

49. $\sin^4 u + \cos^4 u = (\sin^2 u + \cos^2 u)^2 - 2 \sin^2 u \cos^2 u = 1 - \frac{1}{2} \sin^2 2u = 1 - \frac{1}{2} \cdot (1 - \cos 4u)/2 = \frac{3}{4} + \frac{1}{4} \cos 4u$

51. $\cos^2 x + \cos^2 2x + \cos^2 3x = \dfrac{1 + \cos 2x}{2} + \cos^2 2x + \dfrac{1 + \cos 6x}{2} = 1 + \frac{1}{2}(\cos 2x + \cos 6x) + \cos^2 2x =$

$1 + \cos 4x \cos 2x + \cos^2 2x = 1 + \cos 2x(\cos 4x + \cos 2x) = 1 + \cos 2x(2 \cos 3x \cos x) = 1 + 2 \cos x \cos 2x \cos 3x$

53. Since $\alpha + \beta + \gamma = 180°$, $2\gamma = 360° - 2\alpha - 2\beta$. Thus $\sin 2\alpha + \sin 2\beta + \sin 2\gamma = \sin 2\alpha + \sin 2\beta - \sin(2\alpha + 2\beta) = $
$\sin 2\alpha + \sin 2\beta - \sin 2\alpha \cos 2\beta - \cos 2\alpha \sin 2\beta = \sin 2\alpha(1 - \cos 2\beta) + \sin 2\beta(1 - \cos 2\alpha) = 2 \sin \alpha \cos \alpha(2 \sin^2 \beta) +$
$2 \sin \beta \cos \beta(2 \sin^2 \alpha) = 4 \sin \alpha \sin \beta(\cos \alpha \sin \beta + \sin \alpha \cos \beta) = 4 \sin \alpha \sin \beta \sin(\alpha + \beta) = 4 \sin \alpha \sin \beta \sin \gamma$.

55. $(\frac{7}{9}, 4\sqrt{2}/9)$

PROBLEM SET 3-4 (Page 107)

1. $\pi/3$ **3.** $\pi/4$ **5.** 0 **7.** $\pi/3$ **9.** $2\pi/3$ **11.** $\pi/4$ **13.** $.2200$ **15.** $-.2200$ **17.** $.2037$
19. (a) $.7938$ (b) 1.9545 **21.** $.3486; 2.7930$ **23.** $1.2803; 4.4219$ **25.** $\frac{2}{3}$ **27.** 10 **29.** $\pi/3$
31. $\pi/4$ **33.** $\frac{3}{5}$ **35.** $2/\sqrt{5}$ **37.** $\frac{1}{3}$ **39.** $2\pi/3$ **41.** $.9666$ **43.** $.4508$ **45.** 2.2913 **47.** $\frac{24}{25}$
49. $\frac{7}{25}$ **51.** $\frac{56}{65}$ **53.** $(6 + \sqrt{35})/12 \approx .993$ **55.** $\tan(\sin^{-1} x) = \sin(\sin^{-1} x)/\cos(\sin^{-1} x) = x/\sqrt{1 - x^2}$
57. $\tan(2 \tan^{-1} x) = 2 \tan(\tan^{-1} x)/[1 - \tan^2(\tan^{-1} x)] = 2x/(1 - x^2)$ **59.** $\cos(2 \sec^{-1} x) = \cos[2 \cos^{-1}(1/x)] = 2/x^2 - 1$
61. (a) $-\pi/3$ (b) $-\pi/3$ (c) $2\pi/3$ **63.** (a) 43 (b) $\frac{12}{13}$ (c) $7\sqrt{2}/10$ (d) $(4 - 6\sqrt{2})/15$
65. (a) $\pm\sqrt{7}/4$ (b) $\pm.9$ (c) $\frac{11}{6}$ (d) $1; 2$ **67.** (a) $\pi/2$ (b) $\pi/4$ (c) $-\pi/4$ (d) $-\pi/2$ (e) π (f) $\pi/4$
69. (a) $\sin^{-1}(x/5)$ (b) $\tan^{-1}(x/3)$ (c) $\sin^{-1}(3/x)$ (d) $\tan^{-1}(3/x) - \tan^{-1}(1/x)$
71. (a) $.6435011$ (b) $-.3046927$ (c) $.6435011$ (d) 2.6905658 **73.** Show that the tangent of both sides is $120/119$.
75. (a) $\theta = \tan^{-1}(6/b) - \tan^{-1}(2/b)$ (b) $22.83°$ (c) $2\sqrt{3}$

PROBLEM SET 3-5 (Page 115)

1. $\{0, \pi\}$ **3.** $\{3\pi/2\}$ **5.** No solution. **7.** $\{5\pi/6, 7\pi/6\}$ **9.** $\{\pi/4, 3\pi/4, 5\pi/4, 7\pi/4\}$
11. $\{\pi/4, 2\pi/3, 3\pi/4, 4\pi/3\}$ **13.** $\{0, \pi, 3\pi/2\}$ **15.** $\{0, \pi/3, \pi, 4\pi/3\}$ **17.** $\{\pi/3, \pi, 5\pi/3\}$
19. $\{.3649, 1.2059, 3.5065, 4.3475\}$ **21.** $\{0, \pi/2\}$ **23.** $\{\pi/6, \pi/2\}$ **25.** $\{0\}$
27. $\{\pi/6 + 2k\pi, 5\pi/6 + 2k\pi: k$ is an integer$\}$ **29.** $\{k\pi: k$ is an integer$\}$ **31.** $\{\pi/6 + k\pi, 5\pi/6 + k\pi: k$ is an integer$\}$

33. $\{0, \pi/2, \pi, 3\pi/2\}$ **35.** $\{\pi/8, 5\pi/8, 9\pi/8, 13\pi/8\}$ **37.** $\{3\pi/8, 7\pi/8, 11\pi/8, 15\pi/8\}$
39. $\{0, \pi/6, 5\pi/6, \pi\}$ **41.** $\{.9553, 2.1863, 4.0969, 5.3279\}$ **3.** $\{\pi/4, 5\pi/4\}$ **45.** $\{0, \pi/6, 5\pi/6, \pi, 7\pi/6, 11\pi/6\}$
47. $\{.3076, 2.8340\}$ **49.** $\{2\pi/3, 4\pi/3\}$ **51.** $\{2\pi/3, 5\pi/6, 5\pi/3, 11\pi/6\}$ **53.** $\{3\pi/2, 5.6397\}$
55. $\{\pi/4, 3\pi/4, 5\pi/4, 7\pi/4\}$ **57.** (a) 15 inches (b) $\tan \theta = \frac{2}{3}$ (c) 33.7° **59.** (a) 26.6° (b) 10.3°
61. $\{k\pi/3, 2\pi/3 + 2k\pi, 4\pi/3 + 2k\pi: k \text{ is an integer}\}$ **63.** $\{\pi/6, \pi/3, 2\pi/3, 5\pi/6\}$

CHAPTER 3. REVIEW PROBLEM SET (Page 118)

1. (a) $\cot \theta \cos \theta = \cos^2 \theta/\sin \theta = (1 - \sin^2 \theta)/\sin \theta = 1/\sin \theta - \sin \theta = \csc \theta - \sin \theta$
(b) $(\cos x \tan^2 x)(\sec x - 1)/(\sec x + 1)(\sec x - 1) = \cos x \tan^2 x (\sec x - 1)/\tan^2 x = 1 - \cos x$
2. (a) $-\sin^3 x/(1 - \sin^2 x)$ (b) $1 - \sin x$ **3.** (a) $\cos 45° = \sqrt{2}/2$ (b) $\sin 90° = 1$ (c) $\cos 45° = \sqrt{2}/2$
4. (a) $24/25 = .96$ (b) $3/\sqrt{10} \approx .95$ **5.** (a) $\sin 2t \cos t - \cos 2t \sin t = \sin(2t - t) = \sin t$
(b) $\sec 2t + \tan 2t = (1 + \sin 2t)/\cos 2t = (\cos t + \sin t)^2/(\cos^2 t - \sin^2 t) = (\cos t + \sin t)/(\cos t - \sin t)$
(c) $\cos(\alpha + \beta)/\cos \alpha \cos \beta = \cos \alpha \cos \beta/\cos \alpha \cos \beta - \sin \alpha \sin \beta/\cos \alpha \cos \beta = 1 - \tan \alpha \tan \beta = \tan \alpha(\cot \alpha - \tan \beta)$
6. (a) $\{5\pi/6, 7\pi/6\}$ (b) $\{0, \pi, 7\pi/6, 11\pi/6\}$ (c) $\{0\}$ (d) $\{\pi/2, \pi\}$ (e) $\{\pi/6, 5\pi/6, 3\pi/2\}$
7. $-\pi/2 \le t \le \pi/2; 0 \le t \le \pi; -\pi/2 < t < \pi/2$ **8.** (a) $-\pi/3$ (b) $5\pi/6$ (c) $-\pi/3$ (d) 6 (e) π (f) $\sqrt{5}/3$
(g) $-.02$ (h) $120/169$ **9.** See the graph in the text on page 105. **10.** -1.57
11. $\tan(\arctan \frac{1}{2} + \arctan \frac{1}{3}) = (\frac{1}{2} + \frac{1}{3})/(1 - \frac{1}{6}) = 1$. Since $\arctan \frac{1}{2}$ and $\arctan \frac{1}{3}$ are between 0 and $\pi/2$, their sum cannot be in quadrant III and so must equal $\pi/4$.

PROBLEM SET 4-1 (Page 125)

1. $\gamma = 55.5°; b \approx 20.9; c \approx 17.4$ **3.** $\beta = 56°; a = c \approx 53$ **5.** $\beta \approx 42°; \gamma \approx 23°; c \approx 20$
7. $\beta \approx 18°; \gamma \approx 132°; c \approx 12$ **9.** Two triangles: $\beta_1 \approx 53°, \gamma_1 \approx 97°, c_1 \approx 9.9; \beta_2 \approx 127°, \gamma_2 \approx 23°, c_2 \approx 3.9$
11. 93.7 meters **13.** 44.7° **15.** 192.8 **17.** 265.3 **19.** 78.4° **21.** 694.6 square feet
23. 1769 feet **25.** 40 **27.** $6r^2 \sin \phi(\cos \phi + \sqrt{3} \sin \phi)$

PROBLEM SET 4-2 (Page 130)

1. $a \approx 12.5; \beta \approx 76°; \gamma \approx 44°$ **3.** $c \approx 15.6; \alpha \approx 26°; \beta \approx 34°$ **5.** $\alpha \approx 44.4°; \beta \approx 57.1°; \gamma \approx 78.5°$
7. $\alpha \approx 30.6°; \beta \approx 52.9°; \gamma \approx 96.5°$ **9.** 98.8 meters **11.** 24 miles **13.** 106°
15. $s = 6, A = \sqrt{6 \cdot 3 \cdot 2 \cdot 1} = 6$ **17.** 18.63 **19.** 41.68° **21.** 42.60 miles
23. $\cos^{-1}(\frac{3}{4}) \approx 41.41°; \frac{1}{2}(\sqrt{3} + 3\sqrt{7})$ **27.** $\sqrt{15}$

PROBLEM SET 4-3 (Page 134)

1.

3.

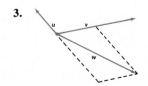

5. $\mathbf{w} = \frac{1}{2}(\mathbf{u} + \mathbf{v})$ **7.** $\|\mathbf{w}\| = 1$

9. $10\sqrt{2 + \sqrt{2}} \approx 18.48$ **11.** 243.7 kilometers; S43.9°W **13.** .026 hours **15.** 479 miles per hour; N2.68°E
17. 15.9; S7.5°W **19.** 163.7, 118.9 **21.** **23. 0**

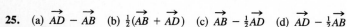

25. (a) $\overrightarrow{AD} - \overrightarrow{AB}$ (b) $\frac{1}{2}(\overrightarrow{AB} + \overrightarrow{AD})$ (c) $\overrightarrow{AB} - \frac{1}{2}\overrightarrow{AD}$ (d) $\overrightarrow{AD} - \frac{1}{2}\overrightarrow{AB}$
27. $\sqrt{7}/2 \approx 1.32$ miles per hour **29.** N10.33°E **31.** N1.019°W; 654.9 miles per hour **33.** 651.3 pounds
35. $\alpha = 65.38°$; $w = 146.8$ pounds

PROBLEM SET 4-4 (Page 141)

1. $4\mathbf{i} - 24\mathbf{j}$; -33; $-33/65$ **3.** $3\mathbf{i} + \mathbf{j}$; 10; $2/\sqrt{5}$ **5.** 101.385° **7.** $5\mathbf{i} + 2\mathbf{j}$; $4\mathbf{i} - 3\mathbf{j}$; 14
9. $-4\mathbf{i} - 5\mathbf{j}$; $-6\mathbf{i} + 5\mathbf{j}$; -1 **11.** $-5\mathbf{i} + 5\sqrt{3}\mathbf{j}$ **13.** $\frac{4}{3}$ **15.** $\frac{3}{5}\mathbf{i} - \frac{4}{3}\mathbf{j}$ **17.** $-4\mathbf{i} + 10\mathbf{j}$
19. $\mathbf{u} \cdot \mathbf{u} = a^2 + b^2 = \|\mathbf{u}\|^2$ **21.** 7 **23.** $(118/169)(5\mathbf{i} + 12\mathbf{j})$ **25.** $-56/5$ **27.** 100
29. $325\sqrt{2}$ dyne-centimeters **31.** $\sqrt{34}$ **33.** $\pm(\frac{4}{5}\mathbf{i} + \frac{3}{5}\mathbf{j})$ **35.** $\frac{9}{25}\mathbf{i} - \frac{12}{25}\mathbf{j}$; 84.09°
37. $|\mathbf{u} \cdot \mathbf{v}| = \|\mathbf{u}\| \|\mathbf{v}\| |\cos \theta| \leq \|\mathbf{u}\| \|\mathbf{v}\|$; $\theta = 0°$ or $\theta = 180°$ **39.** $\|\mathbf{u}\| = \|\mathbf{v}\|$ **41.** $(-1 + \sqrt{5})/4$

PROBLEM SET 4-5 (Page 148)

1. (a)

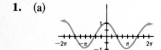

(b)

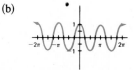

(c)

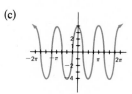

(d)

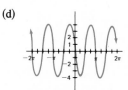

3. (a) π, 4; 0   (b) 2π, 3, $-\pi/8$ (c) $\pi/2$; 1; $-\pi/32$ (d) $2\pi/3$; 3; $\pi/6$

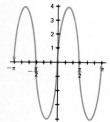

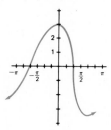

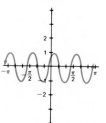

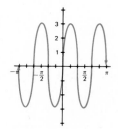

5. 4π, 2; $\pi/8$ **7.**   **9.**   **11.**

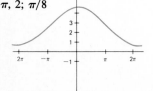

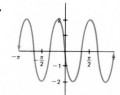

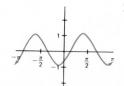

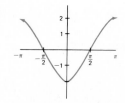

13. $(5 \cos 4t, 5 \sin 4t)$ **15.** $5 \cos 4t, -8 + 5 \sin 4t$
17. (a) $2\pi/5$; 1; 0   (b) 4π; $\frac{3}{2}$; 0   (c) $\pi/2$; 2; $\pi/4$ (d) $2\pi/3$; 4; $-\pi/4$ **19.** 4 feet; after 1.5 seconds
21. $\sin t + \sqrt{25 - \cos^2 t}$ **23.** 156; 55   **25.** (a) 0   (b) 0   (c) 53.53   **27.**

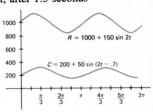

29. (a) $2\sqrt{2} \sin 2t - 2\sqrt{2} \cos 2t$ (b) $-\frac{3}{2}\sqrt{3} \sin 3t + \frac{3}{2} \cos 3t$ **33.** (a) $5 \sin[2t + \tan^{-1}(\frac{4}{3})]$ (b) $2\sqrt{3} \sin(4t + 11\pi/6)$
35. $\sqrt{2}$; $-\sqrt{2}$

PROBLEM SET 4-6 (Page 156)

1-11.

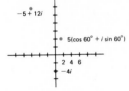

13. $\sqrt{13}$; $\sqrt{13}$; 5; 4; 1; 2   **15.** $0 - 4i$ **17.** $-\sqrt{2} - \sqrt{2}i$ **19.** $4(\cos \pi + i \sin \pi)$

21. $5(\cos 270° + i \sin 270°)$ **23.** $2\sqrt{2}(\cos 315° + i \sin 315°)$ **25.** $4(\cos \pi/6 + i \sin\pi/6)$
27. $6.403(\cos .6747 + i \sin .6747)$ **29.** $6(\cos 210° + i \sin 210°)$ **31.** $\frac{3}{2}(\cos 125° + i \sin 125°)$
33. $\frac{2}{3}(\cos 70° + i \sin 70°)$ **35.** $2(\cos 305° + i \sin 305°)$ **37.** $16 + 0i$ **39.** $-2 + 2\sqrt{3}i$
41. $16(\cos 60° + i \sin 60°)$ **43.** $1(\cos 120° + i \sin 120°)$ **45.** $16(\cos 270° + i \sin 270°)$
47.   (a) $|-5 + 12i| = 13$ (b) $|-4i| = 4$ (c) $|5(\cos 60° + i \sin 60°)| = 5$

49. (a) $12(\cos 0° + i \sin 0°)$ (b) $2(\cos 135° + i \sin 135°)$ (c) $3(\cos 270° + i \sin 270°)$
(d) $4(\cos 300° + i \sin 300°)$ (e) $8(\cos 30° + i \sin 30°)$ (f) $2(\cos 315° + i \sin 315°)$
51. (a) $12(\cos 160° + i \sin 160°)$ (b) $3(\cos 40° + i \sin 40°)$ (c) $\cos 45° + i \sin 45°$
53. (a) $12 + 5i$, $-12 + 5i$ (b) $\pm(4\sqrt{2} + 4\sqrt{2}i)$
55. (a) $r^3(\cos 3\theta + i \sin 3\theta)$ (b) $r[\cos(-\theta) + i \sin(-\theta)]$ (c) $r^2(\cos 0 + i \sin 0)$ (d) $\frac{1}{r}[\cos(-\theta) + i \sin(-\theta)]$
(e) $r^{-2}[\cos(-2\theta) + i \sin(-2\theta)]$ (f) $r[\cos(\theta + \pi) + i \sin(\theta + \pi)]$
57. (a) The distance between U and V in the complex plane.   (b) The angle from the positive x-axis to the vector from U to V.   **59.** (a) -2^8 (b) 0

PROBLEM SET 4-7 (Page 163)

1. $8[\cos(3\pi/4) + i \sin(3\pi/4)]$ **3.** $125(\cos 66° + i \sin 66°)$ **5.** $16(\cos 0° + i \sin 0°)$ **7.** $1 + 0i$
9. $-16\sqrt{3} + 16i$

11. $5(\cos 15° + i \sin 15°)$;
$5(\cos 135° + i \sin 135°)$;
$5(\cos 255° + i \sin 255°)$

13. $2[\cos(\pi/12) + i \sin(\pi/12)]$; $2[\cos(5\pi/12) + i \sin(5\pi/12)]$;
$2[\cos(9\pi/12) + i \sin(9\pi/12)]$; $2[\cos(13\pi/12) + i \sin(13\pi/12)]$;
$2[\cos(17\pi/12) + i \sin(17\pi/12)]$; $2[\cos(21\pi/12) + i \sin(21\pi/12)]$

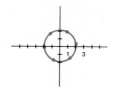

15. $\sqrt{2}(\cos 28° + i \sin 28°)$; $\sqrt{2}(\cos 118° + i \sin 118°)$
$\sqrt{2}(\cos 208° + i \sin 208°)$; $\sqrt{2}(\cos 298° + i \sin 298°)$

17. $\pm 2; \pm 2i$

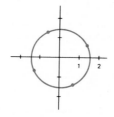

19. $\pm(\sqrt{2} + \sqrt{2}i)$ **21.** $\pm(\sqrt{2} + \sqrt{6}i)$

23. $\pm 1; \pm i$ **25.** $\cos(k \cdot 36°) + i \sin(k \cdot 36°)$, $k = 0, 1, \ldots, 9$

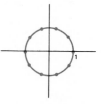

27. (a) $81(\cos 80° + i \sin 80°)$ (b) $90.09(\cos 7.7 + i \sin 7.7)$ (c) $8(\cos 240° + i \sin 240°)$
(d) $16[\cos(5\pi/3) + i \sin(5\pi/3)]$

29. $2(\cos 51° + i \sin 51°)$, $2(\cos 123° + i \sin 123°)$, $2(\cos 195° + i \sin 195°)$, $2(\cos 267° + i \sin 267°)$, $2(\cos 339° + i \sin 339°)$

31. 1, $\sqrt{2}/2 + (\sqrt{2}/2)i$, i, $-\sqrt{2}/2 + (\sqrt{2}/2)i$, -1, $-\sqrt{2}/2 - (\sqrt{2}/2)i$, $-i$, $\sqrt{2}/2 - (\sqrt{2}/2)i$;
Sum $= 0$ and product $= -1$.

33. $\sqrt[5]{2} (\cos 27° + i \sin 27°) \approx 1.0235 + .5215i$

35. *Method 1.* Use the formula $\cos(k\pi/3) + i \sin(k\pi/3)$, $k = 0, 1, 2, 3, 4, 5$. *Method 2.* Write $x^6 - 1 = (x - 1)(x^2 + x + 1)(x + 1)(x^2 - x + 1) = 0$. Both methods give the answers ± 1, $(-1 \pm \sqrt{3}i)/2$, $(1 \pm \sqrt{3}i)/2$.

37. $\pm i$, $\sqrt{2}/2 \pm (\sqrt{2}/2)i$, $-\sqrt{2}/2 \pm (\sqrt{2}/2)i$ **39.** (a) $\sqrt{6}/2 + (\sqrt{2}/2)i$ (b) $\sqrt{2}/2 + (\sqrt{6}/2)i$ **41.** -2^n

CHAPTER 4. REVIEW PROBLEM SET (Page 166)

1. (a) $\beta = 39.1°$; $a \approx 228$; $c \approx 139$ (b) $\alpha \approx 4.0°$; $\beta \approx 154.5°$; $\gamma \approx 21.5°$ (c) $c \approx 13.2$; $\alpha \approx 37.3°$; $\beta \approx 107.7°$
(d) $\gamma \approx 30.7°$; $\alpha \approx 100.7°$; $a \approx 75.5$

2. $x \approx 37.1$; $A \approx 281$ **3.** $6\mathbf{i} + (5\sqrt{3} - 3)\mathbf{j}$

4. (a) 5 (b) 13 (c) -33 (d) $120.5°$ (e) $-33/13$ (f) $(-33/169)(5\mathbf{i} + 12\mathbf{j})$

5. (a) $120\mathbf{i} + 60\mathbf{j}$ (b) $(80\mathbf{i} + 60\mathbf{j}) \cdot (4\mathbf{i} + 2\mathbf{j})/\sqrt{5} \approx 196.8$ foot pounds
6. (a) π; 1; 0 (b) $\pi/2$; 3; 0 (c) $2\pi/3$; 2; $\pi/6$ (d) 4π, 2; -2π
7. (a) (b) (c) (d)

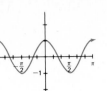
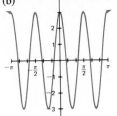
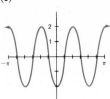
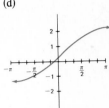

8. $(4\cos(3\pi t/4 + \pi), 4\sin(3\pi t/4 + \pi))$ **9.** (a) $2\pi/5$ seconds (b) At $x = -3$ feet. (c) When $t = \pi/5$ seconds.

10.

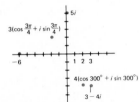

11. (a) 5 (b) 6 (c) 5 (d) 3 (e) 4

12. $-2\sqrt{3} + 2i$ **13.** (a) $3[\cos(\pi/2) + i\sin(\pi/2)]$ (b) $6(\cos\pi + i\sin\pi)$ (c) $\sqrt{2}[\cos(5\pi/4) + i\sin(5\pi/4)]$
(d) $4[\cos(11\pi/6) + i\sin(11\pi/6)]$
14. (a) $32(\cos 145° + i\sin 145°)$ (b) $2(\cos 65° + i\sin 65°)$ (c) $512(\cos 315° + i\sin 315°)$ (d) $4096(\cos 330° + i\sin 330°)$
15. $2(\cos 20° + i\sin 20°)$; $2(\cos 80° + i\sin 80°)$; $2(\cos 140° + i\sin 140°)$; $2(\cos 200° + i\sin 200°)$; $2(\cos 260° + i\sin 260°)$;
$2(\cos 320° + i\sin 320°)$
16. $\cos 0° + i\sin 0°$; $\cos 72° + i\sin 72°$; $\cos 144° + i\sin 144°$; $\cos 216° + i\sin 216°$; $\cos 288° + i\sin 288°$

PROBLEM SET 5-1 (Page 174)

1. $7^{1/3}$ **3.** $7^{2/3}$ **5.** $7^{-1/3}$ **7.** $7^{-2/3}$ **9.** $7^{4/3}$ **11.** $x^{2/3}$ **13.** $x^{5/2}$ **15.** $(x + y)^{3/2}$
17. $(x^2 + y^2)^{1/2}$ **19.** $\sqrt[3]{16}$ **21.** $1/\sqrt{8^3} = \sqrt{2}/32$ **23.** $\sqrt[4]{x^4 + y^4}$ **25.** $y\sqrt[5]{x^4 y}$ **27.** $\sqrt{\sqrt{x} + \sqrt{y}}$
29. 5 **31.** 4 **33.** $\frac{1}{27}$ **35.** .04 **37.** .000125 **39.** $\frac{1}{5}$ **41.** $\frac{1}{16}$ **43.** $\frac{1}{16}$ **45.** $-6a^2$
47. $8/x^4$ **49.** x^4 **51.** $4y^2/x^4$ **53.** y^9/x^{30} **55.** $(2y^3 - 1)/y$ **57.** $x + y + 2\sqrt{xy}$
59. $(7x + 2)/3(x + 2)^{1/5}$ **61.** $(1 - x^2)/(x^2 + 1)^{2/3}$ **63.** $\sqrt[6]{32}$ **65.** $\sqrt[12]{8x^2}$ **67.** $\sqrt{x}$ **69.** 2.53151
71. 4.6364 **73.** 1.70777 **75.** .0050463
77. **79.** **81.** **83.** (a) $b^{3/5}$ (b) $x^{1/2}$ (c) $(a + b)^{2/3}$

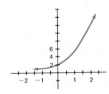

85. (a) 72 (b) $3 \cdot 2^{1/6}$ (c) $a^{17/12}$ (d) $a^{5/6}$ (e) $a^3 + 2 + 1/a^3$ (f) $\dfrac{a}{1 + a^2}$ (g) $a^{1/2}b^{1/12}$ (h) $\dfrac{a^8}{b^{14/3}}$ (i) 8
(j) $4a^{3/2}b^{9/4}$ (k) $3^{3/2}$ (l) $a - b$
87. (a) $-\frac{1}{2}$ (b) -1; 2 (c) All reals. (d) -4; 3 (e) 1; 8 (f) 2

89.

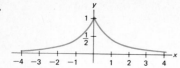

91. The graph of $y = f(x) = a^x$ has the x-axis as a horizontal asymptote. Also f is not a constant function. The graph of a nonconstant polynomial does not have a horizontal asymptote.

PROBLEM SET 5-2 (Page 181)

1. (a) Decays. (b) Grows. (c) Grows. (d) Decays.
3. (a) 4.66095714 (b) 17.00006441 (c) 4801.02063 (d) 9750.87832   **5.** 1480
7. (a) 5.384 billion (b) 6.562 billion (c) 23.772 billion
9. (a) $185.09 (b) $247.60   **11.** (a) $76,035.83 (b) $325,678.40   **13.** $P(1 + r/100)^n$
15. $7401.22   **17.** $7102.09   **19.** $7305.57   **21.** $1000(1 + .08/12)^{120} = $2219.64
23. 8100; 5400; 3600; 2400; 1600   **25.** 800   **27.** (a) 8680; 22,497 (b) About 44 years.
29. (a) $146.93 (b) $148.59 (c) $148.98 (d) $149.18   **31.** (a) About 9 years. (b) About 11 years.
33. (a) $k \approx .0005917$ (b) 10.76 milligrams   **35.** 2270 years   **37.** $320,057,300

PROBLEM SET 5-3 (Page 190)

1. $\log_4 64 = 3$ **3.** $\log_{27} 3 = \frac{1}{3}$ **5.** $\log_4 1 = 0$ **7.** $\log_{125}(1/25) = -\frac{2}{3}$ **9.** $\log_{10} a = \sqrt{3}$
11. $\log_{10} \sqrt{3} = a$ **13.** $5^4 = 625$ **15.** $4^{3/2} = 8$ **17.** $10^{-2} = .01$ **19.** $c^1 = c$ **21.** $c^y = Q$ **23.** 2
25. -1 **27.** $\frac{1}{3}$ **29.** -4 **31.** 0   **33.** $\frac{4}{3}$ **35.** 2   **37.** $\frac{1}{27}$ **39.** -2.9 **41.** 49   **43.** .778
45. 1.204   **47.** $-.602$ **49.** 1.380   **51.** $-.051$ **53.** .699   **55.** 1.5314789   **57.** $.08990511 - 2$
59. 3.9878003   **61.** $\log_{10}[(x + 1)^3(4x + 7)]$ **63.** $\log_2[8x(x + 2)^3/(x + 8)^2]$ **65.** $\log_6(\sqrt{x}\sqrt[3]{x^3 + 3})$
67. 47   **69.** $-\frac{11}{4}$ **71.** $\frac{16}{7}$ **73.** 5; 2 is extraneous.   **75.** 7   **77.** 4.0111687   **79.** 2.0446727
81. (a) 2 (b) $\frac{1}{2}$ (c) 125 (d) 32 (e) 10 (f) 4   **83.** $\frac{8}{9}$
85. (a) 13 (b) -5 (c) No solution. (d) 20 (e) 3 (f) 16
87. (a) $y = x/(x - 1)$ (b) $y = \frac{1}{2}(a^x + a^{-x})$ **89.** $x = 10^c$, $\log_2 x = c \log_2 10$, $\log_2 x = (\log_{10} x)(\log_2 10)$
91. By Problem 90 $\log_a a = \log_b a \cdot \log_a b$, or $1 = \log_b a \cdot \log_a b$. Therefore, $\log_a b = 1/\log_b a$
93.

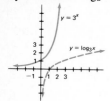

95.

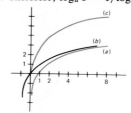

PROBLEM SET 5-4 (Page 198)

1. 1   **3.** 0   **5.** $\frac{1}{2}$ **7.** -3 **9.** 3.5   **11.** $-.2$ **13.** -7.5 **15.** 4.787   **17.** 6.537   **19.** .182
21. 9.1   **23.** .9   **25.** 90   **27.** 1.4609379   **29.** -2.0635682 **31.** -1.8411881 **33.** 50833303
35. 11.818915   **37.** 61.146135   **39.** 8.3311375   **41.** .8824969   **43.** 915.98   **45.** About 2.73.
47. About -0.737.   **49.** About 6.84   **51.** Approximately 6.12 years.   **53.** Approximately 4.71 years.
55. (a) 5^{10} (b) 9^{10} (c) 10^{20} (d) 10^{1000} **57.**

59. $y = ba^x$; $a \approx 1.5$, $b \approx 64$

61. $y = bx^a$; $a \approx 4$, $b \approx 12$ **63.** (a) 4.2 (b) 4 (c) $\frac{1}{2}$ **65.** (a) 7 (b) 12.25 (c) $-\frac{1}{2}$ (d) 0 (e) 125 (f) $\frac{1}{3}$
67. (a) $-.349$ (b) $-.823$ (c) $.633, -3.633$ (d) 2.166 (e) 4.560; .219 (f) $e^e \approx 15.154$ (g) $e \approx 2.718$
(h) $e \approx 2.718$ (i) $\pm\sqrt{e + 5} \approx \pm 2.778$
69. $(\ln 2)/3 \approx .231$ years   **71.** $(\ln 2)/240 \approx .00289$
73.

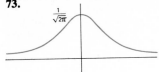

75. $e^{\pi/e - 1} > 1 + \pi/e - 1$, $e^{\pi/e}/e > \pi/e$, $e^{\pi/e} > \pi$, $e^\pi > \pi^e$

77. (a) $100(1 + .01)^{120} \approx \$330.04$ (b) $100(1 + .12/365)^{3650} \approx \$331.95$ (c) $100(1 + .12/(365)(24))^{(3650)(24)} \approx \$332.01$
(d) $100e^{(.12)(10)} \approx \$332.01$

PROBLEM SET 5-5 (Page 206)

1. 4   **3.** -2 **5.** $\frac{11}{2}$ **7.** -15 **9.** 10,000   **11.** .01   **13.** $10^{3/2}$ **15.** $10^{-3/4}$ **17.** .6355
19. 2.1987   **21.** $.5172 - 2$ **23.** 5.7505   **25.** 8.9652   **27.** 32.8   **29.** .0101   **31.** 3.98×10^8
33. 166   **35.** .838   **37.** .7191   **39.** 3.8593   **41.** $.0913 - 3$ **43.** 7.075   **45.** 8184   **47.** .03985
49. (a) $\frac{5}{4}$ (b) $-\frac{4}{3}$ (c) $\frac{5}{8}$ (d) -3 **51.** (a) 2.6926 (b) $.6726 - 2$ (c) 856.4 (d) .001861
53. (a) .0035703 (b) .0000845   **55.** $.2932 - 3$ **57.** .2123   **59.** 16   **61.** About 972.5 miles.

PROBLEM SET 5-6 (Page 210)

1. 128   **3.** .0959   **5.** .0208   **7.** 7.12×10^7 **9.** 3.50   **11.** .983   **13.** 6.05   **15.** 4762
17. 6.143   **19.** 3.530×10^{-6} **21.** 8.90   **23.** 18.2   **25.** 5.19   **27.** $-.5984$ **29.** 2.24
31. (a) .3495 (b) 100.7 (c) .8274   **33.** (a) $\frac{2}{9}$ (b) 2 (c) 1 (d) 3   **35.** About 4.395 hours from now.
37. (a) About 6.17 billion. (b) About the year 2018.

CHAPTER 5. REVIEW PROBLEM SET (Page 213)

1. (a) $-2y^2z^{2/3}$ (b) $2^{5/4}x^{1/2}y^2$ (c) $2(5)^{1/6}$ **2.** (a) 12 (b) 4
3. (a) $125a^3$ (b) $1/a^{1/2}$ (c) $1/5^{7/4}$ (d) $3y^{13/6}/x^6$ (e) $x - 2x^{1/2}y^{1/2} + y$ (f) $2^{7/6}$
4.

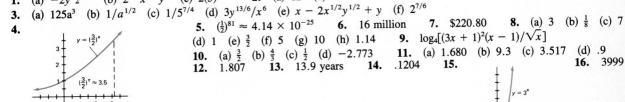

5. $(\frac{1}{2})^{81} \approx 4.14 \times 10^{-25}$ **6.** 16 million   **7.** \$220.80   **8.** (a) 3 (b) $\frac{1}{8}$ (c) 7
(d) 1 (e) $\frac{3}{2}$ (f) 5 (g) 10 (h) 1.14   **9.** $\log_4[(3x + 1)^2(x - 1)/\sqrt{x}]$
10. (a) $\frac{3}{2}$ (b) $\frac{4}{3}$ (c) $\frac{1}{2}$ (d) -2.773 **11.** (a) 1.680 (b) 9.3 (c) 3.517 (d) .9
12. 1.807   **13.** 13.9 years   **14.** .1204   **15.**   **16.** 3999

PROBLEM SET 6-1 (Page 219)

1. Upward.　　　　**3.** To the right.　　　**5.** To the left.　　　**7.** $p = 2$

9. $p = \frac{1}{8}$

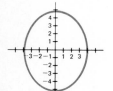

11. $p = \frac{1}{2}$

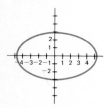

13. $p = \frac{9}{16}$

15. $(-\frac{1}{2}, 1)$; $(\frac{1}{2}, 1)$　　**17.** $3x^2 = 2y$　**19.** $x^2 = -12y$　**21.** $y^2 = 4x$　**23.** $(-\frac{1}{5}, 0)$; $x = \frac{1}{5}$　**25.** $4p$
27. 2.5 feet　　**29.** 225 meters　　**31.** $y^2 = 4(p + r)x$　**33.** $8\sqrt{3}p$

PROBLEM SET 6-2 (Page 225)

1. Vertical; 8, $2\sqrt{7}$.　　**3.** Horizontal; 12, $4\sqrt{5}$.　　**5.** Horizontal; 2, $\frac{4}{3}$.　　**7.** Vertical; $2/k$, $1/k$.
9. Horizontal ellipse; $a = 5$, $b = 3$, $c = 4$.　　　**11.** Vertical ellipse; $a = 2$, $b = 1$ $c = \sqrt{3}$.

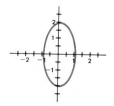

13. $x^2/16 + y^2/25 = 1$; $e = \frac{3}{5}$　　　　**15.** $x^2/49 + y^2/40 = 1$; $e = \frac{3}{7}$

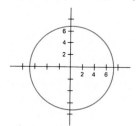

17. $x^2/49 + y^2/4$; $e = 3\sqrt{5}/7$　　　　**19.** $x^2/81 + y^2/9 = 1$; $e = 2\sqrt{2}/3$

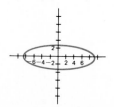

21. $a = 5, b = 2, c = \sqrt{21}, e = \sqrt{21}/5$ **23.** $x^2/128 + y^2/144 = 1$ **25.** $5\sqrt{3}$ feet

27. 20,000 miles, $4000\sqrt{21}$ miles   **29.** $\pi\sqrt{77}$ **31.** (a) 80π square feet   (b) 176π square feet

33. (a) $\overline{PR} + \overline{RQ} = 2a$; $\overline{PR'} + \overline{R'Q} > 2a$ since R' is outside the ellipse.

(b) Let Q' be the mirror image of Q about the line l. Show that $\angle Q'RR' = \alpha$.

PROBLEM SET 6-3 (Page 231)

1. Horizontal; $a = 4, b = 6, c = 2\sqrt{13}$.   **3.** Vertical; $a = 3, b = 4, c = 5$.

5. Horizontal; $a = \frac{1}{2}, b = \frac{1}{4}, c = \sqrt{5}/4$.   **7.** Horizontal; $a = 2, b = 4, c = 2\sqrt{5}$.

9. Horizontal hyperbola: $a = 5, b = 3, c = \sqrt{34}$.   **11.** Vertical hyperbola; $a = 8, b = 6, c = 10$.

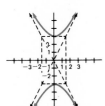

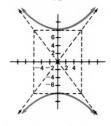

13. $-x^2/\frac{9}{4} + y^2/9 = 1$ **15.** $x^2/1 - y^2/15 = 1$

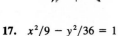

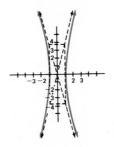

17. $x^2/9 - y^2/36 = 1$

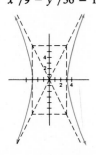

19. $-x^2/28 + y^2/36 = 1$

21. $x^2/16 - y^2/128 = 1$

23. $\frac{32}{3}$ **25.** $\sqrt{2}$ **27.** $5\sqrt{3}$

29. $-x^2/3,630,000 + y^2/1,210,000 = 1$

PROBLEM SET 6-4 (Page 238)

1. $u^2 + 2v^2 = 2$; ellipse.   **3.** $u^2 + v^2 = 1$; circle   **5.** $u^2 = 4v$; parabola.   **7.**

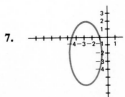

9.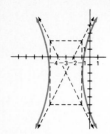
11.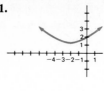
13.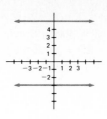

15. $(x + 2)^2 + (y - 1)^2 = 5$; circle: center $(-2,1)$, radius $\sqrt{5}$.

17. $(x - 2)^2/\frac{26}{4} + (y - 4)^2/26 = 1$; vertical ellipse; center $(2, 4)$.　　**19.** $4(x - 2)^2 + (y - 4)^2 = 0$; point: $(2, 4)$.

21. $(x - 2)^2 = -\frac{1}{4}(y - 24)$; vertical parabola: vertex $(2, 24)$.

23. $4(x - 2)^2 - 9(y - 1)^2 = 0$; two lines intersecting at $(2, 1)$.　　**25.**

27. $-(x - 1)^2/4 + (y + 3)^2/9 = 1$; 6　　**29.** Focus; $(21/20, 1)$; directrix: $x = -29/20$.

31. (a) 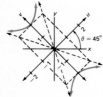　　(b)

33. $a < 0$ (hyperbola); $a = 0$ (parabola); $a > 0$, $a \neq 1$ (ellipse); $a = 1$ (circle).

35. $-4(x - 4) = (y - 5)^2$　　**37.** $x^2/8 + (y - 2)^2/4 = 1$　　**39.** $-5(x - 3)^2/36 + 5(y + 4)^2/144 = 1$

41. (a) $y = x^2 - x$　(b) $x = \frac{1}{4}y^2 - y$　(c) $x^2 + y^2 - 5x - 5y = 0$　　**43.** $x^2 + (y + \frac{7}{3})^2 = \frac{625}{9}$

PROBLEM SET 6-5 (Page 245)

1. $v = 0$　　**3.** $4u^2 + v^2 = 16$　　**5.** $u^2 + 2uv + v^2 - 8u + 8v = 0$　　**7.** $u^2 + 3v^2 = 8$　　**9.** $-2u^2 + 3v^2 = 6$

11. 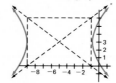　　$3x^2 + 10xy + 3y^2 + 8 = 0$; $x = (\sqrt{2}/2)(u - v)$ and $y = (\sqrt{2}/2)(u + v)$; $-u^2 + v^2/4 = 1$

13. $4x^2 - 3xy = 18$; $x = (1/\sqrt{10})(u - 3v)$ and $y = (1/\sqrt{10})(3u + v)$; $-u^2/36 + v^2/4 = 1$

15. 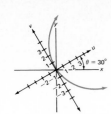 $x^2 - 2\sqrt{3}xy + 3y^2 - 12\sqrt{3}x - 12y = 0$; $x = \frac{1}{2}(\sqrt{3}u - v)$ and $y = \frac{1}{2}(u + \sqrt{3}v)$; $v^2 = 6u$

17. 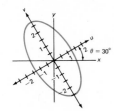 $13x^2 + 6\sqrt{3}xy + 7y^2 - 32 = 0$; $x = \frac{1}{2}(\sqrt{3}u - v)$ and $y = \frac{1}{2}(u + \sqrt{3}v)$; $u^2/2 + v^2/8 = 1$

19. 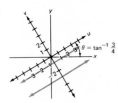 $9x^2 - 24xy + 16y^2 - 60x + 80y + 75 = 0$; $x = \frac{1}{5}(4u - 3v)$ and $y = \frac{1}{5}(3u + 4v)$; $v^2 + 4v + 3 = 0$

21. $u = (5 - 3\sqrt{3})/2$; $v = (-5\sqrt{3} - 3)/2$ **23.** $u = 4$; $v = -2$ **25.** $u = 5$; $v = 0$

27. $x^2 + 2\sqrt{3}xy + 3y^2 = 8(-\sqrt{3}x + y)$ **29.** $u^2/2 + v^2/10 = 1$; ellipse.   **31.** $(u - 4)^2 + v^2 = 16$

33. $u^2/5 - v^2/21 = 1$; hyperbola.   **35.** Equation transforms to $u = d$, a line $|d|$ units from the origin.   **37.** 3

PROBLEM SET 6-6 (Page 253)

1–11. 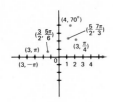   **13.** $(2\sqrt{2}, 2\sqrt{2})$ **15.** $(-3, 0)$ **17.** $(-5, -5\sqrt{3})$ **19.** $(\sqrt{2}, -\sqrt{2})$

21. $(4, 0)$ **23.** $(2, \pi)$ **25.** $(2\sqrt{2}, \pi/4)$ **27.** $(2\sqrt{2}, 3\pi/4)$ **29.** $(2, -\pi/3)$ **31.** $(2\sqrt{3}, 11\pi/6)$

33. 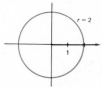   **35.**   **37.** 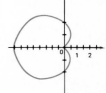   **39.**

41.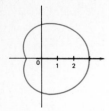

43. $r = 2$ **45.** $r = \tan \theta \sec \theta$ **47.** $y = 2x$ **49.** $(x^2 + y^2)^{3/2} = x^2 - y^2$

51.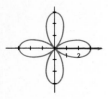

53.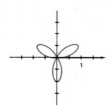

55.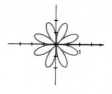

57. (a) $3y - 2x = 5$; a line.   (b) $(x - 2)^2 + (y + 3)^2 = 13$; a circle.

59.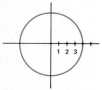

61.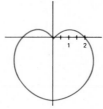

63.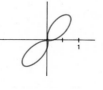

65. (a) 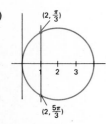 $(2, \pi/3), (2, 5\pi/3)$ (b) 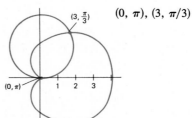 $(0, \pi), (3, \pi/3)$

67. Use the law of cosines; $4\sqrt{5}$ **69.** $\frac{1}{2}(\beta - \alpha)(b - a)(b + a)$

71.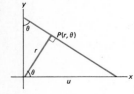

Since $\cos \theta = \dfrac{r}{u}$ and $\sin \theta = \dfrac{u}{4}$, it follows that

$r = u \cos \theta = 4\sin \theta \cos \theta = 2 \sin 2\theta.$
Now compare with Example B.

PROBLEM SET 6-7 (Page 260)

1. $r \cos \theta = 4$ **3.** $r \cos \theta = -3$ **5.**

$x = 6$

7.

$x + \sqrt{3}y = 8$

9.

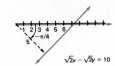

$\sqrt{2}x - \sqrt{2}y = 10$

11. $r = 8 \cos \theta, \ x^2 + y^2 = 8x$

13. $r = 10 \cos (\theta - \pi/3); \ x^2 + y^2 = 5x + 5\sqrt{3}\,y$ **15.** Ellipse; $e = \frac{2}{3}; \ x = 6$.   **17.** Hyperbola; $e = 2; \ x = \frac{5}{4}$.
19. Parabola; $e = 1; \ x = -7$.   **21.** Ellipse; $e = \frac{2}{3}; \ x = -\frac{1}{2}$.   **23.** Parabola; $e = 1; \ y = 5$.
25. Ellipse; $e = \frac{1}{2}; \ y = -6$.   **27.** Hyperbola; $e = \frac{5}{4}; \ y = \frac{8}{3}$.

29.

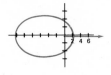

31.

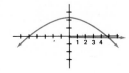

33.

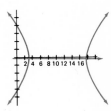

35. (a) $r = 3 \csc \theta$ (b) $r = 3$ (c) $r = -18 \cos \theta$ (d) $r = 6\sqrt{2} \cos (\theta - \pi/4)$
37. (a) $r = 2/(1 - \cos \theta)$ (b) $r = 4/(1 + \sin \theta)$ **39.** $(x + 3y)(x - 2y) = 0$ **41.** $d = 8, \ e = \frac{1}{2}; \frac{32}{3}, \ 16\sqrt{3}/3$
43. 8   **45.** (a)   (b) $2a + a \cos \theta + 2a + a \cos(\theta + \pi) = 4a$

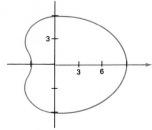

PROBLEM SET 6-8 (Page 268)

1. $x = 2t; \ y = -3t$ **3.** $x = 1 + 3t; \ y = 2 - 7t$ **5.** $-2; 1$ **7.** $2x + 3y = 17$ **9.** $2y = x^2 + 3x + 2$
11. $x^2 + y^2 = 4$ **13.** $x^2/4 + y^2/9 = 1$ **15.** $27y = x^3 - 3x^2 + 3x - 1$ **17.**

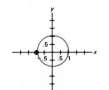

19.

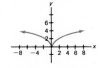

21.

23.

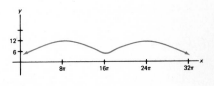

25. (a) $y = -x^2/768 + x/\sqrt{3}$ (b) 4 (c) $256\sqrt{3}$ (d) 64
27. (a) $3x + 2y = 17$ (b) $y = 4 \cos(x/3)$ (c) $x^2/4 - y^2/9 = 1$ (d) $8x + (y + 1)^3 = 8$ (e) $x = (x + y)^6 + 2(x + y)^3$

29. Show that $x^2 + y^2 = 4$ in each case. Verify that the parameter interval gives the same quarter circle in each case.

31. $(x - 2)^2/9 + (y - 1)^2/16 = 1$

33. (a) $y = (\tan \alpha)x - 16x^2/(v_0^2 \cos^2 \alpha)$ (b) $(v_0 \sin \alpha)/16$ (c) $(v_0 \cos \alpha)(v_0 \sin \alpha)/16 = (v_0^2 \sin 2\alpha)/32$ (d) $\pi/4$

35. Eliminate t to get $-x + 9y = 40$. Endpoints are $(-4, 4)$ and $(5, 5)$. Parameter interval $\pi/2$ to π gives same segment traced in the reverse direction.

CHAPTER 6. REVIEW PROBLEM SET (Page 273)

1. (a) xii (b) iii (c) viii (d) ii (e) vii (f) vi (g) v (h) x (i) ix (j) xi (k) x

2. Focus $(-2, -\frac{11}{4})$; vertex $(-2, -3)$. **3.** $x^2 = \frac{4}{3}y$ **4.** $(x - 2)^2/9 + (y - 1)^2/25 = 1$

5.

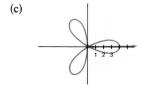

$e = \sqrt{3}/2$

6. $-x^2/9 + y^2/9 = 1$ **7.** $(3 - \sqrt{34}, -1); (3 + \sqrt{34}, -1)$

8. $u^2/8 + v^2/\frac{56}{3} = 1; 4\sqrt{2}$

9.

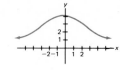

10. (a) $(-3/2, 3\sqrt{3}/2)$ (b) $(0, 2)$ (c) $(-3\sqrt{3}, -3)$

11. (a) $(5, 0)$ (b) $(4, 7\pi/4)$ (c) $(4, 5\pi/6)$

12. (a)   (b)   (c)

13. $r^2 = 4 \sec \theta \csc \theta; (x^2 + y^2)^{3/2} = 2xy$ **14.** (a) iii (b) vii (c) ii (d) v (e) iii (f) ii (g) iv (h) vi

15. (a) $x^2 + y^2 = 25$ (b) $r = 5$ (c) $x = 5 \cos \theta; y = 5 \sin \theta$ **16.**

Index
of Teaser Problems

Index
of Names and Subjects

GEOMETRY

Triangles

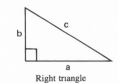

Right triangle

Pythagorean Theorem

$a^2 + b^2 = c^2$

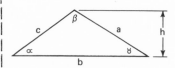

Any triangle

Angles $\quad \alpha + \beta + \gamma = 180°$

Area $\quad A = \frac{1}{2}bh$

Circles

Circumference $\quad C = 2\pi r$

Area $\quad A = \pi r^2$

Cylinders

Surface area $\quad S = 2\pi r^2 + 2\pi rh$

Volume $\quad V = \pi r^2 h$

Cones

Surface area $\quad S = \pi r^2 + \pi r \sqrt{r^2 + h^2}$

Volume $\quad V = \frac{1}{3}\pi r^2 h$

Spheres

Surface area $\quad S = 4\pi r^2$

Volume $\quad V = \frac{4}{3}\pi r^3$

Conversions

1 inch = 2.54 centimeters
1 liter = 1000 cubic centimeters
1 kilogram = 2.20 pounds
1 kilometer = .62 miles
1 liter = 1.057 quarts
1 pound = 453.6 grams
π radians = 180 degrees

FORMULA CARD

to accompany

**ALGEBRA AND TRIGONOMETRY, 3rd ed.
COLLEGE ALGEBRA, 3rd ed.
PRECALCULUS MATHEMATICS, 2nd ed.
PLANE TRIGONOMETRY, 2nd ed.**

Walter Fleming and Dale Varberg
PRENTICE HALL, Englewood Cliffs, N. J. 07632

ALGEBRA

Exponents

$a^m a^n = a^{m+n}$

$(a^m)^n = a^{mn}$

$\dfrac{a^m}{a^n} = a^{m-n}$

$(ab)^n = a^n b^n$

$\left(\dfrac{a}{b}\right)^n = \dfrac{a^n}{b^n}$

Radicals

$(\sqrt[n]{a})^n = a$

$\sqrt[n]{a^n} = a \quad \text{if} \quad a \geq 0$

$\sqrt[n]{ab} = \sqrt[n]{a}\,\sqrt[n]{b}$

$\sqrt[n]{\dfrac{a}{b}} = \dfrac{\sqrt[n]{a}}{\sqrt[n]{b}}$

Logarithms

$\log_a MN = \log_a M + \log_a N \qquad \log_a (M/N) = \log_a M - \log_a N$

$$\log_a (N^P) = P \log_a N$$

Quadratic Formula

Solutions to $ax^2 + bx + c = 0 \quad$ are $\quad x = \dfrac{-b \pm \sqrt{b^2 - 4ac}}{2a}$

Factoring Formulas

$$x^2 - y^2 = (x - y)(x + y)$$
$$x^2 + 2xy + y^2 = (x + y)^2$$
$$x^2 - 2xy + y^2 = (x - y)^2$$
$$x^3 - y^3 = (x - y)(x^2 + xy + y^2)$$
$$x^3 + y^3 = (x + y)(x^2 - xy + y^2)$$
$$x^3 + 3x^2 y + 3xy^2 + y^3 = (x + y)^3$$

Binomial Formula

$$(x + y)^n = {}_nC_0 x^n y^0 + {}_nC_1 x^{n-1} y^1 + \cdots + {}_nC_{n-1} x^1 y^{n-1} + {}_nC_n x^0 y^n$$

$${}_nC_r = \frac{n!}{(n - r)! \, r!} = \frac{n(n - 1) \cdots (n - r + 1)}{r(r - 1) \cdots 3 \cdot 2 \cdot 1}$$

$${}_nC_0 = {}_nC_n = 1$$

TRIGONOMETRY

Basic Identities

$$\tan t = \frac{\sin t}{\cos t} \qquad \cot t = \frac{\cos t}{\sin t} \qquad \cot t = \frac{1}{\tan t}$$

$$\sec t = \frac{1}{\cos t} \qquad \csc t = \frac{1}{\sin t} \qquad \sin^2 t + \cos^2 t = 1$$

$$1 + \tan^2 t = \sec^2 t \qquad\qquad 1 + \cot^2 t = \csc^2 t$$

Confunction Identities

$$\sin\left(\frac{\pi}{2} - t\right) = \cos t \qquad \cos\left(\frac{\pi}{2} - t\right) = \sin t \qquad \tan\left(\frac{\pi}{2} - t\right) = \cot t$$

Odd-even Identities

$$\sin(-t) = -\sin t \qquad \cos(-t) = \cos t \qquad \tan(-t) = -\tan t$$

Addition Formulas

$$\sin(s + t) = \sin s \cos t + \cos s \sin t \qquad \sin(s - t) = \sin s \cos t - \cos s \sin t$$

$$\cos(s + t) = \cos s \cos t - \sin s \sin t \qquad \cos(s - t) = \cos s \cos t + \sin s \sin t$$

$$\tan(s + t) = \frac{\tan s + \tan t}{1 - \tan s \tan t} \qquad \tan(s - t) = \frac{\tan s - \tan t}{1 + \tan s \tan t}$$

Double Angle Formulas

$$\sin 2t = 2 \sin t \cos t$$

$$\tan 2t = \frac{2 \tan t}{1 - \tan^2 t}$$

$$\cos 2t = \cos^2 t - \sin^2 t = 1 - 2 \sin^2 t = 2 \cos^2 t - 1$$

Half Angle Formulas

$$\sin\frac{t}{2} = \pm\sqrt{\frac{1 - \cos t}{2}} \qquad \cos\frac{t}{2} = \pm\sqrt{\frac{1 + \cos t}{2}} \qquad \tan\frac{t}{2} = \frac{1 - \cos t}{\sin t}$$

Product Formulas

$$2 \sin s \cos t = \sin(s + t) + \sin(s - t)$$

$$2 \cos s \cos t = \cos(s + t) + \cos(s - t)$$

$$2 \cos s \sin t = \sin(s + t) - \sin(s - t)$$

$$2 \sin s \sin t = \cos(s - t) - \cos(s + t)$$

Factoring Formulas

$$\sin s + \sin t = 2 \cos\frac{s - t}{2} \sin\frac{s + t}{2}$$

$$\cos s + \cos t = 2 \cos\frac{s + t}{2} \cos\frac{s - t}{2}$$

$$\sin s - \sin t = 2 \cos\frac{s + t}{2} \sin\frac{s - t}{2}$$

$$\cos s - \cos t = -2 \sin\frac{s + t}{2} \sin\frac{s - t}{2}$$

Laws of Sines and Cosines

$$\frac{\sin \alpha}{a} = \frac{\sin \beta}{b} = \frac{\sin \gamma}{c}$$

$$a^2 = b^2 + c^2 - 2bc \cos \alpha$$

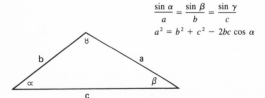

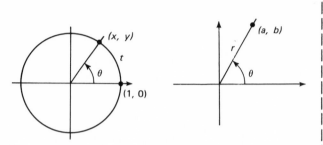

$$\sin t = \sin \theta = y = \frac{b}{r}$$

$$\cos t = \cos \theta = x = \frac{a}{r}$$

$$\tan t = \tan \theta = \frac{y}{x} = \frac{b}{a}$$

$$\cot t = \cot \theta = \frac{x}{y} = \frac{a}{b}$$

Graphs

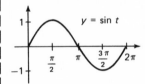

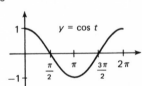

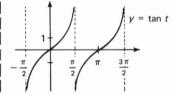

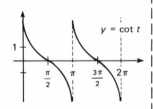

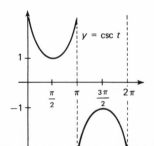

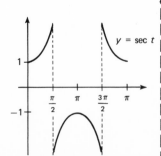

GRAPHS OF TRIGONOMETRIC FUNCTIONS

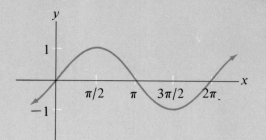

$y = \sin x$

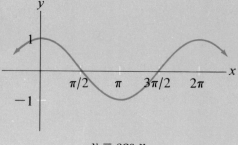

$y = \cos x$

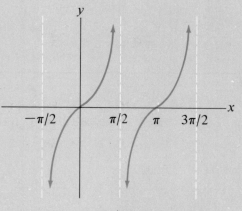

$y = \tan x$

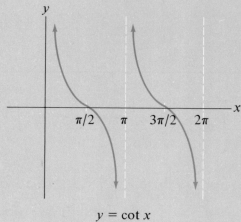

$y = \cot x$

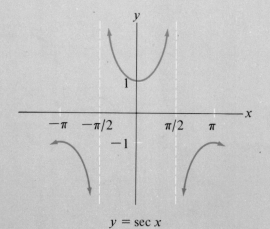

$y = \sec x$

$y = \csc x$